Representations of Finite-Dimensional Algebras

Springer
Berlin
Heidelberg
New York
Barcelona
Budapest
Hong Kong
London
Milan
Paris
Santa Clara
Singapore
Tokyo

P. Gabriel A. V. Roiter

Representations of Finite-Dimensional Algebras

With 98 Figures

Springer

Title of the Russian edition:
Itogi nauki i tekhniki, Sovremennye problemy matematiki,
Fundamental'nye napravleniya, Vol. 73, Algebra 8
Publisher VINITI, Moscow

Second Printing 1997 of the First Edition 1992, which was originally
published as Algebra VIII,
Volume 73 of the Encyclopaedia of Mathematical Sciences.

Die Deutsche Bibliothek – CIP-Einheitsaufnahme

Gabriel, Peter: Representations of finite dimensional algebras / Peter Gabriel; Andrei V. Roiter.
Ed.: A. I. Kostrikin; I. R. Shafarevich. With contributions by B. Keller.
2. print. – Berlin; Heidelberg; New York; Barcelona; Budapest; Hongkong; London; Mailand; Paris;
Santa Clara; Singapur; Tokio: Springer, 1997
ISBN 3-540-62990-4

Mathematics Subject Classification (1991):
16A64, 16A46, 14L30, 11E12, 15A21, 17B67

ISBN 3-540-62990-4 Springer-Verlag Berlin Heidelberg New York

Printed in Germany

Typesetting: Asco Trade Typesetting Ltd., Hong Kong

SPIN: 10629351
41/3143-5 4 3 2 1 0 – Printed on acid-free paper.

List of Editors, Authors and Translators

Editor-in-Chief

R.V. Gamkrelidze, Russian Academy of Sciences, Steklov Mathematical Institute, ul. Gubkina 8, 117966 Moscow, Institute for Scientific Information (VINITI), ul. Usievicha 20 a, 125219 Moscow, Russia; e-mail: gam@ipsun.ras.ru

Consulting Editors

A. I. Kostrikin, Steklov Mathematical Institute, ul. Gubkina 8, 117966 Moscow, Russia

I. R. Shafarevich, Steklov Mathematical Institute, ul. Gubkina 8, 117966 Moscow, Russia

Authors

P. Gabriel, Mathematisches Institut der Universität Zürich, Rämistraße 74, CH-8001 Zürich, Switzerland

A. V. Roiter, Institute of mathematics, Ukrainian Academy of Sciences, ul. Repina 3, 252601 Kiev, Ukrainia

B. Keller (Chapter 12), Université de Paris VII, U.E.R. de Mathématiques, Tour 45/55, 5e étage, 2 place Jussieu, F-75251 Paris Cedex 05, France

Representations of Finite-Dimensional Algebras

P. Gabriel, A.V. Roiter

Contents

Foreword

This monograph aims at a general outline of old and new results on representations of finite-dimensional algebras. In a theory which developed rapidly during the last two decades, the lack of textbooks is the main impediment for novices. We therefore paid special attention to the foundations and included proofs for statements which are elementary, serve comprehension or are scarcely available. In this manner we try to lead the reader up to a point where he can find his way in the original literature.

The bounds imposed upon the volume prevent us from endeavouring after completeness. Thus we centered the discourse round the fairly complete theory of finitely represented algebras, only touching tame denizens and omitting for instance questions of degeneration, bocses, Riedtmann's investigation of self-injective algebras ....

First examples of problems and an introduction to terminology are presented in the Sect. 1 and 2, which should be accessible to "everybody". Readers searching for a deeper insight may look at Sect. 3 and 7, which still are introductory. To complete this sketchy view, we included an "organigram" and propose a short bibliography which points to characteristic representatives of the main branches of development and consists of the references [108], [62], [134], [79], [52], [63], [110], [91], [19], [39], [57], [8], [129], [119], [81], [126], [43], [32], [34], [115], [22], [14], [132], [12], [26], [5], [40], [70], [37] and [124]. The organigram

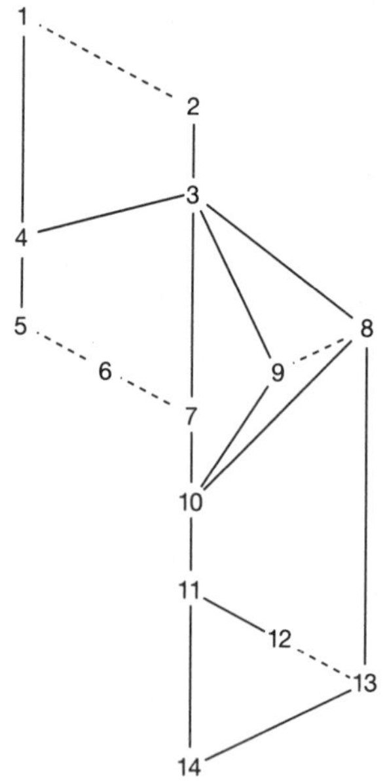

Interdependence of chapters

passes in silence over essential hidden links, for instance between Sect. 5 and Sect. 10–14, which may be used in the proofs omitted here. On the other hand, the short bibliography clearly has a personal component, since other representatives might have been selected. Additional bibliographic information is given at the end of each chapter. A detailed bibliography up to 1984, assembled by V. Dlab, is published in Springer Lecture Notes 831 (1980) and 1178 (1986).

In a quickly growing theory, the terminology happens to be accidental, needless to say. To attain a coherent language, we therefore parted with some words, introduced few others and generally avoided names related to persons. Parallel terminology is indicated in the remarks at the end of each chapter.

The monograph would not have been completed without the assistance of L.A. Nazarova, with whom all points have been discussed. For comments and corrections we are also indebted to T. Brüstle, Deng Bangming, E. Dieterich, T. Guidon, E. Gut, U. Hassler, H.J. von Höhne, B. Keller, J.A. de la Peña, V.V. Sergejchuk and D. Vossieck. Finally, we like to thank Mrs. I. Verdier, whose endurance and unselfish motivation brought the work to its logistic conclusion.

During the completion of the monograph, the authors benefited from large support by the Ukrainian Academy of Sciences and the Schweizerischer Nationalfonds.

1. Matrix Problems

Many problems of representation-theory can be formulated either with matrices or in basis-independent synthetic terms. Synthetic points of view often seem preferable. However, thanks to its visual descriptiveness, the matricial language also partakes of a series of advantages which are essential when the examined spaces carry natural bases. Such situations are typical of present-day representation-theory.

In the present section, k denotes an algebraically closed field.

1.1. Since (finite) matrices describe linear maps between the spaces k^n, $n \in \mathbb{N}$, we admit matrices with no row or no column.[1] By $k^{m \times n}$ we denote the space of all $m \times n$-matrices, by $\mathbb{M}$ the union of the spaces $k^{m \times n}$ when (m, n) ranges over $\mathbb{N} \times \mathbb{N}$. We further write ${}'\mathbb{M}$ and $\mathbb{M}'$ for the subsets of $\mathbb{M}$ formed by the matrices with linearly independent rows and columns respectively. Finally, we set ${}'\mathbb{M}' = {}'\mathbb{M} \cap \mathbb{M}'$.

The following permutations of $k^{m \times n}$ are called *elementary row-transformations*:

a) The replacement of the j-th row $A_{j.}$ of $A \in k^{m \times n}$ by $A_{j.} + aA_{i.}$, where $i \neq j$ and $a \in k$. The resulting matrix is $(\mathbb{1} + aE^{ji})A$, where $\mathbb{1} = \mathbb{1}_m \in k^{m \times m}$ denotes the unit matrix and E^{ji} the matrix whose unique non-zero entry has index (j, i) and equals 1.

b) The multiplication of the i-th row of A by $a \in k \setminus \{0\}$. The resulting matrix is $(\mathbb{1} + (a - 1)E^{ii})A$.

c) The transposition of the i-th and j-th rows of A, $i \neq j$. The new matrix then is $P^{ij}A$, where $P^{ij} = \mathbb{1} - E^{ii} - E^{jj} + E^{ij} + E^{ji}$.

It is well known that the matrices $\mathbb{1} + aE^{ji}$, $\mathbb{1} + (a - 1)E^{ii}$ and P^{ij} considered above "redundantly" generate the linear group $GL_m = k^{m \times m} \cap {}'\mathbb{M}'$. Finite compositions of elementary row-transformations therefore coincide with multiplications from the left by invertible matrices.

Elementary column-transformations are defined in a similar way. Their finite compositions coincide with the multiplications from the right by invertible matrices.

It is notorious that each matrix $A \in k^{m \times n}$ can be reduced by means of row and column transformations to the block form of Fig. 1, where the symbols 0 denote null matrices and r the rank of A: In fact, using row and column transformations of the form c), we can transfer any non-zero entry of A to the left upper corner. Using b), we then reduce this entry to 1. Using a), we annihilate the other entries of the first row and of the first column ...

$$\text{I} \qquad \left[\begin{array}{c|c} \mathbb{1}_r & 0 \\ \hline 0 & 0 \end{array}\right]$$

Fig. 1

In case $A \in \mathbb{M}'$, the right blocks of the reduced form are matrices without column! The reduction can then be performed with row-transformations only. In case $A \in {}'\mathbb{M}$, column-transformations suffice.

Despite its evidence and simplicity, or maybe thanks to them, the above reduction is very useful, and repeated use of it leads to rather profound results.

1.2. Besides the reduction of 1.1, more complicated matrix problems arise in various topics. Most famous are the classification of conjugacy classes of square matrices and the simultaneous reduction of pairs of rectangular matrices which will be treated in 1.7 and 1.8.

Here we start with simpler problems whose solution does not differ essentially from that of 1.1. Let us first subdivide a matrix $A = [A_1 \mid A_2]$ into two blocks by means of a vertical line. Let us on one side allow all row-transformations and on the other all column-transformations within each block. But we refuse column-transformations from left to right or right to left across the vertical line. In other words, we have two matrices $A_1 \in k^{m \times n_1}$, $A_2 \in k^{m \times n_2}$ and accept to replace them by $\bar{A}_1 = SA_1T_1^{-1}$ and $\bar{A}_2 = SA_2T_2^{-1}$, where $S, T_1, T_2 \in {}'\mathbb{M}'$. What are the "simplest possible" forms for $\bar{A}_1$ and $\bar{A}_2$?

Without taking care of A_2, we first reduce A_1 to the form I of 1.1 and thus get a new matrix

$$B = [B_1 \mid B_2] = \left[\begin{array}{cc|c} \mathbb{1} & 0 & B_2^1 \\ \hline 0 & 0 & B_2^2 \end{array}\right]$$

Without spoiling the form of B_1, we can then subject B_2 to the following transformations:

a) Arbitrary column-transformations.

b) Row-transformations of B_2^1; as a result, the block $\mathbb{1}$ is spoiled but can be restored with column-transformations of the left stripe of B_1.

c) Row-transformations of B_2^2.

d) Additions of rows of B_2^2 (multiplied by scalars) to rows of B_2^1.

In this way, the original matrix A, partitioned by a vertical line, provides a new matrix B_2 which is partitioned by a horizontal line. The difference is that now additions across the limiting line are permitted in one direction, from below to above but not reversely.

We now reduce B_2^2 to the form I and construct a stripe of zeros in B_2^1 using d), thus transforming B_2 into C_2 (Fig. 2).

$$C_2 = \left[\begin{array}{c|c} 0 & \\ \hline \mathbb{1} & 0 \\ \hline 0 & 0 \end{array}\right]$$

Fig. 2

The non-completed block obviously allows arbitrary elementary transformations and can be reduced to the form I. Thus A is finally reduced to the block form II. We remark that the entries of II are 0 or 1 and that each column contains at most one entry 1, each row at most 2.

$$\text{II} \qquad \bar{A} = \left[\begin{array}{ccc|ccc} 1 & 0 & 0 & 0 & 1 & 0 \\ 0 & 1 & 0 & 0 & 0 & 0 \\ 0 & 0 & 0 & 1 & 0 & 0 \\ 0 & 0 & 0 & 0 & 0 & 0 \end{array}\right]$$

Fig. 3

The investigation of matrices $A = [A_1 \mid A_2 \mid A_3]$ partitioned into three vertical stripes, where elementary column-transformations are allowed only within each stripe, is alike. Successively transforming blocks into form I, we can finally reduce A to the form III, where some blocks may be empty and the non-completed ones are null. Again the entries are 0 or 1.

III

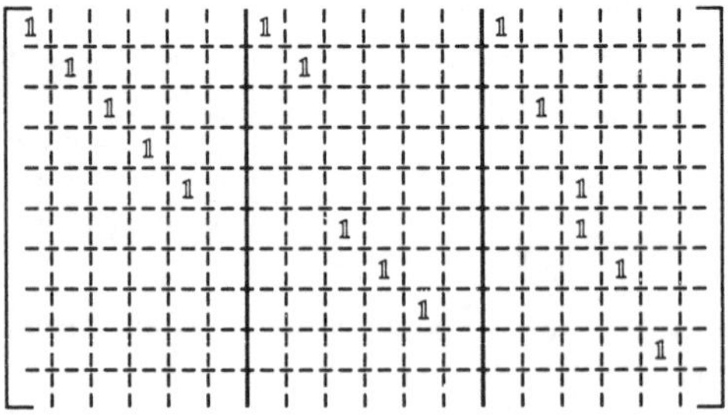

Fig. 4

However, the situation changes for matrices partitioned into 4 vertical stripes when, as above, column-transformations are allowed only within each stripe. In case $\lambda_1 \neq \lambda_2$, it is easy to see that the matrix A_{λ_1} of form IV cannot be transformed into A_{λ_2}.

$$\text{IV} \qquad A_\lambda = \left[\begin{array}{c|c|c|c} 1 & 0 & 1 & \lambda \\ 0 & 1 & 1 & 1 \end{array}\right], \qquad \lambda \in k.$$

Finally, we consider a matrix $A = [A_1 \mid A_2 \mid \ldots \mid A_n]$ partitioned into n vertical stripes. Besides the transformations allowed in the cases above, we now also accept additions of columns of the stripe i to columns of the stripe j if $i < j$. In this case, we have more transformations at our disposal and can reduce A to the form V: We first reduce A_1 to the form I. By adding multiples of the columns of the first stripe to the other columns of A, we then annihilate all the entries of the upper horizontal stripe ...

V 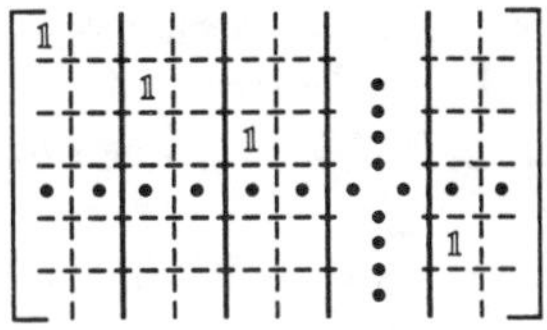

Fig. 5

1.3. Let us draw conclusions and first aim at a generalization: We examine matrices partitioned into some vertical stripes. We allow arbitrary row-transformations of the whole matrix and arbitrary column-transformations of each stripe. Moreover, for some pairs of stripes, we accept arbitrary additions of columns from one stripe to the other, but not reversely (otherwise, the two stripes could equally well be united to a single one). Allowing additions from a stripe A_1 to some A_2 and from A_2 to some A_3, we ipso facto allow additions from A_1 to A_3. Therefore, the relation describing the permitted additions between stripes is a partial order.

So we naturally meet with representations of *posets*[2]: *A representation of the poset* $\mathscr{P} = \{p_1, p_2, \ldots, p_t\}$ *with dimension-vector* $d = [d_0\ d_1\ \ldots\ d_t]^T \in \mathbb{N}^{1+t}$ is by definition a pair (d, A) where $A \in k^{d_0 \times \bar{d}}$ and $\bar{d} = d_1 + \cdots + d_t$. The datum of d provides a partition of $A = [A_1 \mid A_2 \mid \ldots \mid A_t]$ into t "vertical stripes" $A_i \in k^{d_0 \times d_i}$ and permits us to define the following equivalence relation: Two representations (d, A) and (e, B) are *equivalent* if $d = e$ and if B can be obtained from A by performing the following transformations:

a) Arbitrary row-transformations.

b) Arbitrary column-transformations within each vertical stripe.

c) Additions of columns of stripe i to columns of stripe j if $p_i < p_j$.

From this point of view, we may say that 1.1 was devoted to representations of a poset with one element, whereas the problems arising in connection with the forms II, III and IV of 1.2 concerned representations of *antichains*[2] of cardinality 2, 3 and 4. In the last example, we finally saw that each representation of a finite *chain*[2] is equivalent to a partitioned matrix of form V.

Of course, we may also examine matrices subdivided into vertical *and* horizontal stripes, thus introducing the concept of a representation of a pair of two posets[3]. However, such representations can usually be reduced without difficulty to representations of posets.

Our second observation is that the method sketched above, which consists in transforming submatrices to the form I, provides a procedure to reduce one poset to another, for instance an antichain of cardinality 2 to a chain of cardinality 2 and then to a poset of cardinality 1. Fortunately, in this case the procedure leads to simpler posets. As we shall see in Sect. 5, this happens for a rather broad class of posets. For these posets precisely, the representations admit normal forms whose entries are 0 or 1.

In other cases, the procedure reduces a given problem to a similar one (or even to itself) and decreases the dimensions of the involved representations. Such "selfrepeating" problems can also be solved (see 1.7 and 1.8) and play an important role in our theory.

Finally, the procedure may lead to more complicated problems. This happens for the so-called wild posets, for which no practicable method of classification of the representations is known. For instance, an antichain of cardinality 5 is wild.

1.4. The equivalence classes of representations of $\mathscr{P}$ can be interpreted as orbits. Indeed, the dimension-vector d determines a partition of each $Y \in k^{\bar{d} \times \bar{d}}$ into blocks $\bar{Y}_{ij} \in k^{d_i \times d_j}$, $1 \leqslant i$, $1 \leqslant j$. Thus it defines a subgroup $\bar{G}_d = \{Y \in GL_{\bar{d}}: \bar{Y}_{ij} = 0 \text{ if } p_i \nleqslant p_j\}$ of $GL_{\bar{d}}$ and an action of $G_d = GL_{d_0} \times \bar{G}_d$ on $k^{d_0 \times \bar{d}}$ such that $(X, Y)A = XAY^{-1}$. The orbits of this action coincide with the equivalence classes of representations of $\mathscr{P}$ with dimension-vector d.

This interpretation gives rise to a new generalization: We call *matrix problem of size* $m \times n$ a pair $(G, \mathfrak{M})$ formed by an *underlying set* $\mathfrak{M} \subset k^{m \times n}$ and a group $G \subset GL_m \times GL_n$ such that $XAY^{-1} \in \mathfrak{M}$ whenever $A \in \mathfrak{M}$ and $(X, Y) \in G$. The matrix problem is called *separated* if $G = G_1 \times G_2$, where $G_1 \subset GL_m$ and $G_2 \subset GL_n$; it is called *one-sided* if moreover $G_1 = GL_m$ or $G_2 = GL_n$.

The question raised by the matrix problem $(G, \mathfrak{M})$ is to classify the *orbits* of $\mathfrak{M}$ under the action of G defined by $(X, Y)A = XAY^{-1}$. Full classification in general means producing a fundamental domain or *cross-section*, i.e. a subset $F \subset \mathfrak{M}$ which intersects each orbit in one point.

According to these definitions, the representations of a poset with fixed dimension-vector form the underlying set of a one-sided matrix problem, whereas representations of pairs of posets give rise to separated matrix problems. And square matrices under conjugation provide examples of non-separated matrix-problems. In the examples examined in 1.1 and 1.2, the reduced forms I, II, III and V provide cross-sections of the corresponding matrix problems. As for IV, it only provides a *segment*, i.e. a subset $\mathfrak{M}$ which intersects each orbit in one point at most.

1.5. In the present monograph we restrict the investigation to *linear matrix problems* $(G, \mathfrak{M})$. In such problems, G is the group D' of invertible elements of a subalgebra $D \subset k^{m \times m} \times k^{n \times n}$. The condition implies that D coincides with the linear subspace generated by G and with the Zariski-closure[4] of G in $k^{m \times m} \times k^{n \times n}$. It is satisfied in all the examples listed above. It is not satisfied if $\mathfrak{M} = k^{n \times n}$ and $G = \{(X^T, X^{-1}): X \in GL_n\}$ (classification of bilinear forms). However, this problem and some others can be reduced to linear problems[5].

If $(G, \mathfrak{M})$ is linear and $G = D'$, each $A \in \mathfrak{M}$ gives rise to a subalgebra

$$D_A = \{(X, Y) \in D : XA = AY\}$$

of D. We say that A is *schurian* if $D_A = k(\mathbb{1}_m, \mathbb{1}_n)$.

Proposition. *Let* $(D', \mathfrak{M})$ *be a linear matrix problem of size* $m \times n$ *and* $F \subset \mathfrak{M}$ *a cross-section. Assume further that* $\mathfrak{M}$ *is locally closed and irreducible for the Zariski-topology of* $k^{m \times n}$.

Then we have $1 + \dim \mathfrak{M} \leqslant \dim D + \dim \bar{F}$, *where* $\bar{F}$ *is the closure of* F *in* $k^{m \times n}$. *In particular, if* F *is finite, we have* $1 + \dim \mathfrak{M} \leqslant \dim D$; *the equality* $1 + \dim \mathfrak{M} = \dim D$ *then implies the existence of a schurian element in* $\mathfrak{M}$.

On the other hand, we have $\dim D \leqslant 1 + \dim \mathfrak{M}$ *if* $\mathfrak{M}$ *has a schurian element, and* $\dim D < 1 + \dim \mathfrak{M}$ *if it has two non-equivalent schurian elements.*

The proof[6] of the proposition only uses rudiments of algebraic geometry. The statement will be used in Sect. 5 and Sect. 7.

1.6. Further Examples of Linear Matrix Problems

Example 1. Let R and S be finite-dimensional k-algebras, A and B left modules over R and S with k-dimension n and m respectively, $\mathfrak{M}$ the bimodule[7] $\operatorname{Hom}_k(A, B)$ over S and R. Choosing bases in A and B, we can identify $\mathfrak{M}$ with $k^{m \times n}$ and define D as the image of $S \times R$ in $k^{m \times m} \times k^{n \times n}$.

We point out three particular cases:

a) Given a poset $\mathscr{P} = \{p_1, \ldots, p_m\}$, we construct an algebra S with k-basis (e_{ji}), where $i, j \in \{1, \ldots, m\}$ and $p_i \leqslant p_j$, such that $e_{lj}e_{ji} = e_{li}$ and $e_{lh}e_{ji} = 0$ if $h \neq j$. The algebra S acts on a module B with k-basis $b_1, \ldots, b_m$ such that $e_{ji}b_i = b_j$. We further set $A = k^n$, $R = k^{n \times n}$. By transposition, the elements of $\mathfrak{M}$ then correspond to representations of $\mathscr{P}$ with dimension-vector $[n\ 1\ 1\ \ldots\ 1]^T$ (see (4.1)).

b) $S = B = k \oplus ks \oplus \cdots \oplus ks^{m-1}$, $s^m = 0$. In case $A = k^2$ and $R = k^{2 \times 2}$, the null matrix and the three families of Fig. 6 provide a cross-section (the entries equal to 1 are located in the rows numbered i and $i + p$).

c) In most of the linear matrix problems which we can "solve", the constructed cross-section is *piecewise affine*, i.e. of the form $P \backslash Q$, where P and Q are finite unions of affine subspaces of $k^{m \times n}$. *The following counterexample admits no piecewise affine cross-section: $S = k \oplus ks_1 \oplus ks_2 \oplus ks_3$, $s_i s_j = 0$, $B = ke_1 \oplus ke_2 \oplus ke_3 \oplus kf_1 \oplus kf_2 \oplus kf_3$, $s_1 e_i = f_i$, $s_2 e_1 = f_2$, $s_2 e_2 = f_3$, $s_2 e_3 = -f_1$, $s_3 e_1 = f_3$, $s_3 e_2 = f_1$, $s_3 e_3 = -f_2$, $A = k = R$*

$$\begin{bmatrix} - & - & 0 & 1 & 0 & - & - \\ - & - & 0 & 0 & 0 & - & - \end{bmatrix}^T, \quad \begin{bmatrix} - & - & 0 & 1 & 0 & - & - & 0 & 0 & 0 & 0 & - & - \\ - & - & 0 & 0 & 0 & - & - & 0 & 1 & a_1 & a_2 & - & - \end{bmatrix}^T,$$

$$\begin{array}{ccccc} \uparrow & & \uparrow & & \uparrow \\ 1 \leqslant i \leqslant m & , & 1 \leqslant i & < & i + p \leqslant m < i + 2p, \end{array}$$

$$\begin{bmatrix} - & - & 0 & 1 & 0 & - & - & 0 & 0 & 0 & - - & 0 & 0 & 0 & - & - \\ - & - & 0 & 0 & 0 & - & - & 0 & 1 & a_1 & - - & a_{p-1} & 0 & a_{p+1} & - & - \end{bmatrix}^T$$

$$\begin{array}{ccccc} \uparrow & & \uparrow & & \uparrow \\ 1 \leqslant i & < & i + p & < & i + 2p \leqslant m \end{array}$$

Fig. 6

Example 2. Let R, S be finite-dimensional k-algebras and $\mathfrak{M}$ a bimodule over S and R such that $\dim_k \mathfrak{M} < \infty$ and $\lambda\mu = \mu\lambda$, $\forall \mu \in \mathfrak{M}$, $\forall \lambda \in k$. If $R^\cdot$ and $S^\cdot$ denote the multiplicative groups of R and S, the action $((\sigma, \rho), \mu) \mapsto \sigma\mu\rho^{-1}$ of $S^\cdot \times R^\cdot$ on $\mathfrak{M}$ gives rise to a linear matrix problem[7]: First we choose bases of R and S over k which identify these spaces with some k^r and k^s and provide homomorphisms $R \xrightarrow{\varphi} k^{r\times r}$ and $k^{s\times s} \xleftarrow{\psi} S$ such that $\varphi(\rho)x = \rho x$ and $\psi(\sigma)^T y = y\sigma$ respectively. φ admits a retraction $\alpha \mapsto \alpha 1_R$ of left R-modules and ψ a retraction $\beta \mapsto \beta^T 1_S$ of right S-modules. As a consequence, the map $\mathfrak{M} \to k^{s\times s} \otimes_S \mathfrak{M} \otimes_R k^{r\times r}$, $\mu \mapsto \mathbb{1}_s \otimes \mu \otimes \mathbb{1}_r$ is injective. Since each bimodule over $k^{s\times s}$ and $k^{r\times r}$ is a sum of copies of $k^{s\times r}$, we can choose an isomorphism $k^{s\times s} \otimes_S \mathfrak{M} \otimes_R k^{r\times r} \xrightarrow{\sim} (k^{s\times r})^t$ and identify the elements of $\mathfrak{M}$ with matrices of size $s \times rt$ partitioned into t vertical stripes of size $s \times r$. Finally we set $D = \psi(S) \times \bar{R}$, where $\bar{R}$ is the image of R under the composition $\bar{\varphi}$ of φ with the "diagonal" embedding $k^{r\times r} \to k^{rt\times rt}$.

Example 3. With the notations of Example 2, we now suppose that $R = S$, and we consider the action $(\sigma, \mu) \mapsto \sigma\mu\sigma^{-1}$ of $S^\cdot$ on $\mathfrak{M}$. This action again gives rise to a linear matrix problem: As above, $\mathfrak{M}$ can be embedded into $(k^{s\times s})^t$. But the required subalgebra of $k^{s\times s} \times k^{st\times st}$ now is the image $\bar{S}$ of S under the homomorphism $\sigma \mapsto (\psi(\sigma), \bar{\varphi}(\sigma))$.

Unlike the preceding matrix problems, the present one is not separated in general.

Example 4. Let A be a k-algebra generated by $x_1, \ldots, x_t$ and $\mathfrak{M}$ the subvariety of $k^{m\times mt} \xrightarrow{\sim} (k^{m\times m})^t$ formed by the sequences $X_1, \ldots, X_t$ of $m \times m$-matrices such that $\sum_i \lambda_i X_{i(1)} \ldots X_{i(r_i)} = 0$ for each "relation" $\sum_i \lambda_i x_{i(1)} \ldots x_{i(r_i)} = 0$ between the generators x_j. Let further D be the image of the homomorphism $(\mathbb{1}, \delta)$: $k^{m\times m} \to k^{m\times m} \times k^{mt\times mt}$, where δ: $k^{m\times m} \to k^{mt\times mt}$ denotes the "diagonal" embedding. The orbits of the matrix-problem $(D^\cdot, \mathfrak{M})$ then correspond bijectively to the isomorphism classes of A-modules of dimension m.

1.7. The Theorem of Jordan-Weierstrass. Let us examine in detail the classical problem of reducing a square matrix $A \in k^{m\times m}$ by means of *conjugations* $A \mapsto SAS^{-1}$, where $S \in GL_m$. "Jordan-Weierstrass" states that *A can be reduced by such transformations to a block-form matrix whose non-diagonal blocks vanish, whereas the diagonal ones are Jordan-blocks* (Fig. 7).[8]

The conjugacy classes are the orbits of a linear matrix problem with underlying set $k^{m\times m}$ and algebra

$$\left[\begin{array}{ccc|cc} s & 1 & 0 & 0 & 0 \\ 0 & s & 1 & 0 & 0 \\ 0 & 0 & s & 0 & 0 \\ \hline 0 & 0 & 0 & s & 1 \\ 0 & 0 & 0 & 0 & s \end{array}\right] = s\mathbb{1}_n + J_n \in k^{n\times n}$$

Fig. 7

$$D = \{(S, S) \in k^{m \times m} \times k^{m \times m} : S \in k^{m \times m}\}$$

The reduction of a square matrix A to a matrix formed by diagonal Jordan-blocks and vanishing non-diagonal blocks proceeds as in 1.2. In terms of matrix transformations, A allows arbitrary elementary column transformations coupled with suitable row transformations: a column permutation shifting column i to place j is coupled with a row permutation shifting row i to place j; adding λ times column i to column j is coupled with adding $-\lambda$ times row j to row i; multiplying column i by $\lambda \neq 0$ is coupled with multiplying row i by λ^{-1}.

Our first reduction is to produce zero columns by means of column transformations followed by the coupled row transformations (which do not spoil zero columns). In this way, A can be reduced to the block form of Fig. 8. The elementary transformations of $[0 \,|\, A_1]$ which do not spoil the zero block then give rise to the following permissible transformations of A_1: a) arbitrary row transformations of A_1^1; b) column transformations of A_1 coupled with suited row transformations of the square matrix A_1^2 (we hint at them by hatching the block A_1^2); c) additions of multiples of rows of A_1^2 to rows of A_1^1 (observe the arrow!).

As a result, we can perform arbitrary conjugations on A_1^2 and reduce it to the form $[0 \,|\, A_2]$ where $A_2 \in \mathbb{M}^l$, as we did above for A. Going on with transformations of type c), we produce zeros above A_2, thus reducing A_1 to the form I of Fig. 9. By further use of a) we transform B into $\begin{bmatrix} 0 \\ \mathbb{1} \end{bmatrix}$ and reduce A to the form II.

After this first step, a zero block "splits off" at the upper left corner of II. In a second step, we reduce the lower right square matrix formed by 3×3 blocks, applying to A_2^2 and A_2^1 the recipe used for A_1^2 and A_1^1. Thus we obtain the enlarged form III of Fig. 10, which we transform into form IV by suitable permutations of rows and columns (the non-completed blocks are null).

Now a block $\begin{bmatrix} 0 & \mathbb{1} \\ 0 & 0 \end{bmatrix}$ "splits off" at the upper left corner of IV. In a third step, we apply our confirmed recipe to the blocks A_3^2 and A_3^1. Thus the square matrix

$$[0 \,|\, A_1] = \begin{bmatrix} 0 & A_1^1 \\ \hline 0 & A_1^2 \end{bmatrix} \quad , \mathbb{M}^l \ni A_1 = \begin{bmatrix} A_1^1 \\ \hline A_1^2 \end{bmatrix} \uparrow$$

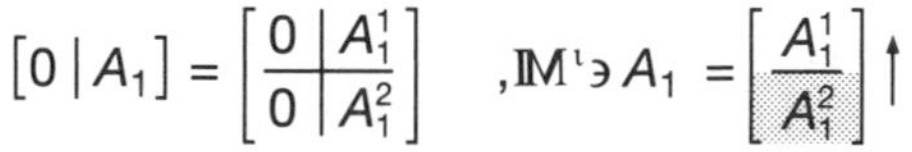

Fig. 8

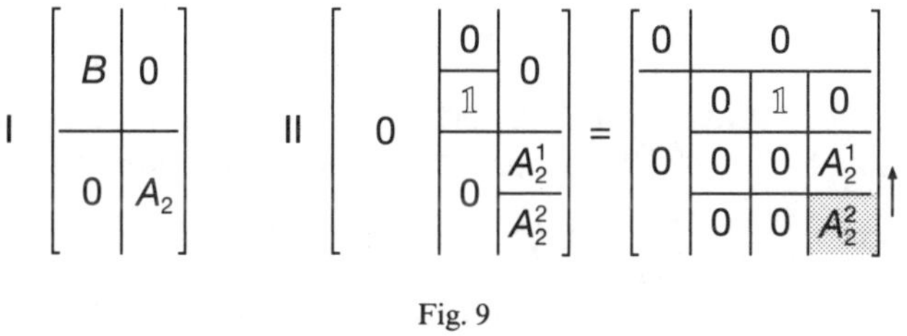

Fig. 9

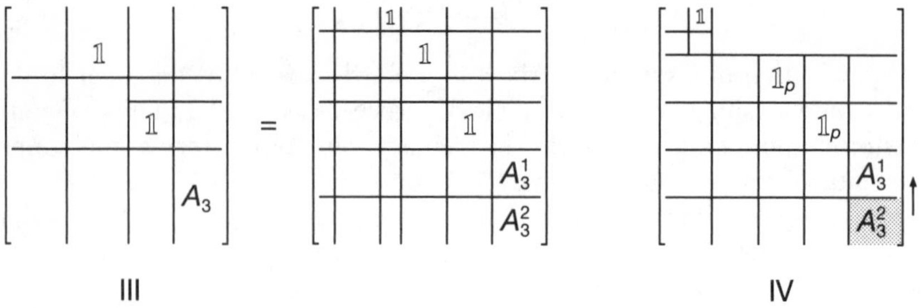

Fig. 10

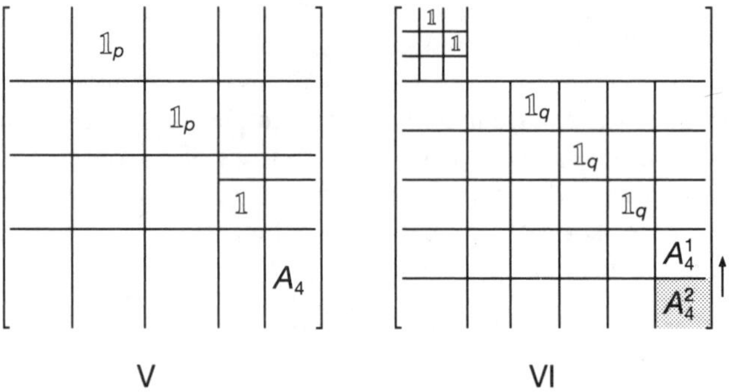

Fig. 11

formed by the 4 × 4 blocks at the lower right corner of IV is first reduced to the enlarged form V of Fig. 11, then by further permutations to form VI.

After the third step, a block $\begin{bmatrix} 0 & \mathbb{1} & 0 \\ 0 & 0 & \mathbb{1} \\ 0 & 0 & 0 \end{bmatrix}$ "splits off" at the upper left corner of VI .... Repeated use of the recipe splits off Kronecker-products[9] $J_1 \otimes \mathbb{1}_{r_1}$, $J_2 \otimes \mathbb{1}_{r_2}$, $J_3 \otimes \mathbb{1}_{r_3}$, ... and produces a sequence $A_1, A_2, A_3, \ldots$ of matrices of $\mathbb{M}'$ of decreasing sizes. Since the sequence cannot strictly decrease for ever, we conclude that A_N^1 is null and that $A_N^2 = A_{N+1}$ is invertible and splits off for some N.

The final reduction of A to the classical normal form is obtained by repeating the whole preceding procedure for $A_{N+1} - s\mathbb{1}$ if A admits a non-zero eigenvalue s ...

1.8. Kronecker's Problem. The underlying set of this linear matrix problem is $k^{m \times 2u}$. The subalgebra of $k^{m \times m} \times k^{2u \times 2u}$ consists of the pairs (X, Y) such that $X \in k^{m \times m}$ and $Y = \begin{bmatrix} Z & 0 \\ 0 & Z \end{bmatrix}$, $Z \in k^{u \times u}$. We interpret a matrix $[A \mid B] \in k^{m \times 2u}$ as a pair of two matrices $A, B \in k^{m \times u}$. Two pairs $[A \mid B]$ and $[A' \mid B']$ are equivalent if $A' = XAZ^{-1}$ and $D' = XBZ^{-1}$ for some $X \in GL_m$, $Z \in GL_u$. A description of the equivalence classes can be obtained as follows:

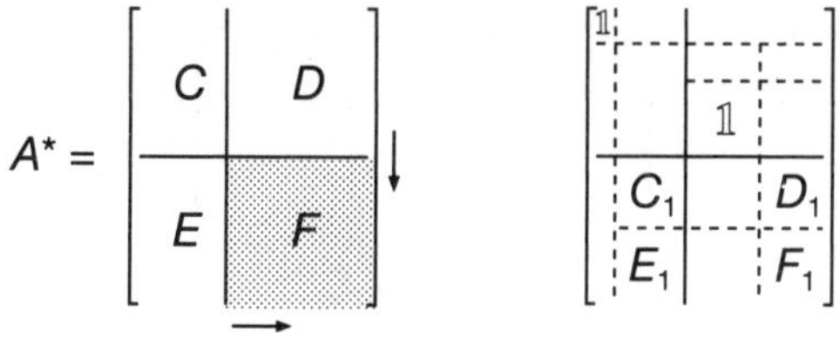

Fig. 12

First we reduce B to the form $\begin{bmatrix} 0 & 0 \\ 0 & \mathbb{1}_r \end{bmatrix}$. The transform A^* of A can be written in the block-form of Fig. 12 where $C \in k^{(m-r)\times(u-r)}$, $D \in k^{(m-r)\times r}$ ... The following matrix transformations fix $\begin{bmatrix} 0 & 0 \\ 0 & \mathbb{1}_r \end{bmatrix}$: a) arbitrary row transformations of $[C \mid D]$ and column transformations of $\left[\frac{C}{E}\right]$; b) arbitrary row transformations $[E \mid F] \mapsto S[E \mid F]$ coupled with column transformations $\left[\frac{D}{F}\right] \mapsto \left[\frac{D}{F}\right] S^{-1}$; c) additions of rows from $[C \mid D]$ to $[E \mid F]$ and of columns from $\left[\frac{C}{E}\right]$ to $\left[\frac{D}{F}\right]$.

To A^* we now apply the following recipe: We first reduce C to the form $\begin{bmatrix} \mathbb{1} & 0 \\ 0 & 0 \end{bmatrix}$ and use c) to annihilate the first rows of D and columns of E. Then we reduce the lower part of the transform of D to the form $\begin{bmatrix} 0 & 0 \\ \mathbb{1} & 0 \end{bmatrix}$ and produce the lower central zero blocks (see Fig. 12, where the non-completed blocks are null). By a permutation of columns we finally reduce A and B to the matrices of Fig. 13.

After this first step, a pair $\left[\begin{array}{c|c} \mathbb{1} & 0 \\ 0 & 0 \end{array}\right]$ splits off at the upper left corner of the pair of Fig. 13. This pair can be further split into a pair $[\mathbb{1} \mid 0]$ and a pair of matrices

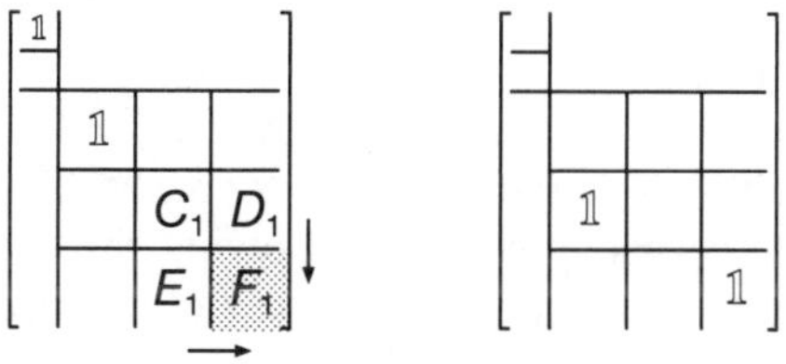

Fig. 13

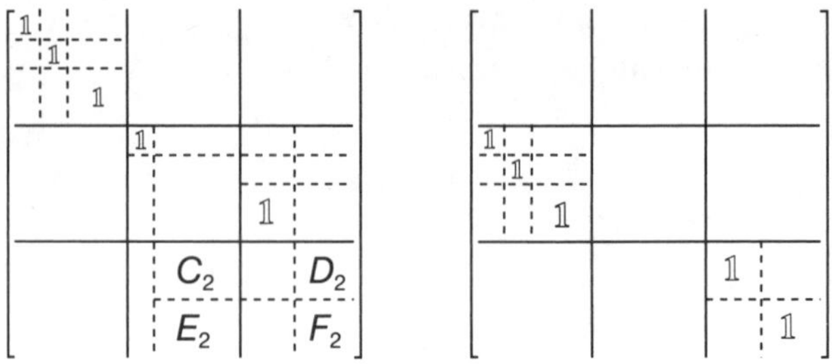

Fig. 14

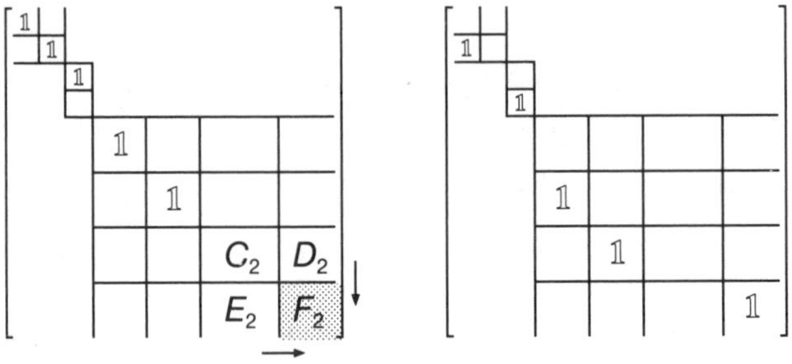

Fig. 15

without column. In a second step, we reduce the pair of right lower matrices formed by 3 × 3 blocks. To this effect, we notice that C_1, D_1, E_1, F_1 allow the same transformations as C, D, E, F. Applying the recipe explained above, we obtain the enlarged pair of Fig. 14 which we transform into the pair of Fig. 15 by permutations of columns and rows.

Now two pairs $\left[\begin{array}{cc|cc} \mathbb{1} & 0 & 0 & 0 \\ 0 & \mathbb{1} & \mathbb{1} & 0 \end{array}\right]$ and $\left[\begin{array}{c|c} \mathbb{1} & 0 \\ 0 & \mathbb{1} \end{array}\right]$ split off at the upper left corner of the pair of Fig. 15. In a third step, we reduce the lower right matrices (formed by 4 × 4 blocks) by applying our recipe to C_2, D_2, E_2, F_2 .... To sum up, repeated use of the recipe splits off pairs of the form $[1\,|\,0] \otimes \mathbb{1}_{r_1}$, $[\,|\,] \otimes \mathbb{1}_{r_2}$, $\left[\begin{array}{cc|cc} 1 & 0 & 0 & 0 \\ 0 & 1 & 1 & 0 \end{array}\right] \otimes \mathbb{1}_{r_3}$, $\left[\begin{array}{c|c} 1 & 0 \\ 0 & 1 \end{array}\right] \otimes \mathbb{1}_{r_4}$, $[\mathbb{1}_3\,|\,J_3^T] \otimes \mathbb{1}_{r_5}$, $\left[\begin{array}{c|c} \mathbb{1}_2 & \text{---} \\ \text{---} & \mathbb{1}_2 \end{array}\right] \otimes \mathbb{1}_{r_6}$, $[\mathbb{1}_4\,|\,J_4^T] \otimes \mathbb{1}_{r_7}$, $\left[\begin{array}{c|c} \mathbb{1}_3 & \text{---} \\ \text{---} & \mathbb{1}_3 \end{array}\right] \otimes \mathbb{1}_{r_8} \ldots$, where — denotes a zero row and $[\,|\,]$ the pair with 1 row and no column.

The preceding algorithm stops after step N if $C_N = 0$ and $D_N = 0$. The pair $[E_N\ F_N\,|\,0\ \mathbb{1}]$ then splits off at the right lower corner. Thus we are reduced to the case of a pair $[\bar{A}\,|\,\bar{B}]$ where $\bar{B} \in {}'\mathbb{M}$. Applying the preceding algorithm to $[\bar{A}^T\,|\,\bar{B}^T]$, we infer that $[\bar{A}\,|\,\bar{B}]$ splits into a pair of matrices without row, pairs of the form $[\mathbb{1}_n\ \ 0\,|\,0\ \ \mathbb{1}_n] \otimes \mathbb{1}_{s_n}$ ($n \geqslant 1$; 0 denotes zero columns) and a pair $[A'\,|\,\mathbb{1}_m]$. Transforming A' by conjugations, we finally attain the following theorem.

$$XAY^{-1} = \left[\begin{array}{ccc:c} A_1 & 0 & 0 & \\ 0 & A_2 & 0 & \\ 0 & 0 & A_3 & \\ \hdashline & & & \end{array}\right] \qquad XBY^{-1} = \left[\begin{array}{ccc:c} B_1 & 0 & 0 & \\ 0 & B_2 & 0 & \\ 0 & 0 & B_3 & \\ \hdashline & & & \end{array}\right]$$

Fig. 16

Theorem.[10] *Given any two matrices $A, B \in k^{m\times u}$, there exist invertible matrices $X \in GL_m$, $Y \in GL_u$ such that* XAY^{-1} and XBY^{-1} *have the block-forms of* Fig. 16, *where the pairs (A_i, B_i) are chosen among the following pairs (U, V)*:

a) $U = \begin{bmatrix} \mathbb{1}_n \\ 0 \end{bmatrix}$ $\quad V = \begin{bmatrix} 0 \\ \mathbb{1}_n \end{bmatrix} \in k^{(n+1)\times n}$ $\quad n \geqslant 0$.

b) $U = \mathbb{1}_n$ $\quad V = J_n^T$ $\quad n \geqslant 1$.

c) $U = s\mathbb{1}_n + J_n$ $\quad V = \mathbb{1}_n$ $\quad n \geqslant 1, s \in k$.

d) $U = [\mathbb{1}_n\ 0]$ $\quad V = [0\ \mathbb{1}_n] \in k^{n\times(n+1)}$ $\quad n \geqslant 0$.

1.9. In some of the linear problems of the present section, we succeeded in reducing the examined matrices to "normal" forms which naturally "split" into smaller entities. We did not discuss the nature of these entities, nor did we touch the question of the unicity of the attained normal forms. This will be done in Sect. 3 within the more general frame described in Sect. 2.

1.10. Remarks and References

1. In our terminology, a matrix A is defined by two numbers $m, n \in \mathbb{N}$ *and* a family of entries A_{ij}, $1 \leqslant i \leqslant m$, $1 \leqslant j \leqslant n$.
2. *poset* = partially ordered set; *chain* = totally ordered poset; *antichain* = poset whose elements are pairwise incomparable.
3. [91, 92, Kleiner, 1972].
4. [140, Shafarevich, 1972, I, §4, n° 1].
5. [56, Gabriel, 1974], [138, Sergejchuk, 1987].
6. The first inequality is due to the surjection $(D'/k') \times \bar{F} \to \mathfrak{M}$ which is induced by the action of D'. For the rest, let us just observe that an element of $\mathfrak{M}$ is schurian if and only if its orbit is identified with D'/k' (compare with 7.3).
7. A *bimodule* over S and R is an abelian group $\mathfrak{M}$ equipped with a left S-module and a right R-module structure such that $s(mr) = (sm)r$ for all $s \in S$, $m \in \mathfrak{M}$, $r \in R$.
8. [147, Weierstrass, 1868], [80, Jordan, 1870]
9. If $A \in k^{m\times n}$ and $B \in k^{p\times q}$, the Kronecker product $A \otimes B \in k^{mp\times nq}$ consists of mn blocks $A_{ij}B \in k^{p\times q}$.
10. For the first solution of this problem raised by Weierstrass, see [95, Kronecker, 1890].

2. Algebras, Modules and Categories

Categories are part of contemporary representation theory. They provide a language as well as objects of investigation. They arise as natural generalizations of algebras as well as of various categories of modules. They are indispensable for the combinatorial descriptions of algebras and modules which we shall produce.

In the present section we introduce the basic categorical notions and describe their relations with classical algebra.

Throughout Sect. 2, k denotes a fixed commutative ring.

2.1. A *k-category* is a category[1] $\mathscr{A}$ whose morphism sets, here denoted by $\mathrm{Hom}(X, Y)$ or $\mathscr{A}(X, Y)$, are endowed with k-module structures such that the composition maps are k-bilinear. A *k-functor* between two k-categories $\mathscr{A}$ and $\mathscr{B}$ is a functor[2] $F: \mathscr{A} \to \mathscr{B}$ whose defining maps $F(X, Y): \mathscr{A}(X, Y) \to \mathscr{B}(FX, FY)$ are k-linear for all $X, Y \in \mathscr{A}$. Unless otherwise stated, the functors between k-categories which we consider in the sequel, and especially the *equivalences*[3] of k-categories, are implicitly supposed to be k-functors.

In most cases the considered k-category will be *svelte*, i.e. its *isoclasses* (=isomorphism[4] classes) will form a set.

Example 1. Each k-algebra[5] A gives rise to a k-category which has one object Ω, the same for all A, and satisfies $\mathrm{Hom}(\Omega, \Omega) = A$. In the sequel, we shall identify k-algebras with the associated k-categories. Homomorphisms of algebras then correspond to k-functors.

Example 2. Let $(D', \mathfrak{M})$ be a linear matrix problem of size $m \times n$ (1.5). Then $\mathfrak{M}$ is the set of objects of a k-category whose morphisms $A \to B$ are furnished by the pairs $(X, Y) \in D$ such that $XA = BY$ $(A, B \in \mathfrak{M} \subset k^{m \times n}, X \in k^{m \times m}, Y \in k^{n \times n})$. The composition is induced by matrix multiplication.

Example 3. In view of Example 1, k-categories generalize k-algebras. The generalization carries over to (two-sided) ideals: An *ideal* $\mathscr{I}$ of a k-category $\mathscr{A}$ is a family of subgroups $\mathscr{I}(X, Y) \subset \mathscr{A}(X, Y)$ such that $f \in \mathscr{I}(X, Y)$ implies $gfe \in \mathscr{I}(W, Z)$ for all $e \in \mathscr{A}(W, X)$ and $g \in \mathscr{A}(Y, Z)$. Each such $\mathscr{I}$ gives rise to a *residue-k-category* $\mathscr{A}/\mathscr{I}$ which has the same objects as $\mathscr{A}$ and satisfies $(\mathscr{A}/\mathscr{I})(X, Y) = \mathscr{A}(X, Y)/\mathscr{I}(X, Y)$ for all $X, Y \in \mathscr{A}$. The canonical projections $\mathscr{A}(X, Y) \to (\mathscr{A}/\mathscr{I})(X, Y)$ give rise to the *projection-functor* $\mathscr{A} \to \mathscr{A}/\mathscr{I}$.

For instance, let $F: \mathscr{A} \to \mathscr{B}$ be a functor which is *quasisurjective*, i.e. such that the maps $F(X, Y): \mathscr{A}(X, Y) \to \mathscr{B}(FX, FY)$ are surjective and that each $Y \in \mathscr{B}$ is isomorphic to some FX. Then F induces an equivalence $\mathscr{A}/\mathscr{I} \xrightarrow{\sim} \mathscr{B}$, where $\mathscr{I}$ denotes the *kernel* of F, i.e. the ideal formed by the morphisms f such that $Ff = 0$.

***Example 4.** Suppose that k is a field. Let $\mathscr{A}$ be the k-category formed by the vector spaces V endowed with an endomorphism φ such that $\varphi^n = 0$ for some fixed $n \geqslant 1$. Let $\mathscr{B}$ be the k-category formed by the vector spaces W equipped with a sequence of subspaces $W(1) \subset W(2) \subset \cdots \subset W(n-1) \subset W$. The functor $F: \mathscr{A} \to \mathscr{B}$ which maps (V, φ) onto the space $W = \mathrm{Ker}\,\varphi$ equipped with the subspaces $W(p) = W \cap \varphi^{n-p}(V)$ is quasisurjective$_*$

Example 5. Let $\mathscr{C}$ be an arbitrary category and $k\mathscr{C}$ its *linearization*: By definition, $k\mathscr{C}$ has the same objects as $\mathscr{C}$, and its morphism-spaces $k\mathscr{C}(X, Y)$

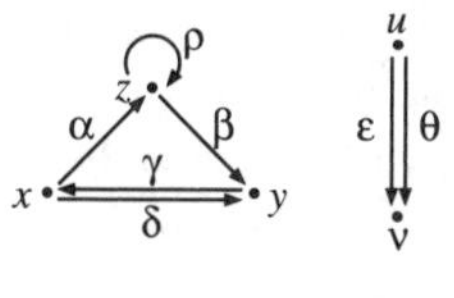

Fig. 1

are the free k-modules with bases $\mathscr{C}(X, Y)$. The definition implies that each functor $E: \mathscr{C} \to \mathscr{A}$ whose range is a k-category uniquely extends to a k-functor $F: k\mathscr{C} \to \mathscr{A}$.

More concretely, let us start with a *poset* $\mathscr{P}$. We identify $\mathscr{P}$ with a category whose objects are the elements of $\mathscr{P}$, whose morphism-sets $\mathscr{P}(x, y)$ consist of one element $(y|x)$ if $y \geqslant x$, or else are empty. The composition is such that $(z|y) \circ (y|x) = (z|x)$.

k-categories of the form $k\mathscr{P}$ provide plenty of instructive examples.

Example 6. Let Q be a *quiver*, i.e. a set of *vertices* connected by *arrows* as illustrated in Fig. 1. We denote by Q_v the set of vertices of Q, by Q_a the set of arrows. Thus we have two maps $t, h: Q_a \to Q_v$ which map an arrow α onto its *tail* $t\alpha$ and its *head* $h\alpha$ ($t\alpha = x$, $h\alpha = z$ in Fig. 1).

Given two vertices x, y of Q, we call *path of length* $d \geqslant 1$ with *origin* x and *terminus* y a sequence $\alpha_d|\ldots|\alpha_2|\alpha_1$ of arrows α_i such that $t\alpha_1 = x$, $t\alpha_{i+1} = h\alpha_i$ and $h\alpha_d = y$. Besides the paths of length $\geqslant 1$, we also consider *identical* paths $\mathbb{1}_x$ of length 0 with origin and terminus $x \in Q_v$. For instance, in the case of Fig. 1, $\beta|\rho|\rho|\alpha$ and $\delta|\gamma|\delta$ are paths with the same origin and the same terminus; we therefore call them *parallel*. Another example of parallel paths is provided by $\mathbb{1}_z$ and $\rho^{|3} = \rho|\rho|\rho$.

The quiver Q gives rise to the *category of paths* $\mathbb{P}Q$ whose set of objects is Q_v, whereas each morphism set $\mathbb{P}Q(x, y)$ consists of the paths with origin x and terminus y. The composition is the juxtaposition of paths.

The linearization $k\mathbb{P}Q$ of $\mathbb{P}Q$ will play an important rôle in the sequel. We call it the *k-category of paths* of Q and simply denote it by kQ. In the very simple case of a quiver Q with one vertex and one *loop* (= arrow whose head and tail coincide), kQ is identified with the algebra of polynomials in one variable.

If $\mathscr{I}$ is the ideal of kQ "generated" by a family of morphisms μ_i, $i \in I$, we shall say that $kQ/\mathscr{I}$ is the *k-category defined by the quiver Q and the relations* $\mu_i = 0$, $i \in I$.

Example 7. Each k-category $\mathscr{A}$ gives rise to a new k-category $|\mathscr{A}$ whose objects are the pairs (X, e) formed by an object $X \in \mathscr{A}$ and an *idempotent* $e \in \mathscr{A}(X, X)$, i.e. an endomorphism such that $e^2 = e$. The morphisms $(X, e) \to (Y, f)$ are identified with the elements of $f\mathscr{A}(X, Y)e \subset \mathscr{A}(X, Y)$.

In practice, only suitable full subcategories of $|\mathscr{A}$ are taken into consideration. For instance, if A is an algebra and $1_A = e_1 + \cdots + e_n$ a decomposition of its unit element into idempotents such that $e_i e_j = 0$ for $i \neq j$, we may consider the full subcategory of $|A$ formed by the pairs (Ω, e_i) (Example 1). More concretely, let A be the algebra of lower triangular $n \times n$ – matrices and $e_i \in A$ the matrix whose unique non-zero entry equals 1 and has index (i, i). Then the full subcategory of $|A$ formed by the (Ω, e_i) is isomorphic to the linearization of the poset $\{1 < 2 < \cdots < n\}$ (Example 5).

Example 8. For each k-category $\mathscr{A}$, we denote by $\oplus\mathscr{A}$ a k-category whose objects are the sequences $(X_1, \ldots, X_m)$ of objects of $\mathscr{A}$ $(m \in \mathbb{N})$. The morphisms $(X_1, \ldots, X_m) \to (Y_1, \ldots, Y_n)$ are identified with the "matrices" $\mu = [\mu_{ji}] \in \bigoplus_{i,j} \mathscr{A}(X_i, Y_j)$. The composition of $\oplus\mathscr{A}$ obeys the rules of matrix multiplication.
For instance, in case $\mathscr{A} = k$, the morphism-spaces of $\oplus k$ are just the matrix spaces $k^{n \times m}$.

Example 9. Let M be a *bimodule* over the algebras S, R such that $\lambda m = m\lambda$ for all $\lambda \in k$, $m \in M$. Then M gives rise to a k-category $\overline{M}$ with 2 objects A, Z and morphism spaces $\overline{M}(A, A) = R$, $\overline{M}(Z, Z) = S$, $\overline{M}(A, Z) = M$ and $\overline{M}(Z, A) = 0$.

Example 10. Let $\mathscr{C}$ be a svelte category and $\mathscr{A}$ a k-category. The category[6] $\mathscr{C}\mathscr{A}$ of the functors from $\mathscr{C}$ to $\mathscr{A}$ then carries a natural k-category structure. We are especially interested in the case where $\mathscr{C}$ is the category of paths of the quiver $\cdot \to \cdot$ or of $\cdot \to \cdot \to \cdot$. In the first case, the objects of $\mathscr{C}\mathscr{A}$ are the morphisms of $\mathscr{A}$; in the second, the objects are the *composable pairs* of morphisms.

Example 11. Let M be as in Example 9. Then the elements of $M = \overline{M}(A, Z)$ are the objects of a full subcategory $\mathscr{A}_M$ of the k-category of morphisms of $\overline{M}$. In fact, if $m, n \in M$, $\mathscr{A}_M(m, n)$ is identified with the set of pairs (s, r) such that $s \in S$, $r \in R$ and $sm = nr$ (confer Example 2).
An important particular case is the following: Let X and Y be modules over an algebra A. The *extensions*[7] of X by Y then form a bimodule $M = \mathrm{Ext}_A(Y, X)$ over $S = \mathrm{Hom}_A(X, X)$ and $R = \mathrm{Hom}_A(Y, Y)$. The associated k-category $\mathscr{A}_M$ is related to the k-category $\mathscr{S}$ of short exact sequences of A-modules of the form $0 \to X \to E \to Y \to 0$ ($\mathscr{S}$ is a full subcategory of the category of composable pairs!). The relation is provided by a functor $F\colon \mathscr{S} \to \mathscr{A}_M$ which maps each sequence onto its equivalence class in $\mathrm{Ext}_A(Y, X)$. The functor F is an *epivalence*, i.e. is quasisurjective and *detects*[8] *isomorphisms*. It therefore induces a bijection between the sets of isoclasses of $\mathscr{S}$ and $\mathscr{A}_M$.

2.2. A (right) *module* M over a k-category $\mathscr{A}$ consists of abelian groups $M(X)$, one for each object $X \in \mathscr{A}$, and of maps $M(Y) \times \mathscr{A}(X, Y) \to M(X)$, $(m, f) \mapsto mf$ which satisfy the usual axioms: $m\mathbb{1}_Y = m$, $m(fg) = (mf)g$, $(m_1 + m_2)f = m_1 f + m_2 f$ and $m(f_1 + f_2) = mf_1 + mf_2$. In particular, the maps $M(X) \times k \to M(X)$, $(m, \lambda) \mapsto m(\lambda \mathbb{1}_X)$ define a k-module structure on each group $M(X)$. The axioms above can also be expressed by saying that the maps $\mathscr{A}(X, Y) \to \mathrm{Hom}_k(M(Y), M(X))$ provided by the multiplications $(m, f) \mapsto mf$ give rise to a k-functor from the *dual*[1] $\mathscr{A}^{op}$ of $\mathscr{A}$ to the k-category of k-modules.
A *morphism* $\mu\colon M \to N$ of modules over $\mathscr{A}$ is given by group homomorphisms $\mu = \mu(Y)\colon M(Y) \to N(Y)$, one for each $Y \in \mathscr{A}$, such that $\mu(mf) = \mu(m)f$

for all $m \in M(Y)$, $f \in \mathscr{A}(Z, Y)$. The condition implies that the maps $\mu(Y)$ are k-linear.

Besides right modules, we will sometimes consider *left modules* over $\mathscr{A}$. A left module N consists of abelian groups $N(X)$ and of maps $\mathscr{A}(X, Y) \times N(X) \to N(Y)$, $(f, n) \mapsto fn$ such that $\mathbb{1}_X n = n$, $(gf)n = g(fn)$, $f(n_1 + n_2) = fn_1 + fn_2$ and $(f_1 + f_2)n = f_1 n + f_2 n$. It provides a k-functor from $\mathscr{A}$ into the k-category of k-modules.

Unless otherwise stated, we will implicitly suppose that the considered modules are right modules.

Example 1. If A is a k-algebra and M a right A-module[9] in the usual sense, we shall set $M(\Omega) = M$ and identify M with a module over the k-category associated with A (2.1, Example 1). Since the objective of this monograph is to study modules over algebras, we start from this concept and will finally return to it. But in the meantime the argumentation will require the investigation of other k-categories.

Example 2. The general notions which enter the classical "Theory of modules", such as "submodule", "factor-module" ... can easily be extended to modules over a k-category $\mathscr{A}$. A generalization of the concept of underlying module of an algebra is obtained as follows: By definition, the *module $X^\wedge$ represented by an object $X \in \mathscr{A}$* is such that $X^\wedge(Y) = \mathscr{A}(Y, X)$ for all $Y \in \mathscr{A}$. If M is an arbitrary module, each $m \in M(X)$ determines a morphism $m^\wedge : X^\wedge \to M$, the unique morphism which maps $\mathbb{1}_X \in X^\wedge(X)$ onto $m \in M(X)$. The thus defined map $m \mapsto m^\wedge$ is the *canonical isomorphism* from $M(X)$ onto the k-module $\mathrm{Hom}(X^\wedge, M)$.

Example 3. If the k-category $\mathscr{A}$ is svelte, the modules over $\mathscr{A}$ and their morphisms again form a k-category which we denote by $\mathrm{Mod}\,\mathscr{A}$. For each $M \in \mathrm{Mod}\,\mathscr{A}$, $M^\wedge$ is then a module over $\mathrm{Mod}\,\mathscr{A}$. Other modules over $\mathrm{Mod}\,\mathscr{A}$ such as $N \mapsto \mathrm{Ext}(N, M)$ also have a "right to recognition." They will be used in Sect. 9 in a very fundamental way.

The k-category $\mathrm{Mod}\,\mathscr{A}$ can be given the following more classical interpretation: Let $\mathscr{B}$ be a *full*[10] subcategory of $\mathscr{A}$ whose objects form a set intersecting each isoclass of $\mathscr{A}$. The usual matrix multiplication then equips the space $\Pi\mathscr{B} := \prod_Y \coprod_X \mathscr{B}(X, Y)$,[11] $X, Y \in \mathscr{B}$, with an algebra structure: $\sum_{Z \in \mathscr{B}} a_{YZ} b_{ZX} = (ab)_{YX} \in \mathscr{B}(X, Y)$ for all $a, b \in \Pi\mathscr{B}$. This structure provides a *fully faithful*[12] functor $\coprod_{\mathscr{B}}: \mathrm{Mod}\,\mathscr{A} \to \mathrm{Mod}\,\Pi\mathscr{B}$, $M \mapsto \coprod_{\mathscr{B}} M := \coprod_{Y \in \mathscr{B}} M(Y)$, where the multiplication by scalars is defined by $(ma)_X = \sum_{Y \in \mathscr{B}} m_Y a_{YX}$.

Besides $\Pi\mathscr{B}$ we may consider the algebra $\Pi(\mathscr{B}^{op})^{op} = \prod_X \coprod_Y \mathscr{B}(X, Y)$, which provides a fully faithful functor ${}_{\mathscr{B}}\coprod : \mathrm{Mod}\,\mathscr{A}^{op} \to \mathrm{Mod}\,\Pi(\mathscr{B}^{op})$. The left module structure on ${}_{\mathscr{B}}\coprod N := \coprod_{X \in \mathscr{B}} N(X)$ is now given by $(an)_Y = \sum_{X \in \mathscr{B}} a_{YX} n_X$. Of course, $\Pi\mathscr{B}$ coincides with $(\Pi\mathscr{B}^{op})^{op}$ if the number of objects of $\mathscr{B}$ is finite. In this case, the functors $\coprod_{\mathscr{B}}$ and ${}_{\mathscr{B}}\coprod$ are equivalences, and we call $\Pi\mathscr{B}$ *the algebra of* $\mathscr{B}$. In the particular case where $\mathscr{A} = \mathscr{B} = kQ$ is the k-category of paths of a quiver Q with finitely many vertices, we also say that $\Pi\mathscr{B}$ is *the algebra of the quiver* Q ....

Example 4. A *representation* of a quiver Q over k consists of k-modules $V(x)$, one for each $x \in Q_v$, and of k-linear maps $V(\alpha)\colon V(x) \to V(y)$, one for each arrow $x \xrightarrow{\alpha} y$. The maps $x \mapsto V(x)$ and $\alpha \mapsto V(\alpha)$ uniquely extend to a k-functor $kQ \to \operatorname{Mod} k$, which we still denote by V. Thus we identify representations of Q with *left* modules over kQ. The identification allows us to transfer the notion of a morphism from left modules to representations: A morphism $f\colon V \to W$ between the representations V and W consists of k-linear maps $f(x)\colon V(x) \to W(x)$, $x \in Q_v$, such that $W(\alpha)f(x) = f(y)V(\alpha)$ for each arrow $x \xrightarrow{\alpha} y$.

For instance, a representation V of $\omega\cdot\!\circlearrowleft\!\alpha$ is a k-module $M = V(\omega)$ together with an endomorphism $a = V(\alpha)$. If k is an algebraically closed field and M has finite dimension, the theorem of Jordan-Weierstrass states that a is represented in some basis of M by a matrix A having the block-form of Fig. 2, where the symbols 0 stand for zero-matrices and the A_i for "Jordan-blocks" $s_i \mathbb{1}_{n_i} + J_{n_i}$ (1.7).

$$A = \begin{bmatrix} A_1 & 0 & 0 & \vdots \\ 0 & A_2 & 0 & \vdots \\ 0 & 0 & A_3 & \vdots \\ \cdots & \cdots & \cdots & \end{bmatrix}$$

Fig. 2

Similarly, if k is a field, a representation of the quiver $\cdot \rightrightarrows \cdot$ is given by two vector spaces U, W and two linear maps $a, b\colon U \rightrightarrows W$. If k is algebraically closed and if the dimensions of U and W are finite, the theorem of Kronecker guarantees the existence of bases of U and W in which a, b are represented by the block-form matrices of 1.8, Fig. 16.

In the general case, if $\mathscr{I}$ is the ideal of kQ generated by morphisms μ_i, $i \in I$, we shall identify representations V of Q satisfying $V(\mu_i) = 0$, $\forall i \in I$, with left modules over $kQ/\mathscr{I}$.

***Example 5.** A linear time-invariant dynamical system[13] is a system of ordinary differential equations $\dot{x}(t) = Bx(t) + Au(t)$, $y(t) = Cx(t)$, where u, x, y are functions of $t \in \mathbb{R}$ with values in $\mathbb{C}^m$, $\mathbb{C}^n$ and $\mathbb{C}^p$ respectively, and $A \in \mathbb{C}^{n\times m}$, $B \in \mathbb{C}^{n\times n}$, $C \in \mathbb{C}^{p\times n}$. This system is associated with the representation $\mathbb{C}^m \xrightarrow{A} \overset{B\,\circlearrowleft}{\mathbb{C}^n} \xrightarrow{C} \mathbb{C}^p$ of the quiver $\cdot \to \overset{\circlearrowleft}{\cdot} \to \cdot$.

Given two linear time-invariant dynamical systems, each isomorphism between the associated representations provides a bijection between the spaces of solutions. A classification of the representations up to isomorphism is desirable therefore. Unfortunately, in the case of $\cdot \to \overset{\circlearrowleft}{\cdot} \to \cdot$, there is no known substitute for the theorems of Jordan-Weierstrass and Kronecker$_*$

Example 6. Let $\mathscr{A}$ be a svelte k-category, $|\mathscr{A}$ the category of idempotents of $\mathscr{A}$ and $E\colon \mathscr{A} \to |\mathscr{A}$, $X \mapsto (X, \mathbb{1}_X)$ the canonical embedding (2.1, Example 7). The restriction-functor $\operatorname{Mod}|\mathscr{A} \to \operatorname{Mod}\mathscr{A}$, $M \mapsto ME$ is an equivalence.

Example 7. Let $\mathscr{A}$ be a svelte k-category and $F: X \mapsto (X)$ the canonical embedding of $\mathscr{A}$ into the category $\oplus\mathscr{A}$ of 2.1, Example 8. The restriction-functor $\operatorname{Mod} \oplus \mathscr{A} \to \operatorname{Mod} \mathscr{A}$, $N \mapsto NF$ is an equivalence.

Example 8. Let M be a bimodule over S and R as in 2.1, Example 9. A module over the k-category $\overline{M}$ is then determined by a triple (V, φ, U), where V is an S-module, U an R-module and $\varphi: V \otimes_S M \to U$ an R-linear map. And a morphism from (V, φ, U) to (V', φ', U') is determined by an S-linear map $f: V \to V'$ and an R-linear map $g: U \to U'$ such that $\varphi'(f \otimes_S M) = g\varphi$.

***Example 9.** Let T be a topological space and $\mathcal{O}$ a presheaf[14] of k-algebras over T. With T and $\mathcal{O}$ we associate a k-category $\mathscr{A}$ whose objects are the open subsets of T, whose morphism-spaces are such that $\mathscr{A}(U, V) \xrightarrow{\sim} \mathcal{O}(U)$ if $U \subset V$ and $\mathscr{A}(U, V) = 0$ if $U \not\subset V$, whose composition is defined by $v \circ \mu = (v|U)\mu$ if $\mu \in \mathscr{A}(U, V)$, $v \in \mathscr{A}(V, W)$ and $U \subset V \subset W$. The k-category $\operatorname{Mod} \mathscr{A}$ is then isomorphic to the k-category of presheaves of modules over $\mathcal{O}_*$

2.3. Additive k-Categories. Let $\mathscr{A}$ be a k-category. In order to fix our terminology, we recall the definition of the (direct) sum of two objects $X, Y \in \mathscr{A}$: First we call *summation* of X and Y in $\mathscr{A}$ a quintuplet (S, i, j, p, q) consisting of an object $S \in \mathscr{A}$ and of morphisms $X \underset{p}{\overset{i}{\rightleftarrows}} S \underset{q}{\overset{j}{\leftrightarrows}} Y$ such that $pi = \mathbb{1}_X$, $qj = \mathbb{1}_Y$ and $ip + jq = \mathbb{1}_S$. Such summations are known to be "unique up to uniquely determined isomorphisms"[15]. Therefore, whenever a summation of X and Y exists, we will suppose that a "canonical" one has been chosen. The object S is then called *the sum*[16] of X and Y in $\mathscr{A}$ and is denoted by $X \oplus Y$; the morphisms p, q are called *projections*, the morphisms i, j *immersions*.

The k-category $\mathscr{A}$ is called *additive* if $X \oplus Y$ exists for all $X, Y \in \mathscr{A}$ and if $\mathscr{A}$ contains a *null* object, i.e. an 0 such that $\mathbb{1}_0 = 0$.

Example 1. The k-category $\operatorname{Mod} \mathscr{B}$ is additive for each svelte k-category $\mathscr{B}$. The sum of two modules X, Y then coincides with the cartesian product $X \times Y$ defined by $(X \times Y)(A) = X(A) \times Y(A)$.

Example 2. In fact, most of the additive k-categories to be considered in the sequel are full[10] subcategories $\mathscr{A}$ of some $\operatorname{Mod} \mathscr{B}$ which contain a null module and are such that $X, Y \in \mathscr{A}$ implies $X \times Y \in \mathscr{A}$.

For instance, if $\mathscr{P}$ is a poset, we call *$\mathscr{P}$-space* a k-module V together with a family of submodules $V(s)$, $s \in \mathscr{P}$, such that $V(s) \subset V(t)$ if $s \leqslant t$. A morphism of $\mathscr{P}$-spaces $V \to W$ is given by a k-linear map $f: V \to W$ such that $f(V(s)) \subset W(s)$ for all $s \in \mathscr{P}$. The category of $\mathscr{P}$-spaces is identified with a full additive subcategory of the category of *left* modules over $k\mathscr{P}^*$ (2.1, Example 5), where the poset $\mathscr{P}^*$ is obtained from $\mathscr{P}$ by addition of a maximum. In the particular case $\mathscr{P} = \{1 < 2 < \cdots < n-1\}$, $\mathscr{P}$-spaces were already considered in 2.1, Example 4.

Example 3. A module over a svelte k-category $\mathscr{B}$ is called *finitely free* if it is isomorphic to a finite sum[17] of modules of the form $X^\wedge$. The finitely free modules form a full additive subcategory of $\operatorname{Mod} \mathscr{B}$. This subcategory is equivalent to the "*additive hull*" $\oplus\mathscr{B}$ of $\mathscr{B}$ (map an object $(X_1, \ldots, X_n)$ of $\oplus\mathscr{B}$ onto $X_1^\wedge \oplus \cdots \oplus X_n^\wedge$; 2.1, Example 8).

Example 4. For the moment, we are primarily interested in the full additive subcategory pro $\mathscr{B}$ of Mod $\mathscr{B}$ formed by the *finitely projective* modules, i.e. by the *summands*[18] of finitely free modules. The category pro $\mathscr{B}$ is equivalent to $|\oplus\mathscr{B}$ (map an object $((X_1, \ldots, X_n), e)$ onto the image of the endomorphism of $X_1^\wedge \oplus \cdots \oplus X_n^\wedge$ associated with the idempotent e; 2.1, Example 7).

The maps $\mathscr{B}(X, Y) \to \mathrm{Hom}(X^\wedge, Y^\wedge)$, $m \mapsto m^\wedge$ give rise to the *canonical embedding* $?^\wedge: \mathscr{B} \to \text{pro } \mathscr{B}$, $X \mapsto X^\wedge$, which is an isomorphism of $\mathscr{B}$ onto a full subcategory of pro $\mathscr{B}$. The induced restriction-functor Mod(pro $\mathscr{B}$) $\to$ Mod $\mathscr{B}$, $M \mapsto M \circ ?^\wedge$ is an equivalence. And the "inverse image map" yields a bijection between the ideals of pro $\mathscr{B}$ and those of $\mathscr{B}$.

Remark We say that an idempotent $e \in \mathscr{B}(Z, Z)$ *splits* if there is an isomorphism $u: Z \tilde{\to} X \oplus Y$ such that[19] $ueu^{-1} = \begin{bmatrix} \mathbb{1} & 0 \\ 0 & 0 \end{bmatrix}$. Idempotents always split in pro $\mathscr{B}$, but not necessarily within the category of finitely free modules. In fact, the canonical embedding $?^\wedge: \mathscr{B} \to \text{pro } \mathscr{B}$ is an equivalence if and only if $\mathscr{B}$ is additive and "*has splitting idempotents.*"

Example 5. A *finite presentation* of a module $M \in$ Mod $\mathscr{B}$, $\mathscr{B}$ as above, is an exact sequence $F_1 \to F_0 \to M \to 0$ of Mod $\mathscr{B}$ such that F_0 and F_1 are finitely free. The *finitely presented* modules, i.e. the modules admitting a finite presentation, form a full additive subcategory mod $\mathscr{B}$ of Mod $\mathscr{B}$. Idempotents of mod $\mathscr{B}$ split.

In compliance with this definition, we say that a representation V of a quiver Q is finitely presented if so is the corresponding module over kQ. We denote by rep Q the category formed by these finitely presented representations.

Example 6. Let $\mathscr{P} = \{p_1, \ldots, p_t\}$ be a poset. We denote by rep $\mathscr{P}$ a k-category whose objects are the representations of $\mathscr{P}$, i.e. the pairs (d, A) formed by a dimension-vector $d \in \mathbb{N}^{1+t}$ and a matrix $A \in k^{d_0 \times \bar{d}}$ (1.3). A morphism $(d, A) \to (e, B)$ is defined by a pair $(X, Y) \in k^{e_0 \times d_0} \times k^{\bar{e} \times \bar{d}}$ such that $XA = BY$ and $\bar{Y}_{ij} = 0$ in case $p_i \nleqslant p_j$, where $\bar{Y}_{ij} \in k^{e_i \times d_j}$ denotes the block with index (i, j) in the natural subdivision of Y $(1 \leqslant i, j \leqslant t$; compare with 1.4).

The k-category rep $\mathscr{P}$ is additive. We illustrate our construction of the direct sum $(d + e, C)$ of two representations (d, A) and (e, B) with an example (Fig. 3).

$$d = [2\ 1\ 1\ 1\ 1]^T \qquad e = [1\ 1\ 1\ 0\ 0]^T \qquad d + e = [3\ 2\ 2\ 1\ 1]^T$$

$$A = \left[\begin{array}{c|c|c|c} a_1 & a_2 & a_3 & a_4 \\ a_1' & a_2' & a_3' & a_4' \end{array}\right] \qquad B = [b_1 \,|\, b_2 \,|\,|\,]$$

$$C = \left[\begin{array}{cc|cc|c|c} a_1 & 0 & a_2 & 0 & a_3 & a_4 \\ a_1' & 0 & a_2' & 0 & a_3' & a_4' \\ 0 & b_1 & 0 & b_2 & 0 & 0 \end{array}\right]$$

Fig. 3

2.4. Isoclasses. Representation theory in its broad acceptation signifies the investigation of additive k-categories. One of the main problems raised by the datum of such a category $\mathscr{A}$ is the description of its isoclasses. If $\mathscr{A}$ is svelte, these isoclasses form a *commutative semigroup*[20] Is $\mathscr{A}$, the sum of the isoclasses of X and of Y being the isoclass of $X \oplus Y$.

Example 1. If k is a field, Is(mod k) is isomorphic to $\mathbb{N}$.

Example 2. If k is an algebraically closed field and Q the quiver $\omega \circlearrowleft^{\alpha}$, Is(rep Q) is isomorphic to the free commutative semigroup generated by the isoclasses of the free module $^\wedge\omega = kQ(\omega, ?)$ and of the representations associated with the various Jordan blocks (2.2, Example 4).

Example 3. If k is algebraically closed and Q the quiver $\cdot \rightrightarrows \cdot$, Is(rep Q) is isomorphic to the free commutative semigroup generated by the isoclasses of the representations associated with the pairs of matrices of Theorem 1.8.

Example 4. Is(mod $\mathbb{Z}$) is isomorphic to the free commutative semigroup generated by the isoclasses of $\mathbb{Z}$ and $\mathbb{Z}/p^n$, where p ranges over the prime numbers and $n = 1, 2, 3, \ldots$.

Example 5. In the preceding cases Is $\mathscr{A}$ is free. It is easy to exhibit examples, where this is not so. For instance, if S is an arbitrary subsemigroup of Is $\mathscr{A}$, we have $S = \text{Is}\, \mathscr{B}$, where $\mathscr{B}$ denotes the full additive subcategory of $\mathscr{A}$ formed by the objects of $\mathscr{A}$ with isoclass in S.

***Example 6.** Another classical example is the following: Let $\mathscr{A}$ be the category of morphisms (2.1, Example 10) of pro $\mathbb{Z}$ and denote by $[0]$, $[n]$, $[\infty]$ the isoclasses of $\mathbb{Z} \to 0$, $\mathbb{Z} \xrightarrow{n?} \mathbb{Z}$, $0 \to \mathbb{Z}$ respectively, where $n = 1, 2, 3, \ldots$. Then Is $\mathscr{A}$ is isomorphic to the commutative semigroup defined by the generators $[0]$, $[n]$, $[\infty]$ and the relations $[m] + [n] = [m \vee n] + [m \wedge n]$, where $m \vee n$ denotes the greatest common divisor and $m \wedge n$ the least common multiple of $m, n \in \{1, 2, 3, \ldots\}$ *

***Example 7.** Let A be a noetherian commutative k-algebra with connected spectrum and Krull-dimension 1. The indecomposables of pro A are then projective modules of rank 1. Each non-zero $M \in$ pro A has the form $M \xrightarrow{\sim} A^{r-1} \oplus P$ where r is the rank of M and P has rank 1. The isoclasses of projectives of rank 1 form a group C for the multiplication induced by $(P, Q) \mapsto P \otimes_A Q$. And Is(pro A) is isomorphic to the quotient of the semigroup $\mathbb{N} \times C$ obtained by shrinking $\{0\} \times C$ to one point *

2.5. Modular Equivalence.[21] Two svelte k-categories $\mathscr{A}$ and $\mathscr{B}$ are called *modularly* (k)-*equivalent* if the associated additive k-categories pro $\mathscr{A}$ and pro $\mathscr{B}$ are k-equivalent. It is easy to see that this holds if and only if Mod $\mathscr{A}$ and Mod $\mathscr{B}$ are k-equivalent. For instance, $\mathscr{A}$ is modularly equivalent to the k-categories $|\mathscr{A}$ (2.1, Example 7), $\oplus\mathscr{A}$ (2.1, Example 8) and pro $\mathscr{A}$.

In our monograph, modular equivalence is intimately related to the notion of an indecomposable object: An object X of a k-category $\mathscr{A}$ will be called absolutely indecomposable or simply *indecomposable* if the endomorphism algebra $\mathscr{A}(X, X)$ has precisely two idempotents, namely 0 and $\mathbb{1}_X \neq 0$. Indecomposability depends only on $\mathscr{A}(X, X)$, not on the surrounding category $\mathscr{A}$. For instance, if $\mathscr{A}$ consists of the free k-modules of rank 0 or $\geqslant 2$, k^2 will not be called indecomposable although it admits no proper sum decomposition within $\mathscr{A}$. But of course, if the idempotents split in $\mathscr{A}$, the indecomposability of $X \in \mathscr{A}$ means that X is not zero and not isomorphic to a sum $Y \oplus Z$ where $Y \neq 0 \neq Z$. In the case of a general k-category $\mathscr{A}$, it means that $X \neq 0$ and that $X^\wedge$ admits no "proper" sum decomposition within pro $\mathscr{A}$.

We are especially interested in the case where $\mathscr{A}$ is additive and each object is isomorphic to a finite sum of indecomposables. In this case, $\mathscr{A}$ is equivalent to $\oplus\mathscr{I}\mathscr{A}$, where *$\mathscr{I}\mathscr{A}$ denotes the full subcategory of $\mathscr{A}$ formed by representatives of the isoclasses of indecomposables.*

Example. Two k-algebras A and B are modularly equivalent if and only if B is isomorphic to the k-algebra $\mathrm{Hom}_A(P, P)$, where P is a finitely projective A-module such that, for some $n \in \mathbb{N}$, $P^n = P \oplus \cdots \oplus P$ admits A as a summand.[21] For instance, A is modularly equivalent to $A^{n \times n}$.

2.6. Remarks and References. For general algebraic or categorical notions we refer to [99, MacLane, 1963] and [141, Shafarevich, 1986].

1. [99, MacLane, chap. I, § 7], [141, Shafarevich, 20].
2. [99, MacLane, chap. I, § 8], [141, Shafarevich, 20].
3. An *equivalence* of categories is a functor $F: \mathscr{C} \to \mathscr{D}$ which admits a *quasi-inverse*, i.e. a functor $E: \mathscr{D} \to \mathscr{C}$ such that EF is isomorphic to $\mathbb{1}_{\mathscr{C}}$ and FE to $\mathbb{1}_{\mathscr{D}}$.
4. An *isoclass* is an equivalence class of objects, two objects X, Y being equivalent or *isomorphic* if there is an *isomorphism* $f: X \xrightarrow{\sim} Y$, i.e. a morphism f admitting an *inverse* $g: Y \to X$ ($gf = \mathbb{1}_X$ and $fg = \mathbb{1}_Y$).
5. Unless otherwise stated, *rings* and *algebras* are supposed to be associative and to have unit elements. A homomorphism of rings or algebras is supposed to map the unit element onto the unit element.
6. The functors from $\mathscr{C}$ to $\mathscr{A}$ give rise to a category if the morphisms between any two functors form a set. This condition is satisfied here because $\mathscr{C}$ is svelte.
7. Two short exact sequences $0 \to X \xrightarrow{i} E \xrightarrow{p} Y \to 0$ and $0 \to X \xrightarrow{j} F \xrightarrow{q} Y \to 0$ are called *equivalent* if $j = ui$ and $p = qu$ for some isomorphism $u: E \xrightarrow{\sim} F$. In our terminology, an *extension* of X by Y is an equivalence class of such sequences. Compare with [99, MacLane, chap. III, § 1], [141, Shafarevich, 12].
8. A functor F *detects* isomorphisms if each morphism u with invertible image Fu is invertible.
9. Of course, we assume that $m1_A = m$ for all $m \in M$.
10. A subcategory $\mathscr{B}$ of $\mathscr{A}$ is called *full* if $\mathscr{B}(X, Y) = \mathscr{A}(X, Y)$ for all $X, Y \in \mathscr{B}$.
11. If $(S_i)_{i \in I}$ is a family of k-modules, the (*direct*) *sum* $\coprod_{i \in I} S_i$ is formed by the families $x \in \prod_{i \in I} S_i$ such that $x_i = 0$ for all but finitely many i.
12. A functor $F: \mathscr{K} \to \mathscr{L}$ is *fully faithful* if it induces an equivalence of $\mathscr{K}$ onto a full subcategory of $\mathscr{L}$.
13. [93, Kraft, 1980/82], [143, Tannenbaum, 1981].
14. [64, Godement, 1958].
15. If (S, i, j, p, q) is a summation, (S, i, j) is a direct sum and (S, p, q) a direct product in the category $\mathscr{A}$. Confer [99, MacLane, chap. 1, § 7], [141, Shafarevich, 20].

16. Our definition is not strictly conformable to the classical terminology. What we call sum, is usually called direct sum, and what is usually called sum of two submodules will be called *supremum*.
17. Finite sums are defined inductively: $X_1 \oplus X_2 \oplus X_3 = (X_1 \oplus X_2) \oplus X_3 \ldots$.
18. Y is a *summand* of X if X is isomorphic to some $Y \oplus Z$.
19. By $\begin{bmatrix} a & b \\ c & d \end{bmatrix}: X \oplus Y \to Z \oplus T$ we denote the morphism $iap + ibq + jcp + jdq$, where p, q are the projections of $X \oplus Y$ onto X, Y and i, j the immersions of Z, T into $Z \oplus T$.
20. A *semigroup* is a set equipped with an associative composition law which admits a neutral element.
21. This is generally called *Morita-equivalence*. See [106, Morita, 1958]. There is no doubt indeed that Morita put the final abstract crown onto the notion. We therefore feel some scruple to depart from a terminology received from pole to pole. On the other hand, we use the concept mainly in the form discovered by R. Brauer, one of those teachers who offered their knowledge to students. See [117, Nesbitt/Scott, 1943].

3. Radical, Decomposition, Aggregates

Our objective in the present section is to delimit a broad class of additive categories whose semigroups of isoclasses are free commutative. The line of demarcation chosen here is based on the work of Fitting[1] and is presented in 3.1–3.3. With finiteness conditions, the class delimited by Fitting is then further reduced to the class of aggregates, which constitute one of the central notions used in the sequel.

Throughout Sect. 3, k denotes a commutative ring, which is supposed to be artinian and local in 3.5–3.7.

3.1. Locular Categories. We call a k-category $\mathscr{A}$ *multilocular*[1] if $\mathscr{A}$ is additive and svelte, and if each object is a finite sum of indecomposables with *local*[2] endomorphism algebras. The condition implies that the associated k-category $\mathscr{L} = \mathscr{I}\mathscr{A}$ (formed by representatives of the isoclasses of indecomposables) is *locular*, i.e. that $\mathscr{L}$ is svelte, that the non-invertible morphisms form an ideal (or, equivalently, that all endomorphism algebras are local) and that distinct objects are not isomorphic.

Locular k-categories provide a natural generalization of local k-algebras. In particular, all projective modules over a locular k-category are free.[3]

Example 1. If the k-category $\mathscr{B}$ is svelte, *the full subcategory* $\text{mod}_f\,\mathscr{B}$ *of* $\text{Mod}\,\mathscr{B}$ *formed by the* $\mathscr{B}$*-modules of finite length*[4] *is multilocular.*[5]

***Example 2.** Let R be a complete local noetherian commutative ring, m its maximal ideal and A an R-algebra whose underlying R-module is noetherian. Then the category $\text{mod}\,A$ of noetherian A-modules is multilocular. This is essentially due to the fact that M/Mm^n has finite length for all $M \in \text{mod}\,A$, $n \in \mathbb{N}$, and that the natural map $\text{Hom}_A(M, N) \to \varprojlim_{n \in \mathbb{N}} \text{Hom}_A(M/Mm^n, N/Nm^n)$ is bijective for all $M, N \in \text{mod}\,A_*$

3.2. The Radical. If the k-category $\mathcal{L}$ is locular, we denote by $\mathcal{R}_{\mathcal{L}}$ the ideal of $\mathcal{L}$ formed by the non-invertible morphisms. Thus we have $\mathcal{R}_{\mathcal{L}}(X, Y) = \mathcal{L}(X, Y)$ if $X, Y \in \mathcal{L}$ and $X \neq Y$, whereas $\mathcal{R}_{\mathcal{L}}(X, X)$ is the maximal ideal of the local algebra $\mathcal{L}(X, X)$. The ideal $\mathcal{R}_{\mathcal{L}}$ provides a description of the *radical* $\mathcal{R}M$ (= intersection of the maximal proper[6] submodules) of an arbitrary $\mathcal{L}$-module M: For each $X \in \mathcal{L}$, $(\mathcal{R}M)(X) = \sum_{Y \in \mathcal{L}} M(Y)\mathcal{R}_{\mathcal{L}}(X, Y)$. In particular, for each $Z \in \mathcal{L}$, $Z^\wedge/\mathcal{R}Z^\wedge$ is a simple $\mathcal{L}$-module such that $(Z^\wedge/\mathcal{R}Z^\wedge)(X) = 0$ if $X \neq Z$ and $(Z^\wedge/\mathcal{R}Z^\wedge)(Z) = \mathcal{L}(Z, Z)/\mathcal{R}_{\mathcal{L}}(Z, Z)$. The maps $Z \mapsto Z^\wedge \mapsto Z^\wedge/\mathcal{R}Z^\wedge$ *provide bijections between the sets of objects of* $\mathcal{L}$, *of isoclasses of indecomposable projective* $\mathcal{L}$*-modules and of isoclasses of simple* $\mathcal{L}$*-modules.*

Let now $\mathcal{A}$ be multilocular and $\mathcal{L} = \mathcal{I}\mathcal{A}$. Since $\mathcal{A}$ is equivalent to the k-category of finitely free $\mathcal{L}$-modules, there is a unique ideal $\mathcal{R}_{\mathcal{A}}$ of $\mathcal{A}$ whose restriction to $\mathcal{L}$ coincides with $\mathcal{R}_{\mathcal{L}}$. More precisely, we have $\mathcal{R}_{\mathcal{A}}(\oplus_i X_i, \oplus_j Y_j) \xrightarrow{\sim} \oplus_{i,j}\mathcal{R}_{\mathcal{L}}(X_i, Y_j)$ if $X_i, Y_j \in \mathcal{L} = \mathcal{I}\mathcal{A}$.

The ideals just considered in the locular and multilocular cases admit a common generalization which rests on the following proposition.

Proposition[6]. *If $f \in \mathcal{A}(X, Y)$ is a morphism of a k-category $\mathcal{A}$, the following statements are equivalent:*

(i) *$f \in L(Y)$ for each maximal (proper) left submodule L of ${}^\wedge X = \mathcal{A}(X, ?)$.*

(ii) *$\mathbb{1}_X + gf$ is invertible for each $g \in \mathcal{A}(Y, X)$.*

(iii) *$f \in R(X)$ for each maximal submodule R of $Y^\wedge = \mathcal{A}(?, Y)$.*

(iv) *$\mathbb{1}_Y + fg$ is invertible for each $g \in \mathcal{A}(Y, X)$.*

Morphisms f satisfying (i)–(iv) are called *radical morphisms*. They constitute an ideal, the *radical*[7] $\mathcal{R}_{\mathcal{A}}$ of $\mathcal{A}$, whose importance lies in the following observation: An ideal $\mathcal{I}$ of $\mathcal{A}$ is contained in $\mathcal{R}_{\mathcal{A}}$ if and only if the canonical projection $P: \mathcal{A} \to \mathcal{A}/\mathcal{I}$ *detects isomorphisms*. It follows that a k-functor $F: \mathcal{A} \to \mathcal{B}$ provides a bijection between the "sets" of isoclasses of $\mathcal{A}$ and $\mathcal{B}$ if it satisfies $\operatorname{Ker} F \subset \mathcal{R}_{\mathcal{A}}$ and induces an equivalence $\mathcal{A}/\operatorname{Ker} F \xrightarrow{\sim} \mathcal{B}$, i.e. if F is an *epivalence* (2.1, Example 11).

Example 1. In 2.1, Example 4, $\mathcal{R}_{\mathcal{A}}$ consists of the morphisms $f: (U, \tau) \to (V, \varphi)$ such that $f(\operatorname{Ker} \tau \cap \operatorname{Im} \tau^{n-p}) \subset \operatorname{Ker} \varphi \cap \operatorname{Im} \varphi^{n-p+1}$ for all $p \in \{1, \ldots, n\}$. The functor F considered there is an epivalence.

Example 2. Let $\mathcal{B}$ be a svelte k-category and $\mathcal{A}$ the full subcategory of $\operatorname{mod}_f \mathcal{B}$ formed by the indecomposables of length $\leqslant d$. We claim that *the radical* $\mathcal{R}_{\mathcal{A}}$ *satisfies* $\mathcal{R}_{\mathcal{A}}^{2^d-1} = 0$[8] (the product $\mathcal{I}\mathcal{J}$ of two ideals is defined by $(\mathcal{I}\mathcal{J})(X, Y) = \sum_{Z \in \mathcal{A}} \mathcal{I}(Z, Y)\mathcal{J}(X, Z)$).

To prove this, we must show that a sequence $U_1 \xrightarrow{f_1} U_2 - \cdots \to U_{n-1} \xrightarrow{f_{n-1}} U_n$ of radical morphisms between indecomposables $U_i \in \mathcal{A}$ satisfies $f_{n-1} \cdots f_2 f_1 = 0$ if $n \geqslant 2^d$: We proceed by induction on the lexicographically ordered pairs (d, δ), where δ is the number of f_i which are neither injective nor surjective. The claim is clear if $d \leqslant 1$. If $d > 1$ and $\delta = 0$, the equality $l(U_i) = d$ implies $l(U_{i-1}) < d$ and $l(U_{i+1}) < d$. It follows that the subsequence formed by the U_i of length $\leqslant d - 1$

has at least 2^{d-1} terms, hence that the morphisms connecting them have composition 0.

Finally, if $d > 1$ and $\delta > 0$, let f_i be neither surjective nor injective. Decomposing $\operatorname{Im} f_i = \oplus_j V_j$ into indecomposables V_j, we obtain $f_i = \sum_j f_j'' f_j'$, where f_j': $U_i \to V_j$ is surjective and f_j'': $V_j \to U_{i+1}$ injective. Since the number of non-injective and non-surjective terms in each sequence $f_1, \ldots, f_j', f_j'', \ldots, f_{n-1}$ is $\delta - 1$, we infer that $f_{n-1} \cdots f_i \cdots f_1 = \sum_j f_{n-1} \cdots f_j'' f_j' \cdots f_1 = 0$.

3.3. Multilocular Categories

Theorem[9]. *If $\mathscr{A}$ is an additive svelte k-category, the following statements are equivalent*:

(i) *$\mathscr{A}$ is multilocular.*

(ii) *$\mathscr{A}/\mathscr{R}_{\mathscr{A}}$ is equivalent to a direct sum*[10] *of categories* mod K_i, *where $(K_i)_{i \in I}$ is a family of division algebras.*

(iii) *For each $f \in \mathscr{A}(X, Y)$, there are isomorphisms u: $X_0 \oplus X_1 \tilde{\to} X$ and v: $Y \tilde{\to} Y_0 \oplus Y_1$ such that $vfu = f_0 \oplus f_1$, where $f_1 \in \mathscr{A}(X_1, Y_1)$ is invertible and $f_0 \in \mathscr{R}_{\mathscr{A}}(X_0, Y_0)$. Moreover, splitting chains stop: If $L_0 \xrightarrow{r_1} L_1 \xrightarrow{r_2} L_2 \to \cdots$ is an infinite sequence of retractions*[11], *the r_n are invertible for large n.*

(iv) *All idempotents split. Moreover, for each $X \in \mathscr{A}$, $\mathscr{A}(X, X)/\mathscr{R}_{\mathscr{A}}(X, X)$ is a semisimple ring whose idempotents are images of idempotents of $\mathscr{A}(X, X)$.*

The basic features of a multilocular category $\mathscr{A}$ derive from statement (ii) above and the fact that the canonical projection P: $\mathscr{A} \to \bar{\mathscr{A}} = \mathscr{A}/\mathscr{R}_{\mathscr{A}}$ detects isomorphisms:

a) Each $X \in \mathscr{A}$ is isomorphic to a sum $\bigoplus_{U \in \mathscr{I}\mathscr{A}} U^{d_U}$, where $d_U = 0$ for almost all U. The exponent d_U is the dimension of the vector space $\bar{\mathscr{A}}(U, X)$ over $\bar{\mathscr{A}}(U, U)$. It is independent of the chosen "decomposition" $\bigoplus_{U \in \mathscr{I}\mathscr{A}} U^{d_U} \tilde{\to} X$ (*unicity of the summands*)[1].

b) Let $V_1, \ldots, V_t \in \mathscr{A}$ be indecomposable, $[v_1 \ \ldots \ v_t]$: $V_1 \oplus \cdots \oplus V_t \tilde{\to} X$ an isomorphism and s: $W \to X$ a *section.*[11] Then there is a sequence $i_1 < \cdots < i_p$ such that $[s \ v_{i_1} \ \ldots \ v_{i_p}]$: $W \oplus V_{i_1} \oplus \cdots \oplus V_{i_p} \to X$ is invertible (*exchange theorem*)[1].

c) Set $E_U = \mathscr{A}(U, U)$ and $\bar{E}_U = \bar{\mathscr{A}}(U, U)$ for each $U \in \mathscr{I}\mathscr{A}$. A morphism f: $\bigoplus_{U \in \mathscr{I}\mathscr{A}} U^{d_U} \to \bigoplus_{U \in \mathscr{I}\mathscr{A}} U^{c_U}$ can be represented by a matrix B subdivided into blocks B_{VU} of size $c_V \times d_U$ with coefficients in $\mathscr{A}(U, V)$. Replacing the entries of the "diagonal" blocks B_{UU} by their residue classes modulo $\mathscr{R}_{E_U}$, we obtain matrices $\bar{B}_{UU} \in \bar{E}_U^{c_U \times d_U}$. The morphism f is radical (resp. invertible, resp. a section, resp. a retraction) if and only if, for each $U \in \mathscr{I}\mathscr{A}$, $\bar{B}_{UU}$ is zero (resp. is invertible, resp. has rank d_U, resp. has rank c_U).

d) Let $\mathscr{L}$ be a locular k-category and $\mathscr{A}$ the k-category of finitely free $\mathscr{L}$-modules. Then each $X \in \mathscr{A}$ is a finite sum of indecomposables $U^\wedge$, where $U \in \mathscr{L}$. Using b), we infer that each summand of X is free and that $\mathscr{A}$ coincides with the k-category pro $\mathscr{L}$ of finitely projective $\mathscr{L}$-modules. In fact, *the "maps" $\mathscr{L} \mapsto$ pro $\mathscr{L}$ and $\mathscr{A} \mapsto \mathscr{I}\mathscr{A}$ provide a bijective correspondence between the isoclasses of locular k-categories and the equivalence classes of multilocular k-categories.*

3.4. Artinian Rings. Let A be an artinian[12] k-algebra and $\mathscr{A} = \text{pro}\, A$ the k-category of finitely projective A-modules, which is multilocular by 3.1, Example 1.

"The" locular k-category $\mathscr{I}\mathscr{A}$ associated with $\mathscr{A}$ can be described as follows: Choose a decomposition $1_A = e_1 + \cdots + e_n$ of the unit element into primitive idempotents e_i such that $e_i e_j = 0$ if $i \neq j$ (compare with 2.1, Example 7; *primitive* here means that e_i admits no further decomposition $e_i = e_i' + e_i''$ into non-zero idempotents such that $e_i' e_i'' = 0 = e_i'' e_i'$). The e_i provide a decomposition $A = e_1 A \oplus \cdots \oplus e_n A$, where $e_i A \in \mathscr{A}$ is indecomposable. The morphisms between these indecomposables can be described easily, since we dispose of bijections $e_j A e_i \xrightarrow{\sim} \text{Hom}_A(e_i A, e_j A)$, $a \mapsto a?$ (left multiplication by a).

Each indecomposable of pro A is isomorphic to a summand of a free module. By 3.3.a) it is therefore isomorphic to some $e_i A$. The only trouble is that we may have $e_i A \xrightarrow{\sim} e_j A$ for $i \neq j$. In order to construct $\mathscr{I}\mathscr{A}$, we have to pick out a convenient subsequence $e_{i_1} A, \ldots, e_{i_p} A$ which contains a representative of each isoclass. Modular equivalence (2.5) $\text{Mod}\, A \xrightarrow{\sim} \text{Mod}\, \mathscr{I}\mathscr{A}$ then maps an A-module M onto the functor $\tilde{M}$: $e_{i_s} A \mapsto M e_{i_s}$, where $M e_{i_s}$ can be identified with $\text{Hom}_A(e_{i_s} A, M)$ (map $m e_{i_s}$ onto $e_{i_s} a \mapsto m e_{i_s} a$).

The unicity of $\mathscr{I}\mathscr{A}$ up to isomorphism here means the following: If $1_A = f_1 + \cdots + f_m$ is another decomposition of 1_A into pairwise annihilating primitive idempotents, then m equals n and there exist a permutation σ and an invertible $u \in A$ such that $f_i A = u e_{\sigma i} A$. From $1_A = \sum_i u e_{\sigma i} u^{-1}$ we then conclude that $f_i = u e_{\sigma i} u^{-1}$ (*conjugacy of idempotent partitions of* 1_A).

Thus $\mathscr{I}\mathscr{A}$ has finitely many objects in the case considered here, and the exponents d_U occurring in a decomposition $\bigoplus_{U \in \mathscr{I}\mathscr{A}} U^{d_U} \xrightarrow{\sim} A$ are all $\geqslant 1$. Since we have $A \xrightarrow{\sim} \mathscr{A}(A, A) \xrightarrow{\sim} \mathscr{A}(\oplus U^{d_U}, \oplus U^{d_U}) \xrightarrow{\sim} \bigoplus_{U,V} \mathscr{A}(V^{d_V}, U^{d_U})$ (identify $a \in A$ with $x \mapsto ax$), the radical $\mathscr{R}_A$ of A is a nilpotent ideal (3.2, Example 2), $A/\mathscr{R}_A$ is semisimple and each idempotent of $A/\mathscr{R}_A$ can be lifted to an idempotent of A (3.3).

In the particular case where k is an *algebraically closed field* and A has *finite dimension* over k, each field $\bar{E}_U$ (3.3) is identified with k, and $A \xrightarrow{\sim} \mathscr{A}(\oplus U^{d_U}, \oplus U^{d_U})$ is the sum of the radical $\mathscr{R}_A \xrightarrow{\sim} (\bigoplus_U \mathscr{R}_{\mathscr{A}}(U, U)^{d_U \times d_U}) \oplus (\bigoplus_{U \neq V} \mathscr{A}(V, U)^{d_U \times d_V})$ and a supplement formed by the matrices B such that $B_{UU} \in k^{d_U \times d_U}$ and $B_{UV} = 0$ if $U \neq V$. In other words, as a vector space over k, *the finite-dimensional algebra A is the (direct) sum of its radical and a semisimple subalgebra isomorphic to* $\prod_U k^{d_U \times d_U}$.

Example 1. If k is a field and $A = k^{n \times n}$ the k-algebra of $n \times n$-matrices, we choose as $e_1, \ldots, e_n$ the diagonal matrices having one entry equal to 1 and the others to 0. The spaces $e_j A e_i \xrightarrow{\sim} \text{Hom}_A(e_i A, e_j A)$ then have dimension 1, and the $e_i A$ are all isomorphic. Therefore, $\mathscr{I}\mathscr{A}$ has one object, say $e_1 A$, with endomorphism algebra $k \xrightarrow{\sim} e_1 A e_1$. A quasi-inverse of the equivalence $\text{Mod}\, k^{n \times n} \xrightarrow{\sim} \text{Mod}\, k$, $M \mapsto M e_1$ is provided by the functor $N \mapsto N^n$, where the elements of N^n are written as rows on which $k^{n \times n}$ acts by matrix multiplication.

Example 2. If k is a field and A the algebra of lower triangular $n \times n$-matrices, we can choose $e_1, \ldots, e_n$ as in Example 1. But now we have $e_j A e_i = 0$ for $j < i$, so that $\mathscr{I}\mathscr{A}$ has n objects $e_1 A, \ldots, e_n A$.

In the general case of a finite-dimensional algebra A, the computational difficulties involved in the determination of $\mathscr{I}\mathscr{A}$ can grow enormous. In Sect. 8 below, we shall taste some flavour of this problem. But the question will not concern us further, because *our purpose is to describe the isoclasses of* $\operatorname{mod}_f A$ *in terms of* $\mathscr{I}\mathscr{A}$, *which will be taken as granted.*

3.5. Aggregates and Spectroids. Although Theorem 3.3 applies to general multilocular categories, we shall mainly apply it to "aggregates". We start here with definitions and some examples. From now onward, *the commutative ground ring k is* supposed to be *artinian and local.*

We call *dimension* of a module over a k-algebra the length of the underlying k-module. We say that a module M over a k-category $\mathscr{A}$ is *pointwise finite* if the dimension of $M(X)$ is finite for each $X \in \mathscr{A}$. We denote by $\operatorname{mod}_{pf} \mathscr{A}$ the full subcategory of $\operatorname{Mod} \mathscr{A}$ formed by the pointwise finite modules.

A k-category $\mathscr{A}$ is called *pointwise finite* if so is $X^\wedge = \mathscr{A}(?, X)$ for each $X \in \mathscr{A}$ or, equivalently, if $\mathscr{A}(X, Y)$ is finite-dimensional for all $X, Y \in \mathscr{A}$. If, moreover, $\mathscr{A}$ has finitely many objects (resp. is locular, resp. is multilocular), we say that $\mathscr{A}$ is *finite* (resp. is a *spectroid*, resp. is an *aggregate*). Among these entities, the spectroids are akin to the schemes of algebraic geometry; therefore, their objects are also called *points.*

Example 1. If $\mathscr{A}$ is a pointwise finite k-category, $\operatorname{pro} \mathscr{A}$ is an aggregate, and the full subcategory $\operatorname{sp} \mathscr{A} = \mathscr{I} \operatorname{pro} \mathscr{A}$ of $\operatorname{pro} \mathscr{A}$ formed by representatives of the isoclasses of indecomposables is a spectroid. The maps $\mathscr{A} \mapsto \operatorname{pro} \mathscr{A}$ and $\mathscr{A} \mapsto \operatorname{sp} \mathscr{A}$ provide bijective correspondences between the modular equivalence classes of pointwise finite k-categories, the equivalence classes of aggregates and the isoclasses of spectroids (2.5, 3.3d). By restriction, the natural embeddings $\mathscr{A} \to \operatorname{pro} \mathscr{A} \leftarrow \operatorname{sp} \mathscr{A}$ induce equivalences $\operatorname{Mod} \mathscr{A} \xleftarrow{\sim} \operatorname{Mod} \operatorname{pro} \mathscr{A} \xrightarrow{\sim} \operatorname{Mod} \operatorname{sp} \mathscr{A}$, $\operatorname{mod}_{pf} \mathscr{A} \xleftarrow{\sim} \operatorname{mod}_{pf} \operatorname{pro} \mathscr{A} \xrightarrow{\sim} \operatorname{mod}_{pf} \operatorname{sp} \mathscr{A} \ldots.$

If $\mathscr{A}$ is an aggregate, $\operatorname{pro} \mathscr{A}$ is equivalent to $\mathscr{A}$, and $\operatorname{sp} \mathscr{A}$ is *isomorphic* to the full subcategory $\mathscr{I}\mathscr{A}$ of $\mathscr{A}$ formed by representatives of the isoclasses of indecomposables. In the sequel, we shall call $\mathscr{I}\mathscr{A}$ a *spectroid* of $\mathscr{A}$.

Example 2. The finite-dimensional modules over a fixed k-algebra form an aggregate. Similarly, if $\mathscr{S}$ is a spectroid, the description of the simple $\mathscr{S}$-modules given in 3.2 shows that an $\mathscr{S}$-module M has a finite Jordan-Hölder series if and only if its (*total*) *dimension* $\dim M = \sum_{X \in \mathscr{S}} \dim M(X)$ is finite. It follows that the category $\operatorname{mod}_f \mathscr{S}$ formed by the $\mathscr{S}$-modules of finite length is an aggregate (3.1, Example 1).

Example 3. The k-category $\operatorname{rep} \mathscr{P}$ formed by the representations of a poset $\mathscr{P} = \{p_1, \ldots, p_t\}$ is an aggregate (see 2.3, Example 6).

Example 4. If $\mathscr{A}$ is an aggregate and $\mathscr{J}$ an ideal of $\mathscr{A}$, then $\mathscr{A}/\mathscr{J}$ is an aggregate. If $\mathscr{J}'$ is the restriction of $\mathscr{J}$ to $\mathscr{I}\mathscr{A}$, the non-zero objects of $(\mathscr{I}\mathscr{A})/\mathscr{J}'$ form a spectroid of $\mathscr{A}/\mathscr{J}$.

Example 5. Let $\mathscr{A}$ be an aggregate and $\mathscr{S}$ a full subcategory of $\mathscr{I}\mathscr{A} \subset \mathscr{A}$. Then $\mathscr{S}$ is a spectroid and is identified with $\mathscr{I}\mathscr{A}_{\mathscr{S}}$, where $\mathscr{A}_{\mathscr{S}}$ denotes the subaggregate of $\mathscr{A}$ formed by the objects whose indecomposable summands are isomorphic to objects of $\mathscr{S}$.

***Example 6.** The coherent sheaves on a complete algebraic scheme over k form an aggregate$_*$

3.6. Pointwise Finite Modules. As we already mentioned in 2.2, Example 3, the general philosophy pervading Sect. 9 below is that k-categories must be investigated together with their modules. In this spirit, we want to examine some elementary properties of modules which derive directly from the constraints imposed upon aggregates and spectroids. Let $\mathscr{S}$ be a *spectroid*: If $M \in \text{mod}_{pf}\, \mathscr{S}$ is pointwise finite, each endomorphism φ of M gives rise to decompositions $M(X) = M(X)_0 \oplus M(X)_1$, $X \in \mathscr{S}$, where $\varphi(X)$ is nilpotent on $M(X)_0$ and bijective on $M(X)_1$. These punctual decompositions provide a global decomposition $M = M_0 \oplus M_1$ such that φ is "pointwise nilpotent" on M_0 and invertible on M_1. In particular, if M is indecomposable, φ must be pointwise nilpotent or invertible. By a classical argument[5], it follows that *the endomorphism algebra of an indecomposable pointwise finite $\mathscr{S}$-module is local.*

Let us now consider decompositions[13] $\coprod_{i \in I} M_i \xrightarrow{\sim} M$ of a fixed $M \in \text{mod}_{pf}\, \mathscr{S}$: For each $X \in \mathscr{S}$, the number of indices i such that $M_i(X) \neq 0$ is $\leqslant \dim M(X)$. It easily follows by "transfinite decomposition" that there are isomorphisms $M \xleftarrow{\sim} \coprod_{i \in I} M_i \xrightarrow{\sim} \prod_{i \in I} M_i$, where each M_i is indecomposable and I may be infinite.

Example 1. Let k be a field and $\mathscr{S}^{op}$ the k-category of paths of the infinite quiver

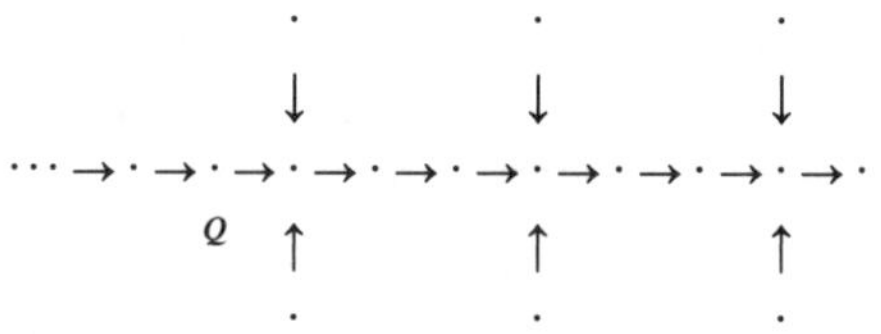

The representation of Q described in Fig. 1 is pointwise finite and indecomposable. Its endomorphism algebra is isomorphic to the algebra $k[[t]]$ of power series in one indeterminate t.

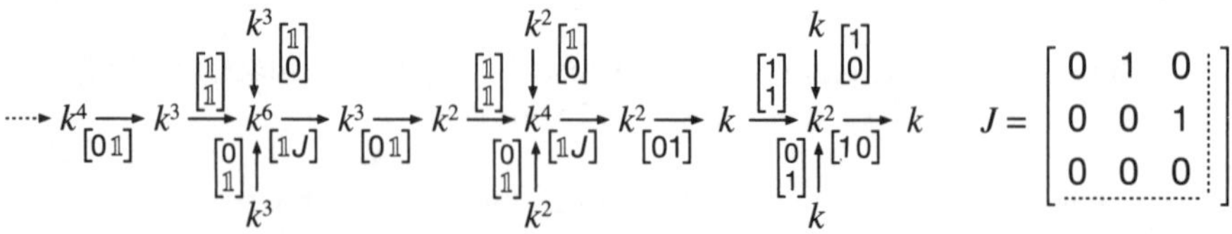

Fig. 1

Example 2. Let $\mathscr{P}$ be an arbitrary *poset* and $\mathscr{S} = k\mathscr{P}$ (2.1, Example 5). Then each *interval* I (= non-empty subset of $\mathscr{P}$ such that $x \leqslant y \leqslant z$, $x \in I$ and $z \in I$ imply $y \in I$) provides an indecomposable $\mathscr{S}$-module k_I: Set $k_I(x) = 0$ if $x \notin I$ and $k_I(y) = k$, $k_I(z|y) = \mathbb{1}_k$ if $y, z \in I$ and $y \leqslant z$.

If k is a field and $\mathscr{P}$ is linearly ordered, each indecomposable of $\text{mod}_{pf}\, k\mathscr{P}$ *is isomorphic to some k_I*: Indeed, suppose that $V \in \text{mod}_{pf}\, \mathscr{S}$ and that $0 \neq v \in V(x)$ for some $x \in \mathscr{P}$. Denote by I the interval formed by the $y \leqslant x$ such that $0 \neq v(x|y) \in V(y)$ and by the $z \geqslant x$ such that $v = w(z|x)$ for some $w \in V(z)$. Let finally $\rho\colon V(x) \to k$ be a linear retraction of $\sigma\colon k \to V(x)$, $\lambda \mapsto \lambda v$. Then we shall construct morphisms $\bar{\sigma}\colon k_I \to V$ and $\bar{\rho}\colon V \to k_I$ such that $\bar{\sigma}(x) = \sigma$, $\bar{\rho}(x) = \rho$. This will imply that $\bar{\rho}\bar{\sigma} = \mathbb{1}$ and that $\bar{\sigma}$ is invertible if V is indecomposable.

In order to construct $\bar{\sigma}$, we have to find vectors $v(y) \in V(y)$, $y \in I$, such that $v(x) = v$ and $v(z)(z|y) = v(y)$ whenever $y, z \in I$ and $y \leqslant z$. The existence of such a family $(v(y))_{y \in I}$ results from a well-known limit argument based on the finite-dimensionality of the spaces $V(y)$. The construction of $\bar{\rho}$ is obtained by duality.

3.7. Projectives and Injectives. In the sequel, we denote by m the maximal ideal of k, by $\bar{k}$ the residue-field k/m, by $\check{k}$ an injective envelope[14] of $\bar{k}$, by $D\colon (\text{mod}\, k)^{op} \to \text{mod}\, k$ the "duality" $N \mapsto DN = \text{Hom}_k(N, \check{k})$, by $\mathscr{S}$ a spectroid.

The existence of D has the following implications for $\text{mod}_{pf}\, \mathscr{S}$:

a) For each $M \in \text{mod}_{pf}\, \mathscr{S}$, the left $\mathscr{S}$-module $DM\colon X \mapsto DM(X)$ is pointwise finite. Conversely, if N belongs to $\text{mod}_{pf}\, \mathscr{S}^{op}$, the right module $DN\colon X \mapsto DN(X)$ lies in $\text{mod}_{pf}\, \mathscr{S}$. We therefore obtain quasi-inverse antiequivalences $\text{mod}_{pf}\, \mathscr{S} \underset{D}{\overset{D}{\leftrightarrows}} \text{mod}_{pf}\, \mathscr{S}^{op}$.

b) Each object $X \in \mathscr{S}$ provides two pointwise finite $\mathscr{S}$-modules, the indecomposable projective $X^\wedge$ and an indecomposable injective $X^\vee = D^\wedge X$ such that $X^\vee(Y) = D\mathscr{S}(X, Y)$. The module $X^\vee$ gives rise to natural isomorphisms $\text{Hom}(M, X^\vee) \xrightarrow{\sim} DM(X)$, $M \in \text{Mod}\, \mathscr{S}$, which are described as follows: For each $f \in \text{Hom}(M, X^\vee)$, the associated $g \in DM(X)$ is the composition $M(X) \xrightarrow{f(X)} X^\vee(X) = D\mathscr{S}(X, X) \xrightarrow{e} \check{k}$, where e denotes the evaluation at $\mathbb{1}_X \in \mathscr{S}(X, X)$. Reversely, if g is given, f maps $m \in M(Y)$ onto the linear form $\mathscr{S}(X, Y) \to M(X) \xrightarrow{g} \check{k}$, $\xi \mapsto m\xi \mapsto g(m\xi)$.

The isomorphisms $\text{Hom}(M, X^\vee) \xrightarrow{\sim} DM(X)$ show that the $\mathscr{S}$-modules $X^\vee$ are even injective within $\text{Mod}\, \mathscr{S}$ and that each $M \in \text{Mod}\, \mathscr{S}$ can be embedded into a product of such injectives.

c) The isomorphism $\text{Hom}(M, X^\vee) \xrightarrow{\sim} D\,\text{Hom}(X^\wedge, M)$[15] which results from $\text{Hom}(M, X^\vee) \xrightarrow{\sim} DM(X)$ and $M(X) \xrightarrow{\sim} \text{Hom}(X^\wedge, M)$ (2.2, Example 2) can also be described as follows: $\mu \in \text{Hom}(M, X^\vee)$ is associated with the linear form $\nu \mapsto \varepsilon(\mu\nu)$ on $\text{Hom}(X^\wedge, M)$, where ε denotes the composition $\text{Hom}(X^\wedge, X^\vee) \xrightarrow{\sim} X^\vee(X) = D\mathscr{S}(X, X) \xrightarrow{e} \check{k}$.

d) Each $X \in \mathscr{S}$ provides a simple right $\mathscr{S}$-module $X^- = X^\wedge/\mathscr{R}X^\wedge$ and a simple left $\mathscr{S}$-module ${}^-X = {}^\wedge X/\mathscr{R}{}^\wedge X$, which satisfy ${}^-X(Y) = 0 = X^-(Y)$ if $X \neq Y \in \mathscr{S}$, whereas ${}^-X(X) = X^-(X)$ is the division algebra $\bar{\mathscr{S}}(X, X) = \mathscr{S}(X, X)/\mathscr{R}_{\mathscr{S}}(X, X)$. These simple modules are coupled in the sense that each non-zero $\varphi \in D\bar{\mathscr{S}}(X, X)$ induces an isomorphism $\varphi^{\cdot}\colon X^- \xrightarrow{\sim} D^-X$ (which maps $1 \in \bar{\mathscr{S}}(X, X)$

$= X^-(X)$ onto $\varphi \in D\overline{\mathscr{S}}(X, X) = (D^- X)(X))$. Since ^-X is the unique simple quotient of $^\wedge X$, it follows that $X^\vee = D^\wedge X$ has a unique simple submodule which is identified with $D^- X$ and isomorphic to X^-. Since each proper submodule of $^\wedge X$ is contained in the kernel of $^\wedge X \to {}^-X$, each non-zero submodule of $X^\vee$ contains $D^- X \xleftarrow{\sim} X^-$. Finally, the composition

$$X^\wedge \xrightarrow{\text{can}} X^- \underset{\sim}{\overset{\varphi^\cdot}{\to}} D^- X \xrightarrow{\text{can}} D^\wedge X = X^\vee$$

coincides with the morphism attached to the form $\mathscr{S}(X, X) \xrightarrow{\text{can}} \overline{\mathscr{S}}(X, X) \xrightarrow{\varphi} \check{k}$ by the canonical isomorphism $D\mathscr{S}(X, X) = X^\vee(X) \xrightarrow{\sim} \operatorname{Hom}(X^\wedge, X^\vee)$.

Example. Suppose that $\mathscr{S} = k\mathscr{P}$, where $\mathscr{P}$ is a poset (3.6, Example 2). For each $s \in \mathscr{P}$, we then have $s^\wedge = k_{s]}$, $s^- = k_{[s]}$ and $s^\vee = k_{[s}$, where $s] = \{t \in \mathscr{P}: t \leqslant s\}$, $[s = \{t \in \mathscr{P}: s \leqslant t\}$ and $[s] = s] \cap [s$.

3.8. Remarks and References

1. [48, Fitting, 1935]. In the case of 3.1, Example 1, the unicity of the summands (3.3a) goes back to [97, Krull, 1925], the exchange theorem to [137, Schmidt, 1929]. Multilocular categories are sometimes called *Krull-Schmidt categories*.
2. A ring is *local* if the non-invertible elements form an ideal.
3. For finitely projective modules this will be proved in 3.3d) below. In the general case, it is easy to extend Kaplansky's original proof which applies to local rings [82, Kaplansky, 1958].
4. [141, Shafarevich, 1986, Chap. 9].
5. We reproduce Fitting's fundamental original proof: One first observes that each endomorphism $\varphi: M \to M$ of $\operatorname{mod}_f \mathscr{B}$ gives rise to the *Fitting-Jordan-decomposition* $M \xrightarrow{\sim} M_0 \oplus M_1$, where M_0, M_1 are φ-stable submodules of M such that φ is nilpotent on M_0 and bijective on M_1: Indeed, since the lengths $l(\operatorname{Ker} \varphi^n)$ cannot grow beyond $l(M)$, we must have $\operatorname{Ker} \varphi^n = \operatorname{Ker} \varphi^{2n}$ for some n. The equality is equivalent to $\operatorname{Ker} \varphi^n \cap \operatorname{Im} \varphi^n = \{0\}$, hence to $M = \operatorname{Ker} \varphi^n \oplus \operatorname{Im} \varphi^n$ since $l(M) = l(\operatorname{Ker} \varphi^n) + l(\operatorname{Im} \varphi^n)$. Thus we can set $M_0 = \operatorname{Ker} \varphi^n$ and $M_1 = \operatorname{Im} \varphi^n$.

Now suppose that $M \in \operatorname{mod}_f \mathscr{B}$ is non-zero and admits no proper sum decomposition. Then φ is either bijective or nilpotent. In the second case, $\varphi\psi$ and $\psi\varphi$ are nilpotent for each $\psi: M \to M$. If ψ is also nilpotent, then so is $\varphi + \psi$: Otherwise, we would have $\mathbb{1}_M = (\varphi + \psi)\chi$ for some χ, hence the contradiction $\mathbb{1}_M = \mathbb{1}_M^n = \sum_r \binom{n}{r} (\varphi\chi)^r (\psi\chi)^{n-r} = 0$ for large n. We conclude that the non-invertible endomorphisms form an ideal of the endomorphism ring E of M, hence that E is local.
6. A submodule L of a module M is *proper* if $L \neq M$. We include a proof of the proposition: The map $L \mapsto L(X)$ provides a bijection between the maximal left submodules of $^\wedge X$ and the maximal left ideals of $\mathscr{A}(X, X)$. The inverse map $I \mapsto L$ is such that $L(Y) = \{f \in \mathscr{A}(X, Y): \forall g \in \mathscr{A}(Y, X), gf \in I\}$. Therefore, (i) is equivalent to saying that $gf \in I$ for each $g \in \mathscr{A}(Y, X)$ and each maximal left ideal I. As it is easy to see, the last statement is also equivalent to the condition (*) saying that $\mathbb{1}_X + gf$ has a left inverse for each g. On the other hand, such a left inverse u satisfies $u = \mathbb{1}_X - ugf$. Thus, if (*) holds, u is also left invertible, hence invertible. This proves (i)$\Leftrightarrow$(*)$\Leftrightarrow$(ii), and (iii)$\Leftrightarrow$(iv) holds by duality.

Now each $g \in \mathscr{A}(Y, X)$ induces a morphism $^\wedge g: {}^\wedge X \to {}^\wedge Y$, which maps the intersection of all maximal left submodules of $^\wedge X$ into the corresponding intersection for $^\wedge Y$. Therefore (i) implies that $fg \in N(Y)$ for each maximal left submodule N of $^\wedge Y$. Since (i) implies (ii), it follows that $\mathbb{1}_Y + fg$ is invertible, hence that (i) implies (iv). By duality, (iii) implies (ii).
7. The radical considered here is also called *Jacobson* radical. See [123, Perlis, 1942] and [77, Jacobson, 1945].
8. [73, Harada/Sai, 1970].
9. We include a *proof*: To prove (i)$\Rightarrow$(ii), let I be the set of isoclasses of objects of $\mathscr{A}$ with local endomorphism algebra, U_i an object in $i \in I$, $\bar{\mathscr{A}}$ the residue-category $\mathscr{A}/\mathscr{R}_{\mathscr{A}}$ and $K_i = \bar{\mathscr{A}}(U_i, U_i)$. The functor $X \mapsto (\bar{\mathscr{A}}(U_i, X))_{i \in I}$ provides the wanted equivalence.

(ii) ⇒ (iii): Let $\bar{f} \in \bar{\mathscr{A}}(X, Y)$ be the image of f. Then there are isomorphisms $\bar{u}: X_0 \oplus X_1 \xrightarrow{\sim} X$ and $\bar{v}: Y \xrightarrow{\sim} Y_0 \oplus Y_1$ in $\bar{\mathscr{A}}$ such that $\bar{v}\bar{f}\bar{u} = 0 \oplus \bar{d}$, where $\bar{d}: X_1 \to Y_1$ is invertible. If u', v' and d are counter-images of $\bar{u}$, $\bar{v}$ and $\bar{d}$ in $\mathscr{A}$, this means that $v'fu'$ is represented by a matrix $\begin{bmatrix} a & b \\ c & d \end{bmatrix}$ where a, b, c are radical. It then suffices to set $u = u'\begin{bmatrix} \mathbb{1} & 0 \\ -d^{-1}c & \mathbb{1} \end{bmatrix}$ and $v = \begin{bmatrix} \mathbb{1} & -bd^{-1} \\ 0 & \mathbb{1} \end{bmatrix} v'$! The last condition of (iii) holds in $\bar{\mathscr{A}}$, hence in $\mathscr{A}$.

(iii) ⇒ (i): The chain condition implies that each $X \in \mathscr{A}$ is isomorphic to some $X_1 \oplus \cdots \oplus X_n$, where the X_i are non-zero and cannot be decomposed further. The first part of (iii) then implies that each $\mathscr{A}(X_i, X_i)$ is local.

(ii) ⇒ (iv): Set $E = \mathscr{A}(X, X)$. Then $\bar{E} = E/\mathscr{R}_E$ is identical with $\bar{\mathscr{A}}(X, X)$, hence semisimple. If $\bar{e} \in \bar{E}$ is idempotent, there is an isomorphism $\bar{u}: X_0 \oplus X_1 \xrightarrow{\sim} X$ of $\bar{\mathscr{A}}$ such that $\bar{u}^{-1}\bar{e}\bar{u} = 0 \oplus \mathbb{1}_{X_1}$. If u is a counter-image of $\bar{u}$ in $\mathscr{A}$, $u(0 \oplus \mathbb{1}_{X_1})u^{-1}$ is an idempotent of E with image $\bar{e}$.

Now let $e \in \mathscr{A}(X, X)$ be idempotent: Then there is an isomorphism $\bar{u}: X_0 \oplus X_1 \xrightarrow{\sim} X$ of $\bar{\mathscr{A}}$ such that $\bar{u}^{-1}\bar{e}\bar{u}$ has the matrix form $\begin{bmatrix} 0 & 0 \\ 0 & \mathbb{1} \end{bmatrix}$. If v is a counter-image of $\bar{u}$ in $\mathscr{A}$, we have $v^{-1}ev = \begin{bmatrix} a & b \\ c & \mathbb{1} + d \end{bmatrix}$, where a, b, c, d are radical. Taking into account that $(v^{-1}ev)^2 = v^{-1}ev$, we verify that

$$\begin{bmatrix} a & b \\ c & \mathbb{1} + d \end{bmatrix}\begin{bmatrix} \mathbb{1} - a & b \\ -c & \mathbb{1} + d \end{bmatrix} = \begin{bmatrix} \mathbb{1} - a & b \\ -c & \mathbb{1} + d \end{bmatrix}\begin{bmatrix} 0 & 0 \\ 0 & \mathbb{1} \end{bmatrix}$$

i.e. that $u^{-1}eu = \begin{bmatrix} 0 & 0 \\ 0 & \mathbb{1} \end{bmatrix}$, where $u = v\begin{bmatrix} \mathbb{1} - a & b \\ -c & \mathbb{1} + d \end{bmatrix}$.

(iv) ⇒ (i): In order to prove that X is a finite sum of objects with local endomorphism algebras, we proceed by inducion on the length $l(\bar{E})$ of $\bar{E}$ (considered as an $\bar{E}$-module). If $l(\bar{E}) = 1$, $\bar{E}$ is a field and E is local. If $l(\bar{E}) > 1$, $\bar{E}$ has a non-trivial idempotent $\bar{e}$, which can be lifted to an idempotent $e \in E$. Since e splits, there is an isomorphism $u: X_0 \oplus X_1 \xrightarrow{\sim} X$ such that $u^{-1}eu = \begin{bmatrix} 0 & 0 \\ 0 & \mathbb{1} \end{bmatrix}$. The endomorphism algebras of X_0 and X_1 are identified with $(\mathbb{1} - e)E(\mathbb{1} - e)$ and eEe, and the residue-algebras modulo the radicals with $(\mathbb{1} - \bar{e})\bar{E}(\mathbb{1} - \bar{e})$ and $\bar{e}\bar{E}\bar{e}$. Their lengths are $< l(\bar{E})$, so that induction applies to X_0 and X_1.

10. If $(\mathscr{B}_i)_{i \in I}$ is a family of additive k-categories, the objects of their *direct sum* $\mathscr{B}$ are the families $(X_i)_{i \in I}$ where $X_i \in \mathscr{B}_i$ and $X_i = 0$ for almost all i. A morphism $(f_i)_{i \in I}$ is a family of morphisms $f_i: X_i \to Y_i$.

11. A morphism $r: X \to Y$ is a *retraction* if it admits a *section*, i.e. a morphism $s: Y \to X$ such that $rs = \mathbb{1}_Y$.

12. *Artinian* here means that $A \in \text{mod}_f\, A$, i.e. that A has finite length as a right module over itself.

13. By $\coprod_{i \in I} M_i$ we denote the *direct sum*, by $\prod_{i \in I} M_i$ the *product* of a finite or infinite family $(M_i)_{i \in I}$ of objects of a category [99, MacLane, 1963, Chap. I, § 7].

14. $\check{k} = k$ if k is a field. In the general case, set $d = \dim(m/m^2)$ and $h = \min\{i: m^i = 0\}$. It is easy to prove that $\dim M \leqslant 1 + d + \cdots + d^{h-1}$ if the k-module M has only one simple submodule. Among these M, $\check{k}$ has maximal dimension. If follows that $\check{k}$ is injective and that the canonical maps $N \to DDN$ are bijective (induction on $\dim N$) [101, Matlis, 1958].

15. According to [116, Nesbitt, 1938], this formula goes back to R. Brauer.

4. Finitely Spaced Modules

The present section is devoted to one-sided linear matrix problems. Besides general statements, we present a method which reduces representations of algebras to such problems. We also introduce an algorithm which will lead us in particular to an efficient characterization of finitely represented posets (5.4).

In all sections except Sect 9, we henceforth suppose for simplicity's sake that k is an algebraically closed field.

4.1. As we have seen earlier, modules over aggregates are natural generalizations of modules over algebras. On the other hand, they also look in some sense like generalizations of posets. Let us explain this point of view, which is essential in the present section.

Consider an aggregate $\mathscr{A}$. By definition, the *maximal dimension* of a module M over $\mathscr{A}$ is the supremum of the *dimension-function* $x \mapsto \dim M(x)$ on a spectroid $\mathscr{I}\mathscr{A}$ of $\mathscr{A}$. If $\mathscr{A}$ admits a *faithful*[1] left module M with maximal dimension 1, $\mathscr{A}$ defines a poset $\mathscr{P}_{\mathscr{A}}$ whose elements are the points of $\mathscr{I}\mathscr{A}$, whose inequalities $x \leqslant y$ mean that $\mathscr{A}(x, y) \neq 0$. It is easy to see that we thus obtain an order relation which determines M up to isomorphism. In fact, the reverse construction works as follows:

Each poset $\mathscr{P}$ gives rise to an aggregate $\mathscr{A} = \oplus k\mathscr{P}$ (the "additive hull" of $k\mathscr{P}$: 2.1, Examples 5 and 8) and to a left module $M = M^{\mathscr{P}}$ over $\mathscr{A}$ such that $M(s_1 \oplus \cdots \oplus s_m) = k^m$ for each object $s_1 \oplus \cdots \oplus s_m \tilde{\rightarrow} (s_1, \ldots, s_m)$ of $\mathscr{A}$ ($s_i \in \mathscr{P}$); the value of M at the "matrix"

$$\mu\colon (s_1, \ldots, s_m) \to (t_1, \ldots, t_n)$$

with entries $\mu_{ij} = \lambda_{ij}(t_i|s_j)$ if $t_i \geqslant s_j$ and $\mu_{ij} = \lambda_{ij} = 0$ otherwise, is the linear map $k^m \to k^n$ with matrix $M(\mu) = [\lambda_{ij}] \in k^{n\times m}$.

Returning to the general case of a pointwise finite left module M over an aggregate $\mathscr{A}$, we denote by M^k the aggregate whose objects are the triples (V, f, X) where $V \in \operatorname{mod} k$, $X \in \mathscr{A}$ and $f \in \operatorname{Hom}_k(V, M(X))$. We call these triples *spaces on* M or *representations of* M. A morphism from (V, f, X) to (V', f', X') is defined by a pair of morphisms $\varphi \in \operatorname{Hom}_k(V, V')$ and $\psi \in \mathscr{A}(X, X')$ such that $f'\varphi = M(\psi)f$.

Of course, if X and V are fixed, each basis of $M(X)$ over k provides a one-sided linear matrix problem $(A' \times GL_n, k^{m\times n})$, where A consists of the matrices describing the action of $\mathscr{A}(X, X)$ on $M(X)$ and $m = \dim M(X)$, $n = \dim V$. The orbits of this matrix problem correspond bijectively to the isoclasses of spaces on M of the form (V, f, X).

In the particular case where $\mathscr{I}\mathscr{A}$ has finitely many points $p_1, \ldots, p_t$ and M is faithful with maximal dimension 1, we can choose basis vectors in the spaces $M(p_i)$ and thus obtain an equivalence

$$(\operatorname{rep} \mathscr{P}_{\mathscr{A}})^{op} \tilde{\rightarrow} M^k, \qquad (d, A) \mapsto (k^{d_0}, A^T, p_1^{d_1} \oplus \cdots \oplus p_t^{d_t}).$$

For similar equivalences in the case of maximal dimension $\geqslant 2$, we refer to the end of Sect. 4. For a generalization of the notions introduced in 4.1 and 4.2, we refer to the remarks.[2]

4.2. Let $\mathscr{L}$ be a finite set of submodules of the left module M over the aggregate $\mathscr{A}$. We say that a space (V, f, X) on M *avoids* $\mathscr{L}$ if $f^{-1}(L(X)) = \{0\}$ for each $L \in \mathscr{L}$ and that M is *finitely $\mathscr{L}$-spaced* (or *$\mathscr{L}$-represented*) if the full subcategory $M^k_{\mathscr{L}}$ of M^k formed by the spaces on M avoiding $\mathscr{L}$ has a finite spectroid. In case $\mathscr{L} = \varnothing$, we then also say that M is *finitely spaced* (or *represented*).

For *example*, let A be a finite-dimensional algebra, $\mathscr{J} \neq 0$ an ideal such that $\mathscr{M}\mathscr{J} = 0$ for some maximal ideal $\mathscr{M}$, and $\bar{A} = A/\mathscr{J}$. Let further M denote the left module $n \mapsto \mathrm{Ext}^1_A(e, n)$ over mod $\bar{A}$, where e is a simple A-module annihilated by $\mathscr{M}$. The module M gives rise to a functor

$$F\colon \mathrm{mod}\, A \to M^k, \qquad m \mapsto (\mathrm{Hom}(e, m/\bar{m}), f, \bar{m}),$$

where $\bar{m}$ is the largest submodule of m annihilated by $\mathscr{J}$ and f the image of the extension $\varepsilon \in \mathrm{Ext}^1_A(m/\bar{m}, \bar{m})$ associated with the canonical sequence

$$0 \to \bar{m} \to m \to m/\bar{m} \to 0$$

under the natural isomorphism

$$\mathrm{Ext}^1_A(m/\bar{m}, \bar{m}) \xrightarrow{\sim} \mathrm{Hom}_k(\mathrm{Hom}_A(e, m/\bar{m}), \mathrm{Ext}^1_A(e, \bar{m}))$$

($\mathscr{M}\mathscr{J} = 0$ implies that $m/\bar{m}$ is semi-simple of type e).

Now we must distinguish between *two* cases: *In case* $\mathscr{J} \subset \mathscr{R}_A$, e is annihilated by $\mathscr{J}$ and gives rise to a proper submodule $L\colon n \mapsto \mathrm{Ext}^1_{\bar{A}}(e, n)$ of M. The functor F then induces an *epivalence* mod $A \to M^k_{\{L\}}$. For instance, if $A = k \oplus kx_1 \oplus \cdots \oplus kx_t$ with $x_i x_j = 0$ for all i, j and $\mathscr{J} = \mathscr{R}_A = kx_1 \oplus \cdots \oplus kx_t$, then $M^k_{\{L\}} = M^k_{\{0\}}$ is equivalent to the category formed by the finite-dimensional representations V of the quiver $a \rightrightarrows z$ with t arrows $\alpha_1, \ldots, \alpha_t$ such that $\bigcap_i \mathrm{Ker}\, V(\alpha_i) = 0$ (see 1.8 for the case $t = 2$).

4.3. In the *second case*, where $\mathscr{J} \not\subset \mathscr{R}_A$ hence $\mathscr{J} \not\subset \mathscr{M}$, the functor $F\colon \mathrm{mod}\, A \to M^k$ is an *equivalence*. Geometry here intrudes on algebra: If $\mathscr{S}$ is a spectroid of A, the simple A-module e then corresponds to some $\mathscr{S}$-module s^-, where $s \in \mathscr{S}$. Our assumptions imply that s is a *sink* of $\mathscr{S}$ ($\mathscr{S}(s, s) = k\mathbb{1}_s$ and $\mathscr{S}(s, t) = 0$ if $t \neq s$) and that the ideal $\mathscr{J}_s$ of $\mathscr{S}$ associated with $\mathscr{J}$ satisfies $\mathscr{J}_s(t, p) = 0$ if $p \neq s$ and $\mathscr{J}_s(t, s) = \mathscr{S}(t, s)$ for all $t \in \mathscr{S}$. The full subcategory $\mathscr{T}$ of $\mathscr{S}$ formed by the points $\neq s$ is identified with a spectroid of $\bar{A}$. The natural embedding mod $\bar{A} \to$ mod A corresponds to the extension $n \mapsto n^\circ$ of $\mathscr{T}$-modules to $\mathscr{S}$ by 0, and $\mathrm{Ext}^1_{\mathscr{S}}(s^-, n^\circ)$ is identified with $\mathrm{Hom}(s^\wedge | \mathscr{T}, n) =: N(n)$. Finally, F corresponds to an *isomorphism of categories*

$$\mathrm{mod}\, \mathscr{S} \xrightarrow{\sim} N^k, \qquad m \mapsto (m(s), f, m|\mathscr{T}),$$

where f is induced by the structure maps $m(s) \otimes \mathscr{S}(t, s) \to m(t)$, $t \in \mathscr{T}$.

For instance, let $\mathscr{P} = \{p_1, \ldots, p_t\}$ be an antichain, $\mathscr{P}^*$ the poset obtained by addition of a largest element s and $\mathscr{S} = k\mathscr{P}^*$, $\mathscr{T} = k\mathscr{P}$. The $\mathscr{T}$-modules $p_i^\vee$ then form a spectroid of mod $\mathscr{T}$. Since $\dim N(p_i^\vee) = \dim D\mathscr{S}(p_i, s) = 1$, we infer that mod $\mathscr{S} \xrightarrow{\sim} N^k$ is equivalent to $M^{\mathscr{P}k}$ and $(\mathrm{rep}\, \mathscr{P})^{op}$.

Remark. Of course, geometry also gives an insight into the first case: Let again $\mathscr{S}$ denote a spectroid of A and let s^- be the simple $\mathscr{S}$-module corresponding to e. For all $p, t \in \mathscr{S}$, set $\mathscr{J}_s(t, p) = 0$ if $p \neq s$ and $\mathscr{J}_s(t, s) = \{v \in \mathscr{R}_{\mathscr{S}}(t, s)$: $\mathscr{R}_{\mathscr{S}}(s, r)v = 0, \forall r \in \mathscr{S}\}$. Then $\mathscr{J} \subset A$ corresponds to a subideal of $\mathscr{J}_s$.

4.4. As we have seen, quite a few problems of representation theory can be formulated in terms of aggregates of the form $M^k_{\mathscr{L}}$. To these we can apply the following reduction:

Starting from a proper submodule N of M which is contained in no $L \in \mathscr{L}$, we consider a new aggregate $\hat{\mathscr{A}} = N^k_{\mathscr{L} \cap N}$, where $\mathscr{L} \cap N = \{L \cap N: L \in \mathscr{L}\}$, and left modules $\hat{P}$ on $\hat{\mathscr{A}}$ associated with some submodules P of M and defined by

$$\hat{P}(W, g, X) = (g(W) + P(X))/g(W) \subset M(X)/g(W) = \hat{M}(W, g, X).$$

With $\hat{\mathscr{L}}$ we denote the set formed by the submodules $\hat{N}$ and $\hat{L}$, $L \in \mathscr{L}$, of $\hat{M}$. Thus we obtain a functor

$$F: M^k_{\mathscr{L}} \to \hat{M}^k_{\hat{\mathscr{L}}}, \qquad (V, f, X) \mapsto (V/V', f'', (V', f', X)),$$

where $V' = f^{-1}(N(X))$ is the inverse image of $N(X)$ under $f: V \to M(X)$, and $f': V' \to N(X)$, $f'': V/V' \to M(X)/f(V')$ are induced by f.

Proposition. *The functor* $F: M^k_{\mathscr{L}} \to \hat{M}^k_{\hat{\mathscr{L}}}$ *is an equivalence if* $\mathscr{L} \neq \emptyset$. *In case* $\mathscr{L} = \emptyset$, *F is an epivalence.*

The proof is straightforward. For the practical application it imports us to translate the proposition into the language of matrix problems. We illustrate the translation with an example: Let us set $\mathscr{P} = \{p_1 < p_2, p_3, p_4\}$, $\mathscr{A} = \oplus k\mathscr{P}$, $M = M^{\mathscr{P}}$ and $\mathscr{L} = \emptyset$. We examine the aggregate rep $\mathscr{P}$, which is dual to $M^{\mathscr{P}k}$. An object of rep $\mathscr{P}$ is given by a matrix A partitioned into 4 vertical stripes $A_i \in k^{d_0 \times d_i}$ such that addition of columns between different stripes is only allowed from A_1 to A_2:

$$A = [\overrightarrow{A_1 \,|\, A_2} \,|\, A_3 \,|\, A_4].$$

The submodule N to be used in our algorithm is defined by $N(p_1) = N(p_2) = 0$ and $N(p_3) = N(p_4) = k$. In order to determine $V' = f^{-1}(N(X))$, where (V, f, X) is the space on M associated with (d, A), we reduce A by row transformations to the form of Fig. 1, where $B_i \in k^{r \times d_i}$ and $r = \text{rank}\,[A_1 \,|\, A_2] = d_0 - \dim V'$. The matrices C_3, C_4 then determine a space on N isomorphic to (V', f', X) which can be decomposed into indecomposables by reducing $[C_3 \,|\, C_4]$ to the form II of 1.1. The result is illustrated by Fig. 2 (non-completed blocks are null).

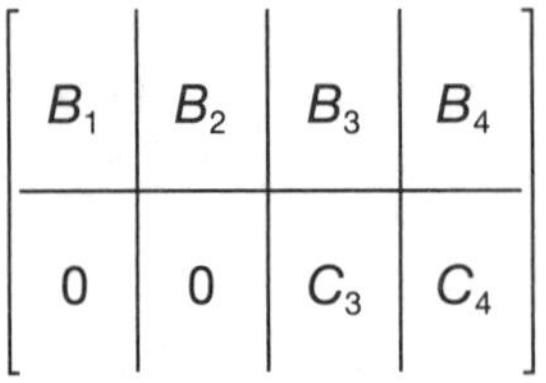

Fig. 1

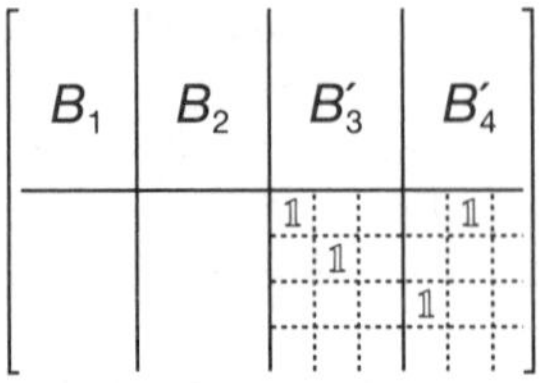

Fig. 2

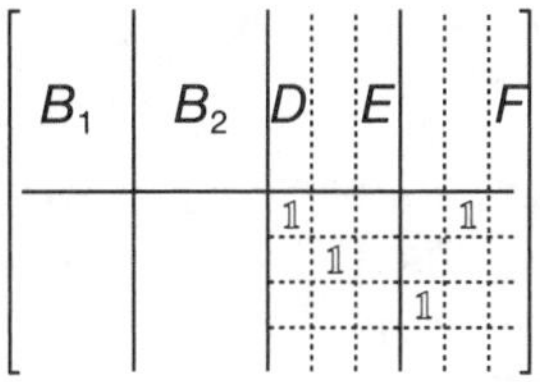

Fig. 3

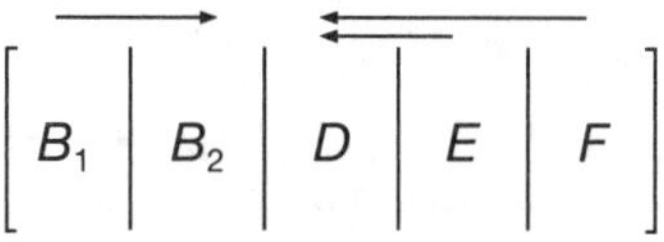

Fig. 4

In a further step we reduce B_3' and B_4' to the forms of Fig. 3 by adding linear combinations of the rows of the lower stripe of Fig. 2 to $[B_1 \mid B_2 \mid B_3' \mid B_4']$. In synthetic terms, this means that we modify the first basis vectors of V in such a way that their images lie in a suitable supplement of $f'(V') = f(V) \cap N(X)$ in $M(X)$. Thus we are reduced to the matrix problems described by Fig. 4, where $[B_1 \mid B_2] \in \mathbb{M}$. These matrix problems coincide with those obtained from $\hat{M}^k_{\{\hat{N}\}}$ by factoring out the annihilator of $\hat{M}$ in $\hat{\mathscr{A}}$ (see 4.5). For their solution, we refer to Sect. 5.

Thus we see how a statement whose categorical formulation is almost trivial can be turned into a quite intricate matricial construction. In fact this construction was elaborated and tested before its abstract reformulation.

4.5. In the situation described in 4.4 when $\mathscr{L} \neq \emptyset$, each N-space (W, g, X) provides an M-space $(W, i \circ g, X)$ where i is an inclusion. Thus we may identify $\hat{\mathscr{A}} = N^k_{\mathscr{L} \cap N}$ with a full subcategory of $M^k_{\mathscr{L}}$. The module $\hat{M}$ on $\hat{\mathscr{A}}$ is then identified with the restriction of the module Cok: $(V, f, X) \mapsto M(X)/f(V)$ on $M^k_{\mathscr{L}}$. In like manner, we can identify each $X \in \mathscr{A}$ with the *trivial M-space*[3] $(0, 0, X)$, and M with Cok$|\mathscr{A}$. In other words, passing from $\mathscr{A}$ to $\hat{\mathscr{A}}$, we stick to our "goal" $M^k_{\mathscr{L}} \xrightarrow{\sim} \hat{M}^k_{\hat{\mathscr{L}}}$ but lessen its distance by extending the known realm of trivial M-spaces.

On the other hand, passing from $\mathscr{L}$ to $\hat{\mathscr{L}}$, we also increase the number of "constraints." This improves our chances provided we can step by step attain the prerequisites of the following lemma.

Lemma.[4] *Suppose that the total dimension $\sum_{s \in \mathscr{I}\mathscr{A}} \dim M(s)$ and the set $\mathscr{M}$ of maximal submodules of M are finite. Then $M^k_{\mathscr{M}}$ has exactly one non-trivial indecomposable up to isomorphism. This has the form $(k, f, \bigoplus_{s \in \Sigma} s)$, where Σ denotes the set of points $s \in \mathscr{I}\mathscr{A}$ such that $M(s) \neq \mathscr{R}M(s)$* $\surd$

Together with 4.4 the preceding lemma provides us with a workable algorithm: Suppose that M has finite total dimension. If $\mathscr{L}$ is any finite set of proper submodules, we start with a proper submodule $N \neq 0$ which is minimal among those contained in no $L \in \mathscr{L}$. According to our lemma, $\mathscr{I}\hat{\mathscr{A}}$ has one point more than $\mathscr{I}\mathscr{A}$: the unique non-trivial indecomposable of $N^k_{\mathscr{L} \cap N}$. In a second step, we then choose a proper submodule $N' \neq 0$ of $\hat{M}$ which is minimal among those contained in no $\hat{L} \in \hat{\mathscr{L}}$, and so on .... Each step adds one point to $\mathscr{I}\mathscr{A}$. *The algorithm stops* and yields all the isoclasses of indecomposable M-spaces avoiding $\mathscr{L}$ *if* M *is finitely* $\mathscr{L}$*-spaced.*

Theorem. *Let* M *be a pointwise finite left module over an aggregate* $\mathscr{A}$ *and* $b \in \mathbb{N}$ *a number such that, for each* $X \in \mathscr{A}$ *satisfying* $\dim M(X) \geqslant b$, *there are only finitely many isoclasses of indecomposable* M*-spaces avoiding* $\mathscr{L}$ *of the form* (V, f, X). *Then the same conclusion holds if* $\dim M(X) < b$.

Proof. *Under the further assumption that the spectroid* $\mathscr{I}\mathscr{A}$ *of* $\mathscr{A}$ *is finite, we first prove that the set* $\mathscr{M}$ *of maximal submodules of* M *is finite*: Indeed, set $\overline{N} = (N + \mathscr{R}M)/\mathscr{R}M$ for each submodule N of M. Then $N \neq M$ implies $\overline{N} \neq \overline{M}$ because $\mathscr{I}\mathscr{A}$ is finite. *If* $\mathscr{M}$ *is not finite*, there is an $r \in \mathscr{I}\mathscr{A}$ such that $\dim \overline{M}(r) \geqslant 2$. We then denote by S the set formed by such an r and by all $t \in \mathscr{I}\mathscr{A}$ such that $\overline{M}(t) \neq \overline{L}(t)$ for some $L \in \mathscr{L}$. For each $s \in S$, we choose a non-zero $\bar{e}_s \in \overline{M}(s)$ such that $\bar{e}_s \notin \overline{L}(s)$ whenever $L \in \mathscr{L}$ and $\overline{L}(s) \neq \overline{M}(s)$. We further choose an $\bar{f} \in \overline{M}(r)$ linearly independent of $\bar{e}_r$ and counter-images $f \in M(r)$, $e_s \in M(s)$ of $\bar{f}$ and of all $\bar{e}_s$. We finally set $\bigoplus_{s \in S} s =: Y \in \mathscr{A}$ and $(\bigoplus_{s \in S} ke_s) \oplus kf =: V \subset M(Y)$. For each $n \geqslant 1$, we thus obtain M-spaces (k^n, g_n^λ, Y^n), $\lambda \in k$, where $g_n^\lambda\colon k^n \to M(Y^n) \tilde{\to} M(Y)^n$ factors through some $k^n \to V^n$ with components $\lambda \mathbb{1}_n + J_n\colon k^n \to (kf)^n \tilde{\to} k^n$ (1.7) and $\mathbb{1}_n\colon k^n \to (ke_s)^n \tilde{\to} k^n, \forall s \in S$. By construction, the M-spaces (k^n, g_n^λ, Y^n) are indecomposable and pairwise not isomorphic (for so are the associated $\overline{M}$-spaces). Furthermore, (k^n, g_n^0, Y^n) avoids $\mathscr{L}$ because each $L \in \mathscr{L}$ satisfies $\bar{e}_s \notin \overline{L}(s)$ for some $s \in S$. We infer that (k^n, g_n^λ, Y^n) avoids L for almost all λ. The wanted *contradiction* then follows for large n from $\dim M(Y^n) = n(\dim M(Y)) \geqslant b$.

In the general case, let $X \in \mathscr{A}$ be such that $d = \dim M(X) < b$. Our objective is to prove by induction on d that the number $v(M, \mathscr{L}, X)$ of isoclasses of indecomposables of the form $(V, f, X) \in M_{\mathscr{L}}^k$ is finite. We present the induction step: By restriction to the indecomposable summands $s \in \mathscr{I}\mathscr{A}$ of X, we are reduced to the case where $\mathscr{I}\mathscr{A}$ and $\mathscr{M}$ are finite. If $L(X) = M(X)$ for some $L \in \mathscr{L}$, $v(M, \mathscr{L}, X) = 0$. Otherwise, we use a second induction on the number n of $N \in \mathscr{M} \backslash \mathscr{L}$ satisfying $N(X) \neq M(X)$. By the lemma above, $v(M, \mathscr{L}, X) = 1$ if $n = 0$. In case $n \geqslant 1$, we choose some $N \in \mathscr{M} \backslash \mathscr{L}$ satisfying $N(X) \neq M(X)$. Then $v(M, \mathscr{L} \cup \{N\}, X)$ is finite by the second induction hypothesis. It is therefore enough to show that there are finitely many isoclasses of indecomposables $(V, f, X) \in M_{\mathscr{L}}^k$ such that $f(V) \cap N(X) \neq 0$. To this end, we use 4.4: The new data $\hat{\mathscr{A}} = N_{\mathscr{L} \cap N}^k$, $\hat{M}$ and $\hat{\mathscr{L}}$ satisfy the assumptions of the theorem for the same bound b. On the other hand, the $\hat{M}$-spaces $(V/V', f'', (V', f', X'))$ associated with (V, f, X) satisfy

$$\dim \hat{M}(V', f', X') = \dim M(X) - \dim f(V) \cap N(X) < d.$$

By the first induction hypothesis, it follows that $v(\hat{M}, \hat{\mathscr{L}}, (V', f', X')) < \infty$ for each (V', f', X'). Hence it remains to prove that (V', f', X') ranges over finitely many isoclasses of $\hat{\mathscr{A}} = N_{\mathscr{L} \cap N}^k$ or, equivalently, that $v(N, \mathscr{L} \cap N, X) < \infty$. By the first induction hypothesis, this follows from the fact that N satisfies the assumptions of the theorem and the inequality $\dim N(X) < d = \dim M(X)$ $\surd$

Corollary.[5] *Let A be a finite-dimensional algebra such that the dimensions of the indecomposable A-modules are bounded. Then A admits only finitely many isoclasses of indecomposable modules.*

This follows immediately from 4.2, 4.3 and our theorem.

4.6. If heaping up constraints may lead to success, we shall experience a case in Sect. 5 where the constraint imposed by Proposition 4.4 can be removed. By way of preparation, we here adduce some preliminary considerations.

First we consider the *annihilator*[6] $\mathcal{J}$ of M in $\mathcal{A}$: Each submodule N of M is identified with a module $\bar{N}$ over $\bar{\mathcal{A}} = \mathcal{A}/\mathcal{J}$, and the canonical "projection" $\mathcal{A} \to \bar{\mathcal{A}}$ induces a quasi-surjective functor $P\colon M^k_{\mathcal{L}} \to \bar{M}^k_{\bar{\mathcal{L}}}$, where $\bar{\mathcal{L}} = \{\bar{L}\colon L \in \mathcal{L}\}$. The points of $\mathcal{I}M^k_{\mathcal{L}}$ are therefore subdivided into two classes: on one side indecomposables annihilated by P, i.e. of the form $(0, 0, s)$ where s is indecomposable in $\mathcal{A}$ and $M(s) = 0$, on the other side points forming a spectroid of $\bar{M}^k_{\bar{\mathcal{L}}}$. The investigation is thus reduced to $\bar{M}^k_{\bar{\mathcal{L}}}$, i.e. to the case of a faithful module. Nonetheless, non-faithful modules cannot be eluded since the construction of 4.4 may transform a faithful M into a non-faithful $\hat{M}$.

Now we define the *support* of M (resp. of a space (V, f, X) on M) as the set of points $s \in \mathcal{I}\mathcal{A}$ where M (resp. (V, f, X)) is *present*; by definition, this means that $M(s) \neq 0$ (resp. that s is a summand of X). Thus, if $s \in \mathcal{I}\mathcal{A}$ does not belong to the support of M, the indecomposable M-spaces present at s are isomorphic to $(0, 0, s)$. If $\dim M(s) = 1$, the indecomposable M-spaces with support $\{s\}$ are isomorphic to $(0, 0, s)$ or to $(M(s), \mathbb{1}, s)$. If moreover the socle of M is isomorphic to ^-s, it follows from the subsequent proposition that, up to isomorphism, $(0, 0, s)$ and $(M(s), \mathbb{1}, s)$ are the only indecomposable M-spaces present at s.

Proposition. *Suppose that, for some $s \in \mathcal{I}\mathcal{A}$, the $\mathcal{A}$-module M contains a submodule S isomorphic to ^-s and that* $\dim M(s) = 1$. *Let further $\mathcal{J}$ denote the ideal of $\mathcal{A}$ generated by $\mathbb{1}_s$, $\bar{\mathcal{A}}$ the quotient $\mathcal{A}/\mathcal{J}$, $\bar{M}$ the $\bar{\mathcal{A}}$-module induced by the $\mathcal{A}$-module M/S and N the annihilator of $\mathcal{J}$ in M. Then the canonical projection $M \to M/S$ induces a quasisurjective functor $M^k_{\{N\}} \to \bar{M}^k$.*

Proof. Each $\bar{M}$-space $(V, \bar{f}, X)$ is isomorphic to the image of some $(V, f, X \oplus s^d) \in M^k_{\{N\}}$, where the first component $V \to M(X)$ of f induces $\bar{f}$ and the second $V \to M(s^d)$ is bijective.

Now let (V, f, X) and (W, g, Y) be two spaces on M. Each morphism between the associated $\bar{M}$-spaces can be obtained from a linear map $\varphi\colon V \to W$ and a morphism $\mu \in \mathcal{A}(X, Y)$ such that $M(\mu)f - g\varphi\colon V \to M(Y)$ factors through $S(Y)$. The problem is to find a $\nu \in \mathcal{J}(X, Y)$ such that $M(\mu - \nu)f = g\varphi$ or, equivalently, $M(\nu)f = M(\mu)f - g\varphi$. To this effect, we set $Y = Y' \oplus s^q$ where Y' has no summand of the form s. Then we have $S(Y) = M(s^q)$, and it suffices to find a $\nu' \in \mathcal{A}(X, s^q)$ such that $M(\nu')f$ is the map induced by $M(\mu)f - g\varphi$. Since $M(\nu')$ factors through $M(X)/N(X)$, into which V is mapped injectively, it is enough to prove that the module structure of M induces a surjection $\mathcal{A}(X, s^q) \to$

$\operatorname{Hom}_k(M(X)/N(X), M(s^q))$. In fact, it induces a surjection $\mathscr{A}(t, s) \to \operatorname{Hom}_k(M(t)/N(t), M(s))$ for each $t \in \mathscr{I}\mathscr{A}$. This is due to the fact that the "transposed" map $M(t)/N(t) \to \operatorname{Hom}_k(\mathscr{A}(t, s), M(s))$ is injective by assumption (each non-zero element of M/N has a non-zero multiple in the socle S of M/N) $\surd$

If M is faithful, the spectroid $\mathscr{I}M^k_{\{N\}}$ contains one indecomposable whose image in $\overline{M}^k$ is zero, namely $(0, 0, s)$. The other points of $\mathscr{I}M^k_{\{N\}}$ provide a spectroid of $\overline{M}^k$, to which the investigation of $M^k_{\{N\}}$ is thus reduced.

4.7. We now turn to the investigation of finitely spaced modules. In the present treatise, the Sect. 4.7–4.11 will be used in 5.8 only. *Until the end of Sect.* 4, *we suppose that the spectroid $\mathscr{I}\mathscr{A}$ of $\mathscr{A}$ is finite and that the pointwise finite left $\mathscr{A}$-module M is faithful.*

Proposition. *If M is finitely spaced and s a point of $\mathscr{I}\mathscr{A}$, then $M(s)$ has a k-basis of the form $m, xm, \ldots, x^{p-1}m$, where $m \in M(s)$, $x \in \mathscr{R}_{\mathscr{A}}(s, s)$ and $x^p = 0$.*

Proof. The isoclasses of M-spaces of the form (k, f, s) correspond bijectively to the *cyclic*[7] submodules of the $\mathscr{A}(s, s)$-module $M(s)$ (with (k, f, s) we associate the submodule $\mathscr{A}(s, s)f(1)$). Accordingly, if M is finitely spaced, $M(s)$ has finitely many cyclic submodules. Since $\mathscr{A}(s, s)$ is local, it follows that $M(s)$ is *uniserial*, i.e. admits a unique Jordan-Hölder series

$$M(s) = M_0 \supset M_1 \supset \cdots \supset M_{p-1} \supset M_p = \{0\}.$$

For each $i \leqslant p - 2$, the ideal $\{a \in \mathscr{R}_{\mathscr{A}}(s, s) : aM_i \subset M_{i+2}\}$ is a hyperplane in $\mathscr{R}_{\mathscr{A}}(s, s)$. We choose $x \in \mathscr{R}_{\mathscr{A}}(s, s)$ outside these hyperplanes and $m \in M(s)$ outside M_1 $\surd$

Our proposition strongly affects the structure of the morphism spaces $\mathscr{A}(s, t)$, $s, t \in \mathscr{I}\mathscr{A}$. Indeed, if M and M_i are as above, let us denote by

$$M(t) = N_0 \supset N_1 \supset \cdots \supset N_{q-1} \supset N_q = \{0\}$$

the Jordan-Hölder series of the $\mathscr{A}(t, t)$-module $M(t)$ and by H_{ji} the space of all $f \in \mathscr{A}(s, t)$ such that $f(M_i) \subset N_j$. The intersections $H_{ji} \cap H_{j-1,0}$ are then linearly ordered by inclusion and provide a filtration

$$\cdots \subset H_{j+1,p-1} \cap H_{j0} \subset H_{j+1,p} \cap H_{j0} = H_{j0} = H_{j0} \cap H_{j-1,0} \subset H_{j1} \cap H_{j-1,0} \subset \cdots$$

of $\mathscr{A}(s, t)$ with smallest term $\{0\} = H_{q0} = H_{q0} \cap H_{q-1,0}$ and largest term $H_{1p} \cap H_{00} = H_{00} = \mathscr{A}(s, t)$. The quotients

$$H_{ji} \cap H_{j-1,0}/H_{j,i-1} \cap H_{j-1,0}$$

of this filtration have dimension $\leqslant 1$, and the indices (j, i) where the dimension is 1 form a "staircase" (Fig. 5): Indeed, if we endow $M(s)$ with the basis of the proposition and choose a similar basis $n, yn, y^2n, \ldots$ for $M(t)$, $\mathscr{A}(s, t)$ can be identified with a subspace of $k^{q \times p}$ which is stable under multiplication by J_q^T and J_p^T (1.7), i.e. under shifts leftwards or downwards.

Like $\mathscr{A}(s, t)$, the space S of all $q \times p$-matrices with vanishing entries above the stairs is equipped with a filtration inherited from $k^{q \times p}$. Annihilating the entries

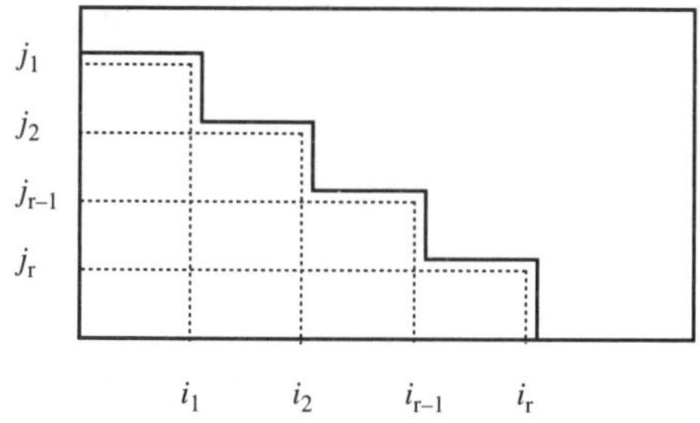

Fig. 5

of the matrices of $\mathscr{A}(s, t)$ which lie above the stairs, we obtain a linear map $a: \mathscr{A}(s, t) \to S$ which is compatible with the filtrations and induces bijections between their quotients (of dimension $\leqslant 1$). Therefore, a itself is bijective, and the "canonical" basis of S is the image of a basis (b^{ji}) of $\mathscr{A}(s, t)$ such that $b^{ji}_{ji} = 1$, whereas the other entries b^{ji}_{vu} vanish if $v \leqslant j$ or if (v, u) is "located below the stairs" (the indexing pair (j, i) runs through the "staircase").

A second "natural" basis of $\mathscr{A}(s, t)$ is formed by the morphisms $y^{f_1}b^{j_1i_1}x^{e_1}$, $y^{f_2}b^{j_2i_2}x^{e_2}$, ... where $f_1 \leqslant q - j_1$, $e_1 < i_1$, $f_2 \leqslant q - j_2$, $e_2 < i_2 - i_1$, ... As a bimodule over $k[y]$ and $k[x]$, $\mathscr{A}(s, t)$ is therefore generated by $b^{j_1i_1}, b^{j_2i_2}, \ldots, b^{j_ri_r}$ and is determined by the entries of these matrices at the lower right corner of Fig. 5. These entries are subjected to obvious polynomial relations expressing that $b^{j_1i_1}x^{i_1}$, $b^{j_2i_2}x^{i_2-i_1}$, ... are linear combinations of the vectors of the second basis.[8]

4.8. The description of subbimodules of $k^{q\times p}$ given in 4.7 raises the question of the classification of all finitely spaced modules. We invest some lines on an approach to this non-solved problem.

Lemma. *If M is finitely spaced, then* $\dim M(s) \leqslant 3$ *for each* $s \in \mathscr{I}\mathscr{A}$.

Proof. Suppose that $\dim M(s) \geqslant 4$. It is enough to prove that M^k then has infinitely many isoclasses of indecomposables of the form (V, f, s^n). Thus we may assume that $\mathscr{I}\mathscr{A}$ has only one point s. In this case, we choose isomorphisms $u_i: \mathscr{R}^{i-1}M/\mathscr{R}^iM \xrightarrow{\sim} {}^-s$, where the $\mathscr{R}^iM$ are the iterated radicals of M and $1 \leqslant i \leqslant 4$. These u_i allow us to construct a functor F from M^k to the category $\mathscr{C}$ formed by the finite-dimensional representations R of Q (Fig. 6) such that $\operatorname{Ker} R(\alpha_i) = 0$ for $i \leqslant 3$: With each M-space (V, f, X), F associates the representation of Fig. 6, where V_i is the inverse image of $(\mathscr{R}^iM)(X)$ under f and f_i the composition of $u_i(X)$ with a map induced by f. The construction implies that each $R \in \mathscr{C}$ is isomorphic to some $F(V, f, X)$ and that (V, f, X) is zero if so is R; it follows that (V, f, X) is indecomposable if so is R and that, like $\mathscr{C}$, M^k has infinitely many isoclasses of indecomposables $\surd$

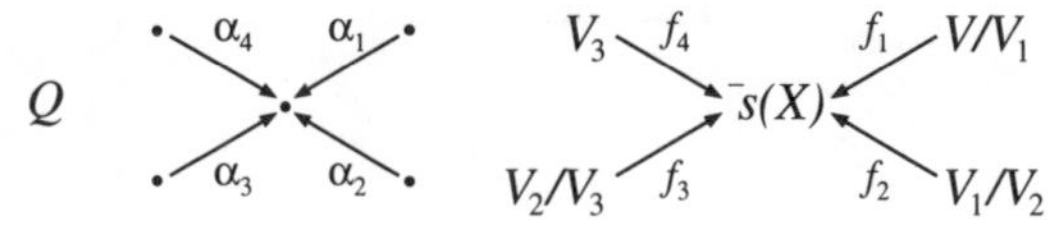

Fig. 6

4.9. The preceding results draw our attention to the disjoint union $\mathscr{C}(M)$ of the sets $\mathscr{C}_s$ formed by the non-zero cyclic $\mathscr{A}(s, s)$-submodules of $M(s)$, $s \in \mathscr{I}\mathscr{A}$. This union is finite if and only if $\mathscr{I}\mathscr{A}$ is finite and each $M(s)$ uniserial. It is naturally equipped with an equivalence relation whose classes are the sets $\mathscr{C}_s$. It further carries an order relation consisting of the pairs $(A, B) \in \mathscr{C}_s \times \mathscr{C}_t$ such that $fA \subset B$, $f\mathscr{R}A \subset \mathscr{R}B$ and $fA \not\subset \mathscr{R}B$ for some $f \in \mathscr{A}(s, t)$. When M is finitely spaced, the equivalence classes $\mathscr{C}_s$ of $\mathscr{C}(M)$ are linearly ordered by this relation.

The structure carried by $\mathscr{C}(M)$ provides us with an informative combinatorial invariant of the pair $(\mathscr{A}, M)$. For instance, let us consider two points $s, t \in \mathscr{I}\mathscr{A}$ such that $M(s)$, $M(t)$ are uniserial of dimension 2. With the notations of 4.7, the poset $\mathscr{C}_s \cup \mathscr{C}_t$ then takes one of the forms of Fig. 7 except possibly for a transposition of s and t (cyclic submodules are represented by generators). In all cases but 2 and 3, the order relation completely determines the bimodules $\mathscr{A}(s, t)$ and $\mathscr{A}(t, s)$ as subspaces of $\mathrm{Hom}_k(M(s), M(t))$ and $\mathrm{Hom}_k(M(t), M(s))$. In the cases 2 and 3, the indetermination concerns $\mathscr{A}(s, t)$ which is either formed by all matrices $\begin{bmatrix} x & 0 \\ y & z \end{bmatrix}$ or by the $\begin{bmatrix} x & 0 \\ y & ax \end{bmatrix}$ where $a \neq 0$ is fixed and $x, y, z \in k$. In the second instance, $\dim \mathscr{A}(s, t) = 2$; we then say that *s affects t*.

Lemma. *Suppose that M is finitely spaced. Then the cases* 9, 10 *and* 11 *of* Fig. 7 *do not occur. Moreover, if s and t both affect or are both affected by some r, then s affects or is affected by t.*

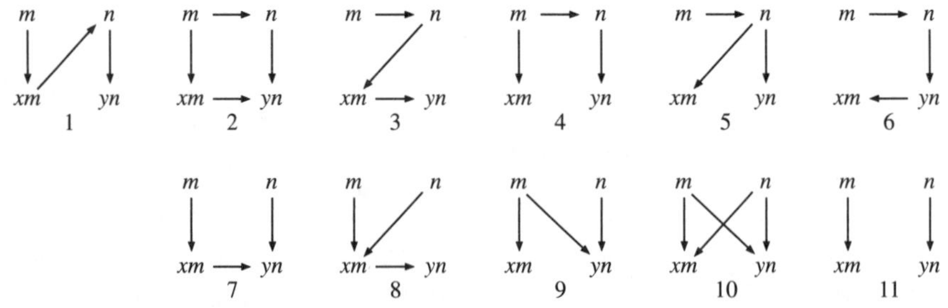

Fig. 7

Proof. The M-spaces $S_\lambda = (k^2, f_\lambda, s \oplus t)$, where f_λ is associated with the matrix

$$\begin{matrix} \begin{bmatrix} 1 & \lambda & 0 & 0 \\ 0 & 0 & 1 & 1 \end{bmatrix}^T, \\ m \quad n \quad xm \quad yn \end{matrix}$$

are pairwise not isomorphic in cases 9, 10 and 11.

Now suppose that s and t affect some r. Since 9–11 is excluded, m must be comparable with n or xm with yn. Suppose for instance that $m < n$: If s did not affect t, there would exist an $f \in \mathscr{A}(s, t)$ such that $fm = n$ and $fxm = 0$. This would imply $\mathscr{A}(r, r)gfm = M(r)$ and $gfxm = 0$ for some $g \in \mathscr{A}(t, r)$. Thus s would not affect r $\surd$

4.10. Let us call *basis* of $(\mathscr{A}, M)$ a pair of two families $(m_{s.})_{s \in \mathscr{I}\mathscr{A}}$ and $(b_{ts.})_{s,t \in \mathscr{I}\mathscr{A}}$, where $m_{s.}$ denotes a basis $m_{s1}, m_{s2}, \ldots$ of the k-space $M(s)$ and $b_{ts.}$ a basis b_{ts1}, $b_{ts2}, \ldots$ of $\mathscr{A}(s, t)$. We say that the basis is *multiplicative* if the following conditions are satisfied: a) For each $s \in \mathscr{I}\mathscr{A}$, the sequence $b_{ss2}, b_{ss3}, \ldots$ is a basis of the radical of $\mathscr{A}(s, s)$, and $b_{ss1} = \mathbb{1}_s$; b) Products of the form $b_{tsj}m_{si}$ are zero or basis vectors m_{tl}; c) $b_{tsl}m_{si} = m_{tp}$, $b_{tsl}m_{sj} = m_{tq}$ and $i \neq j$ imply $p \neq q$.

Proposition. *$(\mathscr{A}, M)$ admits a multiplicative basis provided M is finitely spaced and satisfies* $\dim M(s) \leqslant 2$ *for each* $s \in \mathscr{I}\mathscr{A}$.

Proof. Let us connect two points $s, t \in \mathscr{I}\mathscr{A}$ by an arrow $s \to t$ if s affects t and if there is no r affected by s and affecting t. Affection thus gives birth to a quiver[9] whose vertices are the points of $\mathscr{I}\mathscr{A}$. By 4.9, the connected components of this quiver are vertices or "*strings*" $s_0 \to s_1 \to \cdots \to s_n$, $n \geqslant 1$. For each such string, we choose generators m_i and x_i of $M(s_i)$ and of the radical of $\mathscr{A}(s_i, s_i)$. Accordingly, each $\mathscr{A}(s_{i-1}, s_i)$ contains a unique b_i such that $b_i m_{i-1} = m_i$ and $b_i x_{i-1} m_{i-1} = \lambda_i x_i m_i$ for some $\lambda_i \in k \backslash \{0\}$. The scalars λ_i only depend on $x_0, \ldots, x_n$. They can be reduced to 1 if we replace each x_i by $y_i = \lambda_1 \lambda_2 \ldots \lambda_i x_i$. The thus obtained sequence $y_0, \ldots, y_n$ is called *normal*.

The construction of a multiplicative basis now proceeds as follows: For each $s \in \mathscr{I}\mathscr{A}$, we choose a generator m_s of $M(s)$. We further choose generators x_s of the radicals of $\mathscr{A}(s, s)$ in such a way that $x_{s_0}, \ldots, x_{s_n}$ is normal for each string $s_0 \to \cdots \to s_n$. The required family $(m_{s.})$ is then determined by $m_{s1} = m_s$ and by $m_{s2} = x_s m_s$ if $\dim M(s) = 2$. The bases $b_{ss.}$ consist of the powers of x_s. If s affects t, we choose $b_{ts} \in \mathscr{A}(s, t)$ in such a way that $b_{ts}m_s = m_t$ and $b_{ts}x_s m_s = x_t m_t$ and set $b_{ts1} = b_{ts}$, $b_{ts2} = x_t b_{ts}$. If s does not affect $t \neq s$, $b_{ts.}$ consists of the morphisms $b \in \mathscr{A}(s, t)$ which send one basis vector m_{si} onto some m_{tj} and the others onto 0 $\surd$

4.11. The existence of a multiplicative basis implies that, under the assumptions of 4.10, $(\mathscr{A}, M)$ is determined up to "equivalence" by a purely combinatorial structure which we may grasp as follows: We call *biinvolutive* poset a finite poset $\mathscr{P}$ equipped with two involutions $p \mapsto p^*$ and $(p, q) \mapsto (p, q)^*$ which are defined on $\mathscr{P}$ and $\{(p, q) \in \mathscr{P}^2 : p \leqslant q\}$. We further require that: 1) $p \in \mathscr{P}$ implies $p \leqslant p^*$ or $p^* \leqslant p$; 2) $(p, p)^* = (p^*, p^*)$, $\forall p \in \mathscr{P}$; 3) $(p, q) \neq (p, q)^*$ implies $(p, q)^* =$

(p^*, q^*), $p \neq p^*$ and $q \neq q^*$; 4) $(p, q) \neq (p, q)^*$ and $p < r < q$ imply $(p, r)^* = (p^*, r^*)$ and $(r, q)^* = (r^*, q^*)$.

The involutions equipping $\mathscr{P}$ determine a sub-k-category $\mathscr{S}_{\mathscr{P}}$ of $\oplus k\mathscr{P}$ (4.1) whose objects $\bar{p}$ are associated with the "orbits" $\{p, p^*\}$ of $\mathscr{P}$: $\bar{p} = p \oplus p^* \in \oplus k\mathscr{P}$ if $p < p^*$, whereas $\bar{p} = p \in k\mathscr{P}$ if $p = p^*$. We further set $\mathscr{S}_{\mathscr{P}}(\bar{p}, \bar{q}) = k((q|p) + (q^*|p^*)) \oplus k(q^*|p)$ if $(p, q) \neq (p, q)^*$, $p < p^*$ and $q < q^*$, whereas $\mathscr{S}_{\mathscr{P}}(\bar{p}, \bar{q}) = (\oplus k\mathscr{P})(\bar{p}|\bar{q})$ in the "remaining cases". The k-category $\mathscr{S}_{\mathscr{P}}$ thus defined is a spectroid whose additive hull $\oplus\mathscr{S}_{\mathscr{P}}$ is identified with a subaggregate of $\oplus k\mathscr{P}$. The restriction of the module $M^{\mathscr{P}}$ of 4.1 to $\oplus\mathscr{S}_{\mathscr{P}}$ will be denoted by $N^{\mathscr{P}}$.

If $(\mathscr{A}, M)$ satisfies the assumptions of proposition 4.10, the poset $\mathscr{C}(M)$ of 4.9 carries a natural "biinvolutive structure" such that $M(s)^* = M(s)$ if $\dim M(s) = 1$, $M(s)^* = \mathscr{R}M(s)$ if $\dim M(s) = 2$, $(M(s), M(t))^* = (\mathscr{R}M(s), \mathscr{R}M(t))$ if s affects t, and $(P, Q)^* = (P, Q)$ in all other cases. Moreover, $(\mathscr{A}, M)$ is "equivalent" to $(\oplus\mathscr{S}_{\mathscr{C}(M)}, N^{\mathscr{C}(M)})$.

In the general case, the equivalence $(\text{rep}\,\mathscr{P})^{op} \tilde{\rightarrow} M^{\mathscr{P}k}$ of 4.1 associates the subaggregate $N^{\mathscr{P}k}$ of $M^{\mathscr{P}k}$ with the subcategory $\text{rep}^*\,\mathscr{P}$ of $\text{rep}\,\mathscr{P}$ formed by the *involutive representations* of $\mathscr{P}$, i.e. by the $(d, A) \in \text{rep}\,\mathscr{P}$ such that $d_i = d_j$ whenever $p_i^* = p_j$. A morphism $(d, A) \to (e, B)$ is defined by a pair of matrices $X \in k^{e_0 \times d_0}$, $Y \in k^{\bar{e} \times \bar{d}}$ such that $XA = BY$, $\bar{Y}_{ij} = 0$ if $p_i \nleq p_j$ and $\bar{Y}_{ij} = \bar{Y}_{rs}$ if $(p_i, p_j)^* = (p_r, p_s)$ (for the notations see 2.3, Example 6).

The biinvolutive posets $\mathscr{P}$ giving rise to a finitely spaced $N^{\mathscr{P}}$ will be characterized in 5.8 below.

4.12. Remarks

1. A module M over $\mathscr{A}$ is *faithful* if $M(f) \neq 0$ for each non-zero $f \in \mathscr{A}(p, q)$.
2. Aggregates of the form $M_{\mathscr{L}}^k$ admit the following generalization, which happens to be useful: Let $\mathscr{M}$ be a "bunch" of pointwise finite left modules $M_1, \ldots, M_t$ over an aggregate $\mathscr{A}$ and $\mathscr{L}$ a sequence $\mathscr{L}^1, \ldots, \mathscr{L}^t$, where $\mathscr{L}^i$ is a finite set of submodules of M_i. By $\mathscr{M}_{\mathscr{L}}^k$ we then denote the aggregate whose objects are the *space bunches* on $\mathscr{M}$, i.e. the triples (V, f, X) where X is an object of $\mathscr{A}$, V a "bunch" of spaces $V_1, \ldots, V_t \in \text{mod}\,k$ and f a sequence of maps $f_i \in \text{Hom}_k(V_i, M_i(X))$ such that $f_i^{-1}(L(X)) = \{0\}$, $\forall L \in \mathscr{L}^i$. A morphism $(V, f, X) \to (V', f', X)$ is defined by a sequence of maps $\varphi_i \in \text{Hom}_k(V_i, V_i')$ and a morphism $\psi \in \mathscr{A}(X, X')$ such that $f_i'\varphi_i = M_i(\psi)f_i$, $\forall i$.

Example 1. Set $\mathscr{A} = \text{mod}\,\bar{\mathscr{S}}$, where $\bar{\mathscr{S}}$ is a finite spectroid such that $\mathscr{R}_{\bar{\mathscr{S}}} = 0$, and $\mathscr{L}^i = \varnothing$ for each i. Then $\mathscr{M}^k = \mathscr{M}_{\mathscr{L}}^k$ is equivalent to the category rep Q formed by the finite-dimensional representations of some quiver Q: The vertices of Q are the points of $\bar{\mathscr{S}}$ together with t additional points $s_1', \ldots, s_t'$; the arrows are directed from $s_1', \ldots, s_t'$ towards $\bar{\mathscr{S}}$, the number of arrows $s_i' \to s$ being equal to $\dim M_i(s)$.

Example 2. Let $\mathscr{S}$ be a finite spectroid and $\mathscr{J}$ the ideal of $\mathscr{S}$ which consists of all the radical morphisms v such that $\rho v = 0$ whenever ρ itself is radical. Denote by $s_1, \ldots, s_t$ the points $s \in \mathscr{S}$ such that $\mathscr{J}(?, s) \neq 0$, and set $\bar{\mathscr{S}} = \mathscr{S}/\mathscr{J}$, $\mathscr{A} = \text{mod}\,\bar{\mathscr{S}}$, $M_i = \text{Ext}_{\mathscr{S}}^1(s_i^-, ?)$, $L_i = \text{Ext}_{\bar{\mathscr{S}}}^1(s_i^-, ?)$, $\mathscr{M} = (M_1, \ldots, M_t)$ and $\mathscr{L}^i = \{L_i\}$. Proceeding as in 4.2–4.3 (Remark), we obtain an epivalence $\text{mod}\,\mathscr{S} \to \mathscr{M}_{\mathscr{L}}^k$.

In the particular case where $\mathscr{R}_{\mathscr{S}}^2 = 0$, $\mathscr{M}^k$ is equivalent to the category rep Q described in Example 1, and $\mathscr{M}_{\mathscr{L}}^k$ to the full subcategory formed by the representations without summand of the form $s_i'^-$.

Returning to the *general case*, we can reduce the investigation of $\mathscr{M}_{\mathscr{L}}^k$ step by step to the case of a single module: Denote by $\mathscr{N}$ and $\mathscr{K}$ the sequences $M_1, \ldots, M_{t-1}$ and $\mathscr{L}^1, \ldots, \mathscr{L}^{t-1}$ respectively, by $\bar{P}$ the left module $(W, g, X) \mapsto P(X)$ over $\mathscr{N}_{\mathscr{K}}^k$ which is induced by a left module P over $\mathscr{A}$. Then we obtain in *isomorphism of categories*

$$\mathcal{M}^k_{\mathcal{L}} \tilde{\rightarrow} \overline{M}^k_{t\overline{\mathcal{L}}}, \qquad (V, f, X) \mapsto (V_t, f_t, (V', f', X)),$$

where $\overline{\mathcal{L}} = \{\overline{L}: L \in \mathcal{L}^t\}$, $V' = (V_1, \ldots, V_{t-1})$ and $f' = (f_1, \ldots, f_{t-1})$.

For instance, suppose that $\mathcal{A} = \text{mod } k$, $t = 3$, $M_i(X) = X$ for all X and i, $\mathcal{L}^1 = \mathcal{L}^2 = \{0\}$ and $\mathcal{L}^3 = \varnothing$. Then "the" spectroid of $\mathcal{N}^k_{\mathcal{X}}$ is isomorphic to the k-category defined by the quiver

Q

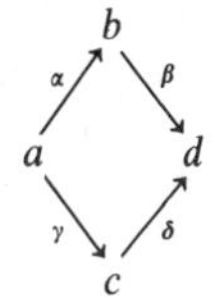

and the relation $\beta\alpha - \delta\gamma = 0$. Moreover, we have $\dim \overline{M}_3(x) = 1$ for each indecomposable $x \in \mathcal{N}^k_{\mathcal{X}}$ and $\overline{M}_3(\mu) \neq 0$ for each morphism $\mu \neq 0$. Thus we can apply 4.1, and infer that $\mathcal{M}^k_{\mathcal{L}}$ is equivalent to $(\text{rep } \mathcal{P})^{op}$, where $\mathcal{P}$ is the poset associated with Q.

3. We simply say *M-space* instead of space on M.

4. [111, Nazarova/Roiter/Gabriel, 1988].

5. [134, Roiter, 1968] Before Roiter's proof, the statement was known as first conjecture of Brauer-Thrall.

6. The *annihilator* of M is the ideal of $\mathcal{A}$ formed by the $f \in \mathcal{A}(p, q)$ such that $M(f) = 0$.

7. *cyclic* = generated by one element.

8. For instance, $k^{3\times 3}$ admits 20 families of subbimodules with the following generators (non-marked entries are null):

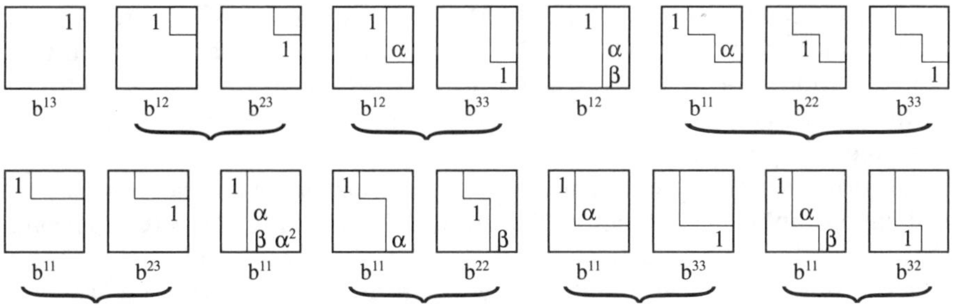

The remaining 10 families are obtained by shifting the preceding ones downwards.

9. This is in fact a subquiver of the quiver of $\mathcal{I}\mathcal{A}$ defined in Sect. 8.

10. As it seems, multiplicative bases should exist in general for finitely spaced modules. We finish Sect. 4 with some considerations in this direction and with the formulation of further simple properties of finitely spaced modules. The proofs are similar to those of 4.8–10.

Throughout (10), we suppose that M *is finitely spaced.* By $w(\mathcal{P})$ we denote the *width* of a poset $\mathcal{P}$ (i.e. the maximum of the cardinalities of its antichains), by $\mathcal{C}^i(M)$ the union of the classes $\mathcal{C}_s \subset \mathcal{C}(M)$ of cardinality $\geqslant i$. In case $f \in \mathcal{I}\mathcal{A}(s, t)$, $r(f)$ denotes the rank of the induced map $M(f)$: $M(s) \rightarrow M(t)$; in case $P \in \mathcal{C}_s$ and $Q \in \mathcal{C}_t$, we write $P \lhd Q$ if there is an $f \in \mathcal{I}\mathcal{A}(s, t)$ such that $fP \subset Q$, $f\mathcal{R}P \subset f\mathcal{R}Q$, $fP \not\subset \mathcal{R}Q$ and $r(f) = 1$.

Lemma 1. $w(\mathcal{C}^i(M))) \leqslant 4 - i$ *if* $1 \leqslant i \leqslant 4$.

Lemma 2. *Each class* $\mathcal{C}_s$ *of cardinality* 2 (*resp.* 3) *contains at least one element* (*resp. two elements*) *comparable with all elements of* $\mathcal{C}^3(M)$ (*resp. of* $\mathcal{C}^2(M)$).

Lemma 3. *Suppose that* $P, Q \in \mathcal{C}^2(M)$ *satisfy* $P < Q$ *and* $P \not\lhd Q$ *and are not comparable to* R. *Then* $R \notin \mathcal{C}^3(M)$.

Lemma 4. *Suppose that* $f \in \mathcal{R}_{\mathcal{I}\mathcal{A}}(s, t)$ *satisfies* $r(f - g) \geqslant r(f)$ *for each* $g \in \mathcal{I}\mathcal{A}(s, t)$ *such that* $r(g) < r(f)$. *Then* $r(f) \leqslant 2$.

Lemma 4 implies Lemma 5 which, together with 4.7, implies Lemma 6 below.

Lemma 5. *Let* $P < Q$ *belong to some* $\mathscr{C}_s$, $s \in \mathcal{I}\mathcal{A}$. *Then* $M \lhd Q$ *follows from* $M < P$ *and* $P \lhd R$ *from* $Q < R$.

Lemma 6. $(\mathcal{A}, M)$ *admits a scalarly multiplicative basis, i.e. a pair of families* $(m_{s.})_{s \in \mathcal{I}\mathcal{A}}$ *and* $(b_{ts.})_{s,t \in \mathcal{I}\mathcal{A}}$ *formed by bases of* $M(s)$ *and* $\mathcal{A}(s, t)$ *which satisfy condition* a) *of* 4.10 *and the following two conditions*: b′) *For all* s, t, i *and* j *there is a scalar* $\lambda \in k$ *and an* l *such that* $b_{tsj}m_{si} = \lambda m_{sl}$; c′) $b_{tsl}m_{si} = \lambda m_{tp}$, $b_{tsl}m_{sj} = \mu m_{tq}$, $i \neq j$ *and* $\lambda, \mu \in k \backslash \{0\}$ *imply* $p \neq q$.

5. Finitely Represented Posets

A poset is called *finitely represented* if it has only finitely many isoclasses of indecomposable representations. Our objective is to characterize such posets and to describe their representations.

Throughout Sect. 5, $\mathscr{P} = \{p_1, \ldots, p_t\}$ denotes a finite poset of cardinality t.

5.1. Since rep $\mathscr{P}$ is antiequivalent to $M^{\mathscr{P}k}$ (4.1), we may transfer the results and the terminology of Sect. 4 from $M^{\mathscr{P}k}$ to rep $\mathscr{P}$. In particular, we say that a representation (d, A) of $\mathscr{P}$ is *present* at p_i if $d_i > 0$. The subset $\mathscr{P}(d, A)$ of $\mathscr{P}$ formed by these p_i is the *support* of (d, A). And $\mathscr{P}$ is called *supportive* if $\mathscr{P} = \mathscr{P}(d, A)$ for some indecomposable (d, A).

We further say that a representation (d, A) is *elementary* if it is indecomposable and has a support of cardinality $|\mathscr{P}(d, A)| \leqslant 2$. In case $|\mathscr{P}(d, A)| = 0$, this means that $d = [1\ 0\ \ldots\ 0]^T$; we then set $(d, A) = (\varnothing, \mathrm{I})$. In case $\mathscr{P}(d, A) = \{p_i\}$, we either have $d_i = 1$ and $d_j = 0$ if $j \neq i$; we then set $(d, A) = (p_i, \mapsto\!\!|)$. Or we have $d_0 = d_i = 1$ and $d_j = 0$ if $j \notin \{0, i\}$; then (d, A) is isomorphic to a representation with matrix $[1]$ which we denote by $(p_i, [1])$. Finally, in case $\mathscr{P}(d, A) = \{p_i, p_j\}$, where $i \neq j$, we have $d_0 = d_i = d_j = 1$ and $d_l = 0$ if $l \notin \{0, i, j\}$; in this case, (d, A) is isomorphic to a representation with matrix $[1\ \ 1]$, which we denote by $(p_i + p_j, [1\ \ 1])$.

A *trichotomy* of $\mathscr{P}$ is a triple $(\mathscr{X}, \mathscr{Y}, \mathscr{Z})$ formed by disjoint subsets with union $\mathscr{P}$ such that the following three conditions are satisfied: a) $\mathscr{X} \neq \varnothing \neq \mathscr{Z}$; b) $x < z$, $\forall x \in \mathscr{X}$, $\forall z \in \mathscr{Z}$; c) the *width* $w(\mathscr{Y})$ (=maximum of the cardinalities of the antichains) of $\mathscr{Y}$ is $\leqslant 1$.

$$\begin{bmatrix} X_1 & Y_1 & Z_1 \\ 0 & Y_2 & Z_2 \\ 0 & 0 & Z_3 \end{bmatrix}$$

Fig. 1

Lemma. *If $(\mathscr{X}, \mathscr{Y}, \mathscr{Z})$ is a trichotomy of $\mathscr{P}$, the support of an indecomposable representation (d, A) of $\mathscr{P}$ is contained in $\mathscr{X} \cup \mathscr{Y}$ or in $\mathscr{Y} \cup \mathscr{Z}$.*

Proof. Use the fang-cheng-method[1], thus reducing A to the block-form of Fig. 1, where the columns of $X_1 \in {}'\mathbb{M}$, $Y_2 \in {}'\mathbb{M}$ and Z_3 are assigned to $\mathscr{X}$, $\mathscr{Y}$ and $\mathscr{Z}$ respectively. Interpret Y_2 as a representation of $\mathscr{Y}$, reduce it to the form V of 1.2, and "kill" the columns of the transform of Y_1 which lie above the ones of V. Finally, kill the transform of Z_1 by adding linear combinations of the columns assigned to $\mathscr{X}$ to the columns assigned to $\mathscr{Z}$. This is possible because the columns of $X_1 \in {}'\mathbb{M}$ generate the corresponding column space.

Our recipe decomposes an arbitrary representation into 2 representations supported by $\mathscr{X} \cup \mathscr{Y}$ and by $\mathscr{Y} \cup \mathscr{Z}$ $\surd$

5.2. For each $p \in \mathscr{P}$, we set $\mathscr{P}^-(p) = \{q \in \mathscr{P} : q < p\}$, $\mathscr{P}^+(p) = \{q \in \mathscr{P} : p < q\}$ and $\mathscr{P}^{\between}(p) = \{q \in \mathscr{P} : q \between p\}$, the expression $q \between p$ meaning that q and p are *incomparable* in $\mathscr{P}$.

Proposition. *If $\mathscr{P}^{\between}(p)$ is a chain, each indecomposable representation of $\mathscr{P}$ present at p is elementary.*

Proof. We apply Lemma 5.1 to $\mathscr{X} = \mathscr{P}^-(p)$, $\mathscr{Y} = \mathscr{P}^{\between}(p)$ and $\mathscr{Z} = \{p\} \cup \mathscr{P}^+(p)$, inferring that an indecomposable (d, A) present at p is supported by $\mathscr{Y} \cup \mathscr{Z} = \mathscr{P}^{\between}(p) \cup \{p\} \cup \mathscr{P}^+(p)$. We then apply Lemma 5.1 to $\mathscr{X}_1 = \{p\}$, $\mathscr{Y}_1 = \mathscr{P}^{\between}(p)$ and $\mathscr{Z}_1 = \mathscr{P}^+(p)$ and infer that (d, A) is supported by $\{p\} \cup \mathscr{P}^{\between}(p)$.

Thus we are reduced to the case where $\mathscr{P} = \{p\} \cup \mathscr{P}^{\between}(p)$. We then proceed by induction on the cardinality $|\mathscr{P}|$. The cases $|\mathscr{P}| \leqslant 2$ are clear by definition. If $|\mathscr{P}| \geqslant 3$, we choose a non-minimal $q \in \mathscr{P}^{\between}(p)$ and apply Lemma 5.1 to $\mathscr{X}_2 = \mathscr{P}^-(q)$, $\mathscr{Y}_2 = \{p\}$ and $\mathscr{Z}_2 = \{q\} \cup \mathscr{P}^+(q)$ $\surd$

Corollary. *In case $w(\mathscr{P}) \leqslant 2$, all indecomposable representations of $\mathscr{P}$ are elementary.*

It follows that *finite posets of width $\leqslant 2$ are finitely represented.* On the other hand, we know that finite posets of width $\geqslant 4$ are not finitely represented (1.2, IV). As we shall see forthwith, finite posets of width 3 may be finitely represented or not.

5.3. Our investigation of finitely represented posets is based on the algorithm which is described in 4.4 and leads to the following device: If a is a maximal point of $\mathscr{P}$, let $\mathscr{P}'_a$ denote the poset obtained from $\mathscr{P}$ by deleting a and adding formal suprema $p \vee q$ for all $p, q \in \mathscr{P}^{\between}(a)$ such that $p \between q$. We endow $\mathscr{P} \setminus \{a\}$ with the order induced by $\mathscr{P}$ and further agree that $s < p \vee q$ if $s \in \mathscr{P} \setminus \{a\}$ satisfies $s \leqslant p$ or $s \leqslant q$, that $p \vee q < s$ if $s \in \mathscr{P} \setminus \{a\}$ satisfies $p \leqslant s$ and $q \leqslant s$, and finally that $p \vee q \leqslant s \vee t$ if $p < s \vee t$ and $q < s \vee t$. We call $\mathscr{P}'_a$ the *derivative* of $\mathscr{P}$ at a.

If $\mathscr{P}^{\between}(a)$ is a chain, $\mathscr{P}'_a$ coincides with $\mathscr{P} \setminus \{a\}$. It follows that finite posets of width $\leqslant 2$ can be reduced to $\varnothing$ by repeated derivations. In the case of width 3, the problem seems knottier since some posets of width 3 can be reduced to $\varnothing$ and some others to posets of width $\geqslant 4$ (Fig. 2)[2].

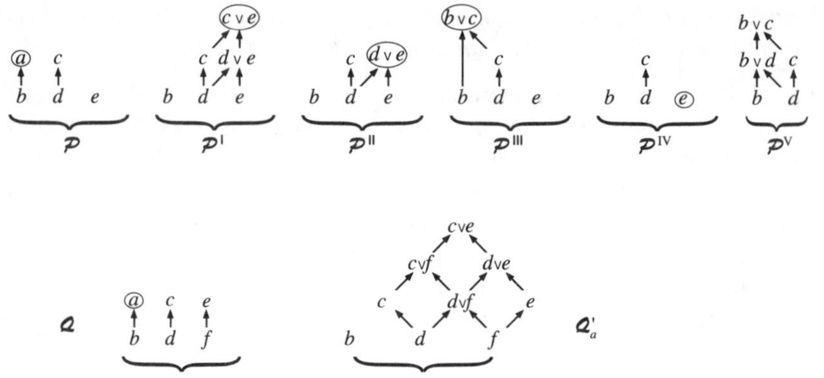

Fig. 2

The same procedure can be applied to the direct sums[3] $\hat{p} \coprod \hat{q} \coprod \hat{r}$, where $\hat{n}$ denotes a chain of cardinality n, and $p \geqslant q \geqslant r \geqslant 1$. We leave it to the reader to convince himself that $\hat{p} \coprod \hat{q} \coprod \hat{r}$ can be reduced to a poset of width $\geqslant 4$ unless $q = 1$ or $r = 1$, $q = 2$ and $p \leqslant 4$.

Theorem.[4] *Let a be a maximal element of $\mathscr{P}$ such that $w(\mathscr{P}^{\between}(a)) \leqslant 2$. Then there exists a quasi-surjective functor F: $\operatorname{rep} \mathscr{P} \to \operatorname{rep} \mathscr{P}'_a$ such that the indecomposables annihilated by F are those isomorphic to $(\varnothing, \mathrm{I})$, $(a, \mapsto\!\!\dashv)$ or $(n, [1])$ where $n \in \mathscr{P}^{\between}(a)$.*

We infer that $\mathscr{P}$ is finitely represented if and only if so is $\mathscr{P}'_a$. If this is the case, the number $\imath(\mathscr{P})$ of isoclasses of indecomposable representations of $\mathscr{P}$ is equal to $\imath(\mathscr{P}'_a) + |\mathscr{P}^{\between}(a)| + 2$, and $\mathscr{P}$ can be reduced to $\varnothing$ by repeated derivations.

Proof. Set $\mathscr{A} = \oplus k\mathscr{P}$ and $M = M^{\mathscr{P}}$ (4.1). Let further N be the submodule of M such that $N(p) = 0$ if $p \leqslant a$ and $N(n) = M(n)$ if $n \in \mathscr{P}^{\between}(a)$. By 4.4, N gives rise to an epivalence $M^k \to \hat{M}^k_{\{\hat{N}\}}$, where $\hat{M}$ is a module over $\hat{\mathscr{A}} = N^k$ which vanishes on the indecomposables $(k, 0, 0)$ and $(k, \mathbb{1}, n)$, $n \in \mathscr{P}^{\between}(a)$, of $\hat{\mathscr{A}}$. The values of $\hat{M}$ at the other points of $\mathscr{I}\hat{\mathscr{A}}$ have dimension 1: In fact, if $\mathscr{J}$ denotes the annihilator of $\hat{M}$ in $\hat{\mathscr{A}}$ and $\mathscr{P}'$ the poset $\mathscr{P}'_a \cup \{a\}$, then $k\mathscr{P}'$ is canonically isomorphic to a spectroid of $\hat{\mathscr{A}}/\mathscr{J}$ (identify $p \in \mathscr{P}$ with $(0, 0, p) \in N^k$ and $p \vee q$ with $(k, [1\ \ 1]^T, p \oplus q) \in N^k$). Considered as a module over $\hat{\mathscr{A}}/\mathscr{J}$, $\hat{M}$ is identified with $M' = M^{\mathscr{P}'}$ (4.1) and $\hat{N}$ with the submodule $N' \subset M'$ such that $N'(p) = 0$ if $p \leqslant a$ and $N'(p) = M'(p) = k$ otherwise. Thus we obtain quasi-surjective functors $M^k \to \hat{M}^k_{\{\hat{N}\}} \to M'^k_{\{N'\}}$ whose composition vanishes at $|\mathscr{P}^{\between}(a)| + 1$ points of $\mathscr{I}M^k$.

To M' and N' we can now apply Proposition 4.6, defining $S' \subset M'$ by $S'(a) = M(a) = k$ and $S'(p) = 0$ if $p \neq a$. The quotient M'/S' is then identified with the module $M^{\mathscr{P}'_a}$ over $\oplus k\mathscr{P}'_a$. The induced functor $M'^k_{\{N'\}} \to M^{\mathscr{P}'_a k}$ is quasi-surjective and vanishes on the image of $(0, 0, a)$. By composition, we finally obtain a quasi-surjective functor $M^{\mathscr{P}k} \to M^{\mathscr{P}'_a k}$ which vanishes on $(k, 0, 0)$, $(0, 0, a)$ and $(k, \mathbb{1}, n)$, $n \in \mathscr{P}^{\between}(a)$ $\surd$

Remarks. 1) In the particular case where $w(\mathcal{P}^{\between}(a)) \leqslant 1$, our theorem also follows from 5.2.

2) The main ingredients of our proof still work when $w(\mathcal{P}^{\between}(a)) \geqslant 3$ except that $\dim \hat{M}(p) \geqslant 2$ then holds for some $p \in \mathcal{I}\hat{\mathcal{A}}$.

3) We leave it to the reader to define the derivative of $\mathcal{P}$ at a minimal element and to prove the dual of the above theorem.

5.4. Theorem 1.[5] *A finite poset $\mathcal{P}$ is infinitely represented if and only if it contains a critical subset, i.e. a subset onto which $\mathcal{P}$ induces an order of one of the five forms displayed in Fig.* 3.[6]

Theorem 2.[5] *Supportive finitely represented posets are isomorphic to posets of Fig.* 4. *The numbers in brackets indicate how many omnipresent*[7] *indecomposables a given poset admits up to isomorphism.*

As a criterion for finite representment and a device for counting indecomposables, the two theorems above provide a good complement to Theorem 5.3. For instance, the poset $\mathcal{P}$ of Fig. 2 contains no critical subset; on the other hand, $\mathcal{P}$ admits 1, 5, 8, 4, 4 and 1 *full*[8] subposets of the form $\varnothing$, $\hat{1}$, $\hat{1} \coprod \hat{1}$, $\hat{1} \coprod \hat{1} \coprod \hat{1}$, $\hat{2} \coprod \hat{1} \coprod \hat{1}$ and $\hat{2} \coprod \hat{2} \coprod \hat{1}$ respectively; finally, no other poset of Fig. 4 is isomorphic to a full subposet of $\mathcal{P}$. It follows that $\mathcal{P}$ is finitely represented and has $1{\cdot}1 + 5{\cdot}2 + 8{\cdot}1 + 4{\cdot}2 + 4{\cdot}1 + 1{\cdot}3 = 34$ isoclasses of indecomposables.

5.5. The two theorems of 5.4 directly result from the following proposition, which will be proved in 5.6 and rests on the lemma below.

Proposition. *If $\mathcal{P}$ is a supportive finite poset and contains no critical subset, then $\mathcal{P}$ is isomorphic to one of the* 14 *posets of Fig.* 4.

Lemma. *If $\mathcal{P}$ is a supportive finite poset of width* 3 *and $a \in \mathcal{P}$ a maximal element, then $\mathcal{P}'_a$ contains a supportive full subposet $\overline{\mathcal{P}} \not\subset \mathcal{P}$ such that $a \neq p \in \mathcal{P} \setminus \overline{\mathcal{P}}$ implies $p \vee q \in \overline{\mathcal{P}}$ for some q.*

Proof. Let us adopt the notations of 4.4 and 5.3, assuming that (V, f, X) is an indecomposable space on M with support $\mathcal{P}$. Then f is injective, and the induced N-space (V', f', X) has no summand isomorphic to $(k, \mathbb{1}, n)$ with $n \in \mathcal{P}^{\between}(a)$:

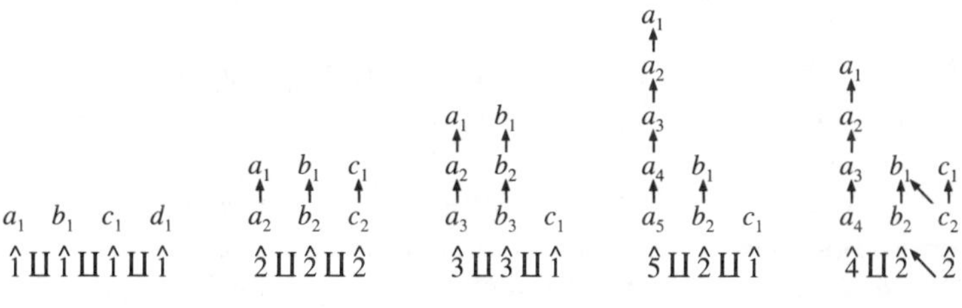

Fig. 3

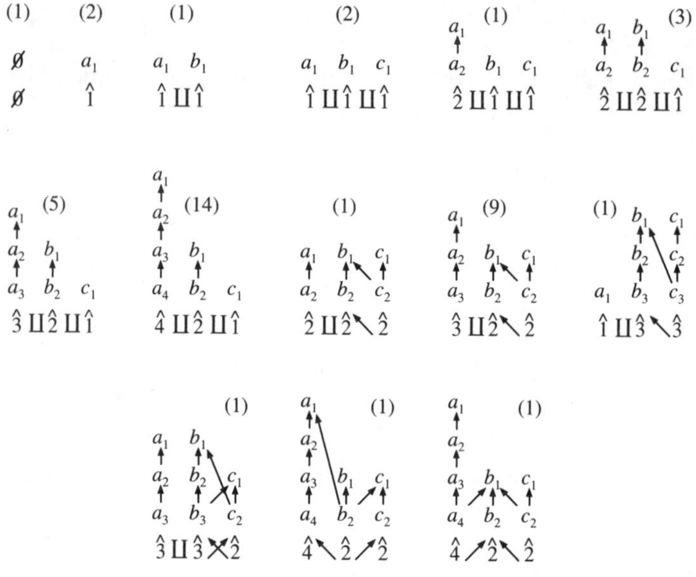

Fig. 4

otherwise, the $\hat{M}$-space associated with (V, f, X) would admit a summand of the form $(0, 0, (k, \mathbb{1}, n))$, and (V, f, X) a summand of the form $(k, \mathbb{1}, n)$.

If (V', f', X) has a summand of the form $(k, [1\ \ 1]^T, p \oplus q)$ with $p, q \in \mathscr{P}^{\between}(a)$, then the $M^{\mathscr{P}'_a}$-space M'/S' associated with (V, f, X) is present at $p \vee q$. Its support can be chosen for $\bar{\mathscr{P}}$.

If (V', f', X) has no summand of the form $(k, [1\ \ 1]^T, p \oplus q)$, the M-space (V, f, X) avoids N. We then set $X \xrightarrow{\sim} Y \oplus a^d$, where Y has no more summand of the form a, and we denote by g the component of f with range $M(Y)$. By Proposition 4.6, (V, g, Y) is an indecomposable space on $M^{\mathscr{P}\setminus\{a\}} \xrightarrow{\sim} M/S$, where $S(a) = k$ and $S(b) = 0$ if $b \in \mathscr{P}\setminus\{a\}$. Considered as a space on $M^{\mathscr{P}}$, (V, g, Y) is indecomposable as well, is present at all points of $\mathscr{P}$ except a and does not avoid N (otherwise it should by 4.6 be isomorphic to (V, f, X)). By the argument above applied to (V, f, X), we now infer that the N-space (V', g', Y) induced by (V, g, Y) has a summand of the form $(k, [1\ \ 1]^T, p \oplus q)$ with $p, q \in \mathscr{P}^{\between}(a)$. For $\bar{\mathscr{P}}$ we then choose the support of the $M^{\mathscr{P}'_a}$-space associated with the $M^{\mathscr{P}}$-space (V, g, Y). $\surd$

5.6. Proof of Proposition 5.5. In view of Corollary 5.2, we may assume that $w(\mathscr{P}) = 3$. By the theorem[9] of Dilworth, $\mathscr{P}$ then is the disjoint union of three chains $\{a_1 > a_2 > \cdots > a_{\bar{l}}\}$, $\{b_1 > b_2 > \cdots > b_{\bar{m}}\}$, $\{c_1 > c_2 > \cdots > c_{\bar{n}}\}$, whose maxima a_1, b_1, c_1 are pairwise incomparable (for instance, $a_1 \geqslant b_1$ would imply that $w(\mathscr{P}^{\between}(a_1)) \leqslant 1$, hence by 5.2 that $\mathscr{P}$ is not supportive).

Since a_1, b_1 and c_1 are the maximal elements of $\mathscr{P}$, the full subposet $\mathscr{P}^+$ of $\mathscr{P}$ formed by the points smaller than one maximal at most has the form

$$\mathscr{P}^+ = \{a_1 > \cdots > a_l, b_1 > \cdots > b_m, c_1 > \cdots > c_n\}$$

and is isomorphic to $\hat{l} \coprod \hat{m} \coprod \hat{n}$ where $l \leqslant \bar{l}$, $m \leqslant \bar{m}$, $n \leqslant \bar{n}$. Since $\mathscr{P}^+$ contains no critical subset, we may suppose that $m = n = 1$ or that $m = 1$, $\min\{l, n\} = 2$ and $\max\{l, n\} \leqslant 4$. This leads us to the case-by-case examination below, throughout which the following lemma will be applied.

Lemma. *Suppose that $\bar{m} = 1$ or $b_2 < a_1$ or $b_2 < c_n$. Then $p \vee q \in \mathscr{P}'_{a_1}$ implies $\{p, q\} \subset \mathscr{P}^+$.*

Proof. Since $n < \bar{n}$ implies $c_{n+1} < a_1$ or $c_{n+1} < b_1$, no $b_1 \vee c_j$ with $j > n$ occurs in $\mathscr{P}'_{a_1}$. In case $\bar{m} = 1$ or $b_2 < a_1$, no $b_i \vee c_j$ with $i \geqslant 2$ occurs in $\mathscr{P}'_{a_1}$. Finally, in case $b_2 < a_1$, we have $b_2 < c_n$, hence $b_i \vee c_j \notin \mathscr{P}'_{a_1}$ for $i \geqslant 2$ and $j \leqslant n$; the same result holds for $j > n$ if $c_{n+1} < a_1$. Otherwise, we have $c_{n+1} < b_1$ and obtain a trichotomy $\mathscr{X} = \{b_2, \ldots, b_{\bar{m}}, c_{n+1}, \ldots, c_{\bar{n}}\}$, $\mathscr{Y} = \{a_1, \ldots, a_{\bar{m}}\}$, $\mathscr{Z} = \{b_1, c_1, \ldots, c_n\}$, which contradicts our assumption that $\mathscr{P}$ is supportive (5.1). $\surd$

1) *Case $m = n = 1$.* By our last lemma, $\mathscr{P}'_{a_1} \backslash \mathscr{P}$ only consists of $b_1 \vee c_1 \in \bar{\mathscr{P}}$ (5.5). Since $\mathscr{P}'^{\,\check{x}}_{a_1}(b_1 \vee c_1)$ is a chain, each indecomposable of $\mathscr{P}'_{a_1}$ with support $\bar{\mathscr{P}}$ is elementary (5.2). If this support is restricted to $\{b_1 \vee c_1\}$, $\mathscr{P}$ equals $\{a_1, b_1, c_1\}$; if it has cardinality 2, $\mathscr{P}$ equals $\{a_1 > a_2, b_1, c_1\}$.

2) *Case $l = n = 2$, $m = 1$.* If $\bar{m} \geqslant 2$, we may further suppose that $b_2 < a_1$. Therefore $\mathscr{P}'_{a_1} \backslash \mathscr{P} = \{b_1 \vee c_1 > b_1 \vee c_2\}$. Since each indecomposable of $\mathscr{P}'_{a_1}$ present at $b_1 \vee c_1$ is elementary (5.2), we have $b_1 \vee c_1 \notin \bar{\mathscr{P}}$, hence $\{a_2, b_1 \vee c_2, c_1\} \subset \bar{\mathscr{P}}^+$, $c_2 \notin \bar{\mathscr{P}}^+$ and $a_3 \notin \bar{\mathscr{P}}^+$ if $\bar{l} \geqslant 3$. Thus $\bar{\mathscr{P}}^+$ is isomorphic to some $\hat{s} \coprod \hat{1} \coprod \hat{1}$, and we may apply to $\bar{\mathscr{P}}$ the results of case 1 above: If $s = 1$ or if $s = 2$ and $b_1 \in \bar{\mathscr{P}} = \bar{\mathscr{P}}^+$, $\mathscr{P}$ equals $\mathscr{P}^+ = \{a_1 > a_2, b_1, c_1 > c_2\}$; if $s = 2$ and $b_2 \in \bar{\mathscr{P}} = \bar{\mathscr{P}}^+$, $\mathscr{P}$ equals $\{a_2 < a_1 > b_2 < b_1, c_1 > c_2\} \tilde{\rightarrow} \hat{2} \coprod \hat{2} \nwarrow \hat{2}$.

3) *Case $l = 3$, $m = 1$, $n = 2$. Here we further suppose that $\mathscr{P}^+$ is not contained in a full subposet of $\mathscr{P}$ isomorphic to* $\hat{3} \coprod \hat{2} \nwarrow \hat{2}$, i.e. that the assumptions of our last lemma are satisfied. Then again $\mathscr{P}'_{a_1} \backslash \mathscr{P} = \{b_1 \vee c_1 > b_2 \vee c_2\}$ and $b_1 \vee c_1 \notin \bar{\mathscr{P}}$, hence $\{a_2 > a_3, b_1 \vee c_2, c_1\} \subset \bar{\mathscr{P}}^+$. Moreover, $\bar{m} \geqslant 2$ implies $b_2 \notin \bar{\mathscr{P}}^+$, or else $\mathscr{P}$ would contain the critical subset $\{a_2 > a_3, b_1 > b_2, c_1 > c_2\}$. If $\bar{\mathscr{P}}^+ = \{a_2 > a_3, b_1 \vee c_2, c_1\}$, we apply the results of case 1 to $\bar{\mathscr{P}}$ and infer that $\bar{\mathscr{P}} = \bar{\mathscr{P}}^+$, hence that $\mathscr{P} = \mathscr{P}^+ = \{a_1 > a_2 > a_3, b_1, c_1 > c_2\}$. If $\bar{\mathscr{P}}^+ = \{a_2 > a_3, b_1 \vee c_1 > b_1, c_1\}$, we apply case 2 and infer that $\bar{\mathscr{P}} = \bar{\mathscr{P}}^+$ and $\mathscr{P} = \mathscr{P}^+$ or that $\bar{\mathscr{P}} \tilde{\rightarrow} \hat{2} \coprod \hat{2} \nwarrow \hat{2}$. In this case, either $\bar{\mathscr{P}} = \bar{\mathscr{P}}^+ \cup \{c_2\}$ and $\mathscr{P} = \mathscr{P}^+$, or $\bar{\mathscr{P}} = \bar{\mathscr{P}}^+ \cup \{c_3\}$, $c_3 \nless b_1$ and $\mathscr{P} = \{a_3 < a_2 < a_1 > c_3 < c_2 < c_1, b_1\} \tilde{\rightarrow} \hat{1} \coprod \hat{3} \nwarrow \hat{3}$.

$3\frac{1}{2}$) *Case $l = 2$, $m = 1$, $n = 3$ and $a_2 \ngtr b_2 \nless c_1$.* Then $\mathscr{P}'_{a_1} \backslash \mathscr{P} = \{b_1 \vee c_1 > b_1 \vee c_2 > b_1 \vee c_3\}$. As above, $b_1 \vee c_1 \notin \bar{\mathscr{P}}$ and $c_1 \in \bar{\mathscr{P}}$. We claim that $b_1 \vee c_2 \notin \bar{\mathscr{P}}$: Otherwise, we have $a_3 \notin \bar{\mathscr{P}}^+$ (if $\bar{l} \geqslant 3$) and $c_2 \notin \bar{\mathscr{P}}^+$, by case 1 $\bar{\mathscr{P}}$ equals $\bar{\mathscr{P}}^+ \tilde{\rightarrow} \hat{1} \coprod \hat{s} \coprod \hat{1}$ where $s \in \{1, 2\}$, and $\bar{\mathscr{P}}$ cannot contain $\{b_1 \vee c_3, b_2\}$.

Therefore, $\bar{\mathscr{P}}^+ \supset \{a_2, b_1 \vee c_3, c_1 > c_2\}$, $a_3 \notin \bar{\mathscr{P}}^+$ if $\bar{l} \geqslant 3$, and $c_3 \notin \bar{\mathscr{P}}^+$. Furthermore, $b_3 \notin \bar{\mathscr{P}}^+$ if $\bar{m} \geqslant 3$, or else $\mathscr{P}$ would contain the critical subset $\{a_2, b_1 > b_2 > b_3, c_1 > c_2 > c_3\}$. Besides a_2, $b_1 \vee c_3$, c_1 and c_2, $\bar{\mathscr{P}}^+$ therefore contains at most b_1 and b_2. If $|\bar{\mathscr{P}}^+| = 4$, case 1 provides the contradiction

$\overline{\mathscr{P}}^+ = \overline{\mathscr{P}} \ni b_2$. If $b_1 \in \overline{\mathscr{P}}^+$ and $b_2 \notin \overline{\mathscr{P}}^+$, case 2 provides the contradiction $\overline{\mathscr{P}} \subset \overline{\mathscr{P}}^+ \cup \{a_3\} \ni b_2$.

If $b_1 \notin \overline{\mathscr{P}}^+$ and $b_2 \in \overline{\mathscr{P}}^+$, case 2 either implies $\overline{\mathscr{P}} = \overline{\mathscr{P}}^+$ and $\mathscr{P} = \{a_2 < a_1 > b_2 < b_1, c_1 > c_2 > c_3\} \tilde{\rightarrow} \hat{3} \coprod \hat{2} \nwarrow \hat{2}$, or $\overline{\mathscr{P}} = \{a_2 > a_3 < b_1 \vee c_3 > b_2, c_1 > c_2\}$ and $\mathscr{P} = \{a_1 > a_2 > a_3 < b_1 > b_2 < a_1, c_1 > c_2 > c_3\} \tilde{\rightarrow} \hat{3} \coprod \hat{3} \nwarrow\!\!\!\nearrow \hat{2}$ (the eventuality $\overline{\mathscr{P}} = \{a_2 > a_3 < c_1 > c_2, b_1 \vee c_3 > b_2\}$ is excluded because $\mathscr{P}$ cannot contain the critical subset $\{a_2 > a_3, b_1 > b_2, c_2 > c_3\}$).

Finally, if $\{b_1, b_2\} \subset \overline{\mathscr{P}}^+$, $\overline{\mathscr{P}}^+ \tilde{\rightarrow} \hat{3} \coprod \hat{2} \coprod \hat{1}$ is not contained in a full subposet of $\overline{\mathscr{P}}$ of the form $\hat{3} \coprod \hat{2} \nwarrow \hat{2}$ (otherwise $\{a_2 > a_3, b_1 > b_2, c_2 > c_3\}$ would be critical in $\mathscr{P}$). Thus we may apply case 3 to $\overline{\mathscr{P}}$ and either obtain $\overline{\mathscr{P}} = \overline{\mathscr{P}}^+$ and $\mathscr{P} = \mathscr{P}^+ \cup \{b_2\} \tilde{\rightarrow} \hat{3} \coprod \hat{2} \nwarrow \hat{2}$, or $\overline{\mathscr{P}} = \overline{\mathscr{P}}^+ \cup \{c_3\}$ and $\mathscr{P} = \mathscr{P}^+ \cup \{b_2\}$, or $\overline{\mathscr{P}} = \overline{\mathscr{P}}^+ \cup \{c_4\}$ and $\mathscr{P} = \{a_2 < a_1 > b_2 < b_1, c_1 > c_2 > c_3 > c_4 < a_1\} \tilde{\rightarrow} \hat{4} \nearrow \hat{2} \nwarrow \hat{2}$.

Thus, case $3\frac{1}{2}$ yields the posets $\hat{3} \coprod \hat{2} \nwarrow \hat{2}$, $\hat{3} \coprod \hat{3} \nwarrow\!\!\!\nearrow \hat{2}$ and $\hat{4} \nearrow \hat{2} \nwarrow \hat{2}$.

4) *Case* $l = 4$, $m = 1$, $n = 2$. If the assumptions of our last lemma were not fulfilled, $\mathscr{P}$ would contain the critical subset $\{a_1 > a_2 > a_3 > a_4, b_1 > b_2 < c_1 > c_2\}$. Hence, $\mathscr{P}'_{a_1} \backslash \mathscr{P} = \{b_1 \vee c_1 > b_2 \vee c_2\}$, $b_1 \vee c_1 \notin \overline{\mathscr{P}}$, $\{a_2 > a_3 > a_4, b_1 \vee c_2, c_1\} \subset \overline{\mathscr{P}}^+$, $c_2 \notin \overline{\mathscr{P}}^+$ and $a_5 \notin \overline{\mathscr{P}}^+$ if $\bar{l} \geq 5$. We further have $b_2 \notin \overline{\mathscr{P}}^+$ if $\overline{m} \geq 2$, or else $\mathscr{P}$ would contain the critical subset $\{a_2 > a_3, b_1 > b_2, c_1 > c_2\}$. Since $\overline{\mathscr{P}}^+ \not\tilde{\rightarrow} \hat{3} \coprod \hat{1} \coprod \hat{1}$ by case 1, we infer that $\overline{\mathscr{P}}^+ = \{a_2 > a_3 > a_4, b_1 \vee c_2 > b_1, c_1\}$.

Now suppose that case 3 applies to $\overline{\mathscr{P}}$. The premise $\overline{\mathscr{P}} \tilde{\rightarrow} \hat{1} \coprod \hat{3} \nwarrow \hat{3}$ then implies the existence of a critical subset $\{a_3 > a_4, b_1 > b_2, c_1 > c_2\}$ in $\mathscr{P}$. Therefore, $\overline{\mathscr{P}} = \overline{\mathscr{P}}^+$ and $\mathscr{P} = \mathscr{P}^+ \tilde{\rightarrow} \hat{4} \coprod \hat{2} \coprod \hat{1}$.

Let us finally suppose that case $3\frac{1}{2}$ applies to $\overline{\mathscr{P}}$. In the subcases $\overline{\mathscr{P}} \tilde{\rightarrow} \hat{3} \coprod \hat{2} \nwarrow \hat{2}$, $\overline{\mathscr{P}} \tilde{\rightarrow} \hat{3} \coprod \hat{3} \nwarrow\!\!\!\nearrow \hat{2}$ and $\overline{\mathscr{P}} \tilde{\rightarrow} \hat{4} \nearrow \hat{2} \nwarrow \hat{2}$, the premise $c_3 \in \overline{\mathscr{P}}$ implies the existence of a critical subset $\{a_2 > a_3 > a_4, b_1, c_1 > c_2 > c_3\}$ in $\mathscr{P}$. In the first subcase, we infer that $\overline{\mathscr{P}} = \overline{\mathscr{P}}^+ \cup \{c_2\}$ and $\mathscr{P} = \mathscr{P}^+ \tilde{\rightarrow} \hat{4} \coprod \hat{2} \coprod \hat{1}$. In the second, we infer that $\overline{\mathscr{P}} = \overline{\mathscr{P}}^+ \cup \{b_2, c_3\}$, where $b_2 < c_1$; since $\mathscr{P}$ has no critical subset of the form $\hat{4} \coprod \hat{2} \nwarrow \hat{2}$, it follows that $a_1 > b_2$ and $\mathscr{P} = \mathscr{P}^+ \cup \{b_2\} \tilde{\rightarrow} \hat{4} \nwarrow \hat{2} \nearrow \hat{2}$. In the third subcase, we finally obtain the contradiction $a_5 < b_1 \vee c_2$ though $a_5 \nless b_1$ and $a_5 \nless c_2$ $\surd$

5.7. We are now prepared to apply Proposition 1.5 to the representations of the poset $\mathscr{P}$. For a given dimension-vector $d \in \mathbb{N}^{1+t}$, the isoclasses of representations of $\mathscr{P}$ are the orbits of a group G_d (1.4) of dimension $d_0^2 + d_1^2 + \cdots + d_t^2 + \sum_{p_i < p_j} d_i d_j$ acting on a space $\mathfrak{M} = k^{d_0 \times \bar{d}}$ of dimension $d_0 d_1 + \cdots + d_0 d_t$. The difference $q_{\mathscr{P}}(d) = \dim G_d - \dim \mathfrak{M} = \sum_{l=0}^{t} d_l^2 + \sum_{p_i < p_j} d_i d_j - \sum_{i=1}^{t} d_0 d_i$ is the value at d of a *unit form*, i.e. of a $\mathbb{Z}$-valued quadratic form which is defined on some $\mathbb{Z}^n$ and has the value 1 at the natural basis vectors of $\mathbb{Z}^n$. Such a unit form is called *weakly positive* if its values on $\mathbb{N}^n \backslash \{0\}$ are >0. The points of $\mathbb{Z}^n$ where it takes the value 1 are the *roots* of the unit form.

Theorem.[10] *A finite poset $\mathscr{P}$ is finitely represented if and only if its unit form $q_{\mathscr{P}}$ is weakly positive. If this is the case, the function mapping each representation onto its dimension-vector provides a bijection between the isoclasses of indecomposable representations and the roots of $q_{\mathscr{P}}$ in $\mathbb{N}^{1+t}$.*

Proof. If $q_{\mathscr{P}}$ is not weakly positive, there is a $d \in \mathbb{N}^{1+t} \setminus \{0\}$ such that $\dim G_d - \dim \mathfrak{M} = q_{\mathscr{P}}(d) \leqslant 0$. It follows by 1.5 that there are infinitely many isoclasses of representations with dimension-vector d, hence infinitely many indecomposables with dimension-vector $\leqslant d$.

If $q_{\mathscr{P}}$ is weakly positive, $\mathscr{P}$ admits no antichain of cardinality 4, say $\mathscr{Q} = \{p_1, p_2, p_3, p_4\}$, or else we would have $q_{\mathscr{P}}(d) = q_{\mathscr{Q}}(d') = \sum_{i=0}^{4} d_i^2 - d_0 \sum_{i=1}^{4} d_i = 0$ for $d = [2\ 1\ 1\ 1\ 1\ 0\ \ldots]^T$ and $d' = [2\ 1\ 1\ 1\ 1]^T$. In the same way, one proves that $q_{\mathscr{Q}}$ is not weakly positive when $\mathscr{Q}$ is one of the four other critical posets of Fig. 3. It follows that $\mathscr{P}$ contains no critical subset, hence that it is finitely represented.

The second statement directly follows from 1.5 and from the fact that the indecomposable representations of a finitely represented poset $\mathscr{P}$ are schurian. To prove this fact, one may examine the 43 omnipresent indecomposables associated with the 14 posets of Fig. 4. We prefer the following *direct proof*, which proceeds by induction on $\iota(\mathscr{P})$ (5.3).

Let (V, f, X) be an indecomposable space on $M = M^{\mathscr{P}}$, where $X = \bigoplus_{p \in \mathscr{P}} p^{n_p}$. We may suppose that $n_p \geqslant 1, \forall p \in \mathscr{P}$, that $\mathscr{P}$ has at least three elements and that $f: V \to M(X)$ is the inclusion of a subspace. Then $\mathscr{P}$ has 3 maximal elements a, b, c (first alinéa of 5.6), and the endomorphisms of (V, f, X) are identified with the endomorphisms ψ of X such that $M(\psi)(V) \subset V$. We suppose that ψ is nilpotent and shall infer that $\psi = 0$.

We write ψ as a matrix $[\psi_{pq}]$, where $\psi_{pq}: q^{n_q} \to p^{n_p}$. *We first claim that* $\psi_{pq} = 0$ *if* $p \in \mathscr{P}$ *is not maximal*: Indeed, assume that $p < a$ and derive $\mathscr{P}$ at a (5.3). If (U, h, Z) is the space on $M^{\mathscr{P}'_a}$ attached to (V, f, X) by the derivation algorithm, we have

$$M^{\mathscr{P}'_a}(Z) = M(X) \Big/ \left(M(a^{n_a}) + V \cap M\left(\bigoplus_{r \between a} r^{n_r} \right) \right).$$

We infer that, in case $\psi_{pq} \neq 0$, ψ would induce a non-zero nilpotent endomorphism of the indecomposable (U, h, Z), thus contradicting our induction hypothesis.

Let us now denote by ξ and η the endomorphisms of X such that $\xi_{aq} = \psi_{aq}$, $\eta_{bq} = \psi_{bq}$, $\eta_{cq} = \psi_{cq}$ for all $q \in \mathscr{P}$, and $\xi_{rq} = 0 = \eta_{rq}$ in all other cases. Thus we have $\psi = \xi + \eta$, and the images of $M(\xi)$ and $M(\eta)$ lie in $M(a^{n_a})$ and $M(b^{n_b} \oplus c^{n_c})$ respectively. Since ψ induces zero on (U, h, Z), the image of $M(\eta)$ lies in $V \cap M(b^{n_b} \oplus c^{n_c}) \subset V$, hence η is identified with the endomorphism of (V, f, X), and so is ξ. Since ψ is nilpotent, so are ξ and η (3.5, Example 3).

Now we claim that $\eta = 0$: Indeed, if we derive $\mathscr{P}$ at b, the representation (U', h', Z') of $\mathscr{P}'_b$ associated with (V, f, X) is such that

$$(*) \quad M^{\mathscr{P}'_b}(Z') = M(X) \Big/ \left(M(b^{n_b}) + V \cap M\left(\bigoplus_{r \between b} r^{n_r} \right) \right).$$

By our induction hypothesis, η induces zero on (U', h', Z'). Hence, the images of the maps $M(\eta_{cq}) = M(\psi_{cq})$ lie in $V \cap M(\bigoplus_{r \between b} r^{n_r})$. On the other hand, $V \cap$

$M(c^{n_c})$ is 0 because (V, f, X) has no summand of the form $(c, [1])$. We infer that $\eta_{cq} = \psi_{cq} = 0$, and similarly $\eta_{bq} = \psi_{bq} = 0$.

Finally, *it remains to show that* $\xi = 0$: In fact, since ξ vanishes on (U', h', Z') (induction hypothesis), (*) shows that Im $M(\xi) \subset V$. Since $V \cap M(a^{n_a}) = 0$, we infer that $\xi = 0$. $\surd$

***5.8.** We finally turn to *biinvolutive posets*, supposing that $\mathscr{P}$ is equipped with two involutions satisfying the conditions 1) – 4) of 4.11. Our objective is to state a criterion *ensuring* that the module $N^{\mathscr{P}}$ over $\oplus\mathscr{S}_{\mathscr{P}}$ defined in 4.11 is finitely spaced. To this effect, *we* may by 4.9 further *suppose that* $\mathscr{P}$ *contains no subset* $\{p, q\}$ *such that* $p > p^*$, $q > q^*$, $p \between q$ *and* $p^* \between q^*$.

We first construct an extended biinvolutive poset $\bar{\mathscr{P}} = \mathscr{P} \cup \{0, 1\}$ by adding to $\mathscr{P}$ a minimum $0 = 0^*$ and a maximum $1 = 1^*$. Then we introduce a new poset $\mathscr{S}t(\mathscr{P})$ whose elements are the *staggered sequences* of $\bar{\mathscr{P}}$, i.e. the sequences $t = (t_0, t_1, \ldots, t_n)$, $n \in \mathbb{N}$, of elements of $\bar{\mathscr{P}}$ such that $\mathscr{P} \ni t_0 \neq t_0^* \between t_1 \neq t_1^* \between \ldots \between t_{n-1} \neq t_{n-1}^*$ and $t_n = t_n^*$; in case $t_n \in \mathscr{P}$, we also require that $t_{n-1}^* \between t_n$. If s and t are staggered sequences with $m + 1$ and $n + 1$ terms respectively, the inequality $s \leqslant t$ is equivalent to the existence of an $l \in \mathbb{N}$ satisfying the inequality $l \leqslant \inf(m, n)$, the condition *a)* below and at least one of the conditions b_1), b_2), b_3):

$$\begin{aligned}
&a)\ s_i \leqslant t_i &&\text{and}\quad (s_i, t_i)^* \neq (s_i, t_i) \quad \text{if } i < l.\\
&b_1)\ s_l \leqslant t_l &&\text{and}\quad (s_l, t_l)^* = (s_l, t_l).\\
&b_2)\ s_{l-1}^* < t_l &&\text{and}\quad (s_{l-1}^*, t_l)^* = (s_{l-1}^*, t_l).\\
&b_3)\ s_l < t_{l-1}^* &&\text{and}\quad (s_l, t_{l-1}^*)^* = (s_l, t_{l-1}^*).
\end{aligned}$$

Theorem.[11] *The module* $N^{\mathscr{P}}$ *attached to a finite biinvolutive poset* $\mathscr{P}$ (4.11) *is finitely spaced if and only if the associated poset* $\mathscr{S}t(\mathscr{P})$ *is finite and finitely represented.*

The raison d'être of the staggered sequences is the following: With each staggered sequence $t = (t_0, \ldots, t_n)$ we associate a space $(V_t, f_t, \bar{t})$ on $N^{\mathscr{P}}$ such that $\bar{t} = \bar{t}_0 \oplus \cdots \oplus \bar{t}_{n-1} \oplus \bar{t}_n$ (4.11) if $t_n \in \mathscr{P}$ and $\bar{t} = \bar{t}_0 \oplus \cdots \oplus \bar{t}_{n-1}$ if $t_n = 0, 1$. The vector space $N^{\mathscr{P}}(\bar{t})$ then has $t_0, t_0^*, \ldots, t_{n-1}, t_{n-1}^*, t_n$ as a basis if $t_n \in \mathscr{P}$ and $t_0, t_0^*, \ldots, t_{n-1}, t_{n-1}^*$ if $t_n = 0, 1$. The map f_t is the inclusion of the subspace

$$V_t = kt_0 \oplus k(t_0^* + t_1) \oplus \cdots \oplus k(t_{n-2}^* + t_{n-1}) \oplus \begin{cases} 0 & \text{if } t_n = 0 \\ k(t_{n-1}^* + t_n) & \text{if } t_n \in \mathscr{P} \\ kt_{n-1}^* & \text{if } t_n = 1. \end{cases}$$

Besides V_t, we consider the hyperplane V_t' generated by the listed basis vectors of V_t except t_0. When t varies, the hyperplanes V_t' give rise to a spectroid $\mathscr{S}t^*(\mathscr{P})$ whose objects are the staggered sequences, whose morphisms $s \to t$ are identified with the morphisms from $(V_s, f_s, \bar{s})$ to $(V_t, f_t, \bar{t})$ which map V_s' into V_t'. Now, if $\mathscr{J}$ denotes the annihilator of the left module $\bar{V}: s \mapsto V_s/V_s'$, the residue-category $\mathscr{S}t^*(\mathscr{P})/\mathscr{J}$ is identified with the spectroid $k\mathscr{S}t(\mathscr{P})$ attached to the poset $\mathscr{S}t(\mathscr{P})$;

and the extension $\oplus\bar{V}$ of $\bar{V}$ to the additive hull $\oplus\mathcal{S}t^*(\mathcal{P})$ is identified with $M^{\mathcal{S}t(\mathcal{P})}$ (4.1 and 4.6). Thus we obtain an epivalence $(\oplus\bar{V})^k \to (M^{\mathcal{S}t(\mathcal{P})})^k$.

On the other hand, we have a natural functor $(\oplus\bar{V})^k \to (N^{\mathcal{P}})^k$ which maps $(W, g, \oplus_i s^i)$ onto $(W', g', \oplus_i \bar{s}^i)$, where the s^i denote staggered sequences, W' the fibre-product of

$$W \xrightarrow{\;g\;} \oplus_i\ V_{s^i}/V'_{s^i} \xleftarrow{\text{can.}} \oplus_i\ V_{s^i}$$

and g' the composition

$$W' \longrightarrow \oplus_i\ V_{s^i} \xrightarrow{\text{incl.}} \oplus_i\ N^{\mathcal{P}}(\bar{s}^i)$$

induced by g. Thus we can relate the representations of the poset $\mathcal{S}t(\mathcal{P})$ with those of the biinvolutive poset $\mathcal{P}$. Unfortunately, this does not provide a bijection between the isoclasses ...[11] *

5.9. Remarks and References

1. Advice repeated 2 000 years before Gauss by Chang Tsang throughout the eighth of his "Nine Books on Arithmetics."
2. We describe a finite poset by its (Hasse-)quiver: An arrow $p \to q$ signifies that $p < q$ and that there is no r satisfying $p < r < q$.
3. The direct sums considered here are those of the category of posets: Thus, $\mathcal{P} \coprod \mathcal{Q}$ is the disjoint union of $\mathcal{P}$ and $\mathcal{Q}$. It induces the original order relation on $\mathcal{P}$ and $\mathcal{Q}$, and each $p \in \mathcal{P}$ is incomparable with each $q \in \mathcal{Q}$.
4. [110, Nazarova/Roiter, 1971/72].
5. [91 and 92, Kleiner, 1971/72].
6. Below each poset we indicate the symbol used for identification.
7. *omnipresent* = present at each point = sincere in Ringel's terminology = faithful in Kleiner's.
8. A subposet of $\mathcal{P}$ is *full* if it carries the order relation induced by $\mathcal{P}$.
9. A finite poset $\mathcal{P}$ of width w is the disjoint union of w chains. This theorem can easily be proved by induction on the cardinality of $\mathcal{P}$: If $\mathcal{P}$ contains only one antichain A of cardinality w, then $\mathcal{P}\backslash\{a\}$ has width $w-1$ for each $a \in A$ and is the disjoint union of $w-1$ chains $C_1, \ldots, C_{w-1}$; hence $\mathcal{P} = C_1 \cup \cdots \cup C_{w-1} \cup \{a\}$. If $\mathcal{P}$ contains exactly two antichains A and B of cardinality w, then it contains a chain $\{a, b\}$ such that $a \in A\backslash B$ and $b \in B\backslash A$; it follows that $\mathcal{P}\backslash\{a, b\}$ had width $w-1$ and is the disjoint union of $w-1$ chains $C_1, \ldots, C_{w-1}$; hence $\mathcal{P} = C_1 \cup \cdots \cup C_{w-1} \cup \{a, b\}$. Finally, if $\mathcal{P}$ has at least 3 antichains of cardinality w, then it has an antichain $\{a_1, \ldots, a_w\}$ which contains a non-maximal and a non-minimal element; it follows that $\mathcal{P}^- := \{p \in \mathcal{P}: \exists i, p \leqslant a_i\} \neq \mathcal{P} \neq \{p \in \mathcal{P}: \exists i, a_i \leqslant p\} =: \mathcal{P}^+$, hence that $\mathcal{P}^-$ and $\mathcal{P}^+$ are the disjoint unions of chains $C_i^- \ni a_i$ and $C_i^+ \ni a_i$ respectively; we infer that $\mathcal{P} = (C_1^- \cup C_1^+) \cup \cdots \cup (C_w^- \cup C_w^+)$.
10. [42, Drozd, 1974].
11. In [113, Nazarova/Roiter, 1991] the necessity is proved, the sufficiency announced. The theorem (and the analogous statement for "weakly completed posets") is proved for posets with one involution in [111, Nazarova/Roiter, 1983]. In this particular case, all indecomposables of $(N^{\mathcal{P}})^k$ come from $(\oplus\bar{V})^k$. Another important particular case, to which the general case can in some sense be reduced, is treated in [112, Nazarova/Roiter, 1988]. In this case, a proof is given, and the indecomposables are described if $N^{\mathcal{P}}$ is finitely spaced; they do not all come from $(\oplus\bar{V})^k$.

6. Roots

In the course of our investigation of posets, we discussed a remarkable interplay of Representation and Number Theory (5.7). The link between them is a special class of integral quadratic forms thus brought into the limelight. The main features of these forms are delineated in the present section.

6.1. In our terminology, a *unit form* is a quadratic form

$$q\colon \mathbb{Z}^n \to \mathbb{Z}, \qquad x \mapsto \sum_{i=1}^{n} x_i^2 + \sum_{i<j} q_{ij} x_i x_j, \qquad q_{ij} \in \mathbb{Z},$$

with value 1 at each natural basis vector.[1] The obvious extension of q to $\mathbb{R}^n$ will be denoted by $\bar{q}$. We further denote by $q(x|y)$ and $\bar{q}(x|y)$ the values of the associated bilinear forms and set $q_{ij} = q_{ji}$ if $i > j$:

$$q(x|y) = q(x+y) - q(x) - q(y) = 2\sum_{i=1}^{n} x_i y_i + \sum_{i \neq j} q_{ij} x_i y_j.$$

We usually depict the unit form q as a *bigraph* with vertices $1, 2, \ldots, n$. Two vertices i, j are linked by $|q_{ij}|$ full edges if $q_{ij} < 0$, by q_{ij} broken edges if $q_{ij} > 0$. As an illustration, we reproduce the bigraphs of the unit forms associated with the critical posets of 5.4 (Fig. 1).

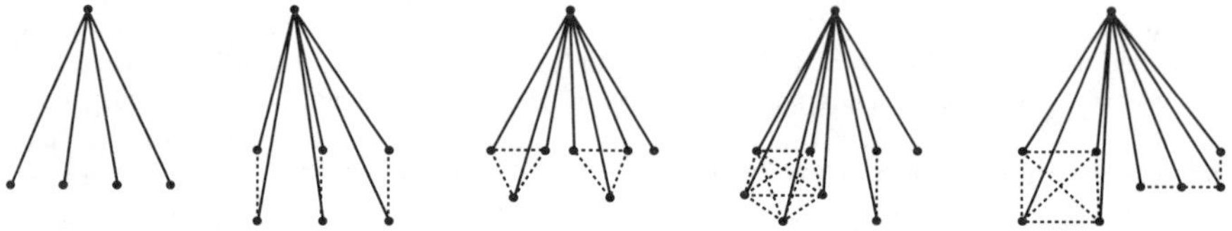

Fig. 1

A *root* of the unit form q is an "integral" vector $x \in \mathbb{Z}^n$ such that $q(x) = 1$. Each sequence $r = (r^1, \ldots, r^m)$ formed by such roots gives rise to a new unit form q_r defined by $q_r(y) = q(y_1 r^1 + \cdots + y_m r^m)$. In particular, if $I = \{i_1 < i_2 < \cdots < i_m\}$ is a subset of $\{1, 2, \ldots, n\}$, we denote by q_I and call *restriction of q to I* the unit form q_r associated with the subsequence $r = (e^{i_1}, \ldots, e^{i_m})$ of the natural basis $(e^1, \ldots, e^n)$ of $\mathbb{Z}^n$. In case $I = \{1, \ldots, n\} \setminus \{i\}$, we set $q_I =: q^i$.

6.2. A unit form q on $\mathbb{Z}^n$ is called *positive* if $0 \neq x \in \mathbb{Z}^n$ implies $q(x) > 0$. As it is easy to prove, this is equivalent to saying that the extension $\bar{q}$ of q to $\mathbb{R}^n$ is positive definite (compare with 6.4b). It implies that the number of roots of q is finite.

Theorem. *If the unit form q on $\mathbb{Z}^n$ is positive, q admits a root basis b (=basis of $\mathbb{Z}^n$ formed by roots b^i of q) such that the connected components of the bigraph of q_b are Dynkin graphs.*

The Dynkin graphs are the bigraphs (without broken edge) depicted in Fig. 2. We include a proof of the theorem, though it is highly classical.[2]

Proof. We first notice that $2 \pm q_{bij} = q(b^i \pm b^j) > 0$ implies $|q_{bij}| \leqslant 1$. This means that the bigraph of q_b admits at most one edge between i and j. If this edge is broken, we set $b'^j = b^j - b^i$ and $b'^l = b^l$ for $l \neq j$. Since $q(b'^j) = q(b^j - b^i) = q(b^j) + q(b^i) - q(b^j|b^i) = 2 - q_{bij} = 1$, we thus obtain a root basis b' which is better than b in the sense that $\pi(b') > \pi(b)$, if $\pi(b)$ denotes the number of roots of q within $\sum \mathbb{N}b = \{\sum_{i=1}^n x_i b^i : x_i \in \mathbb{N}\}$ (notice that $b'^j \in \sum \mathbb{N}b' \supset \sum \mathbb{N}b \not\ni b'^j$).

The argument shows that we can increase $\pi(b)$ whenever the bigraph of q_b contains a broken edge. Since $\pi(b)$ is bounded by the number of roots of q, we infer that there exists a root-basis b whose bigraph has no broken edge. So it remains to convince oneself that a graph which contains no extended Dynkin graph (Fig. 3) is a disjoint sum of Dynkin graphs. The unit form of such a sum is positive (exercise). As for the unit forms of extended Dynkin graphs, they vanish on the vectors of $\mathbb{Z}^n$ whose components are the numbers attached to the vertices of Fig. 3. $\surd$

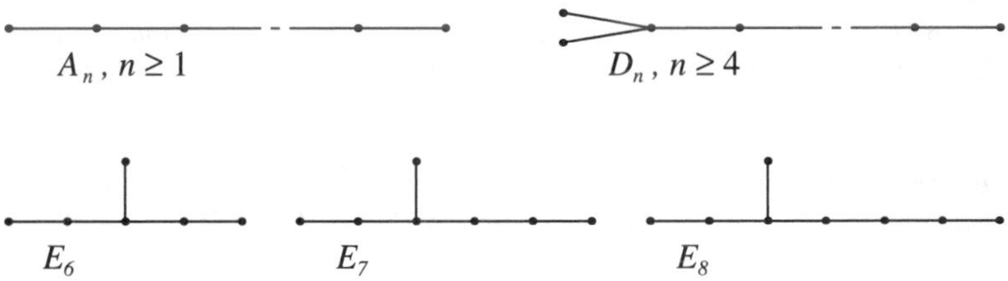

Fig. 2 (n = number of vertices)

6.3. Given two vectors $x, y \in \mathbb{Z}^n$, we write $y \leqslant x$ if $y_i \leqslant x_i$, $\forall i$. Accordingly, x is called *positive* if $0 \leqslant x$. And a unit form q on $\mathbb{Z}^n$ is called *weakly positive* if $0 < x$ (i.e. $0 \leqslant x$ and $0 \neq x$) implies $q(x) > 0$.

The importance of weakly positive forms calls for a more practicable definition. To this effect, we say that a unit form p on $\mathbb{Z}^m$ is *critical* if each restriction p^i is weakly positive though p itself is not. According to this new definition, a necessary and sufficient condition for a unit form q on $\mathbb{Z}^n$ to be weakly positive is that q_I is critical for no subset I of $\{1, \ldots, n\}$.

Superseding one notion by another is sensible here because critical forms can be detected with ease. In fact, they are closely related to the *extended Dynkin graphs* of Fig. 3. If $\tilde{\Delta}$ denotes one of these graphs, each enumeration of the vertices gives rise to a unit form with "bigraph" $\tilde{\Delta}$ (6.1) which we simply denote by $q^{\tilde{\Delta}}$. In case $\tilde{\Delta} = \tilde{D}_5$ for instance, if we number the vertices from the left to the right, we have $q^{\tilde{\Delta}}(x) = \sum_{i=1}^{6} x_i^2 - x_1x_3 - x_2x_3 - x_3x_4 - x_4x_5 - x_4x_6$. In the general case, $q^{\tilde{\Delta}}$ is positive semi-definite. Its null space equals some $\mathbb{Z}\delta^{\tilde{\Delta}}$, where $\delta^{\tilde{\Delta}} \in \mathbb{Z}^n$ is >0 and *omnipresent* ($\delta_i^{\tilde{\Delta}} \neq 0, \forall i$). The components $\delta_i^{\tilde{\Delta}}$ are the numbers attached to the vertices in Fig. 3. In the case $\tilde{\Delta} = \tilde{D}_5$ considered above, $\delta^{\tilde{\Delta}} = [1\ 1\ 2\ 2\ 1\ 1]^T$.

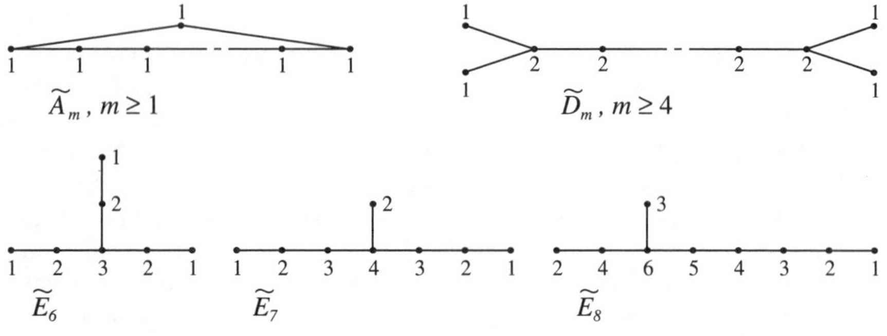

Fig. 3 ($n = m + 1$ = number of vertices)

The above description shows that the unit forms $q^{\tilde{\Delta}}$ are critical. By changes of bases, they can even produce other critical forms. More precisely, let us say that a root basis b of $\mathbb{Z}^{m+1}$ is *clasping* if the coordinates of the *isotropic generator* $\delta^{\tilde{\Delta}}$ in the basis b are >0. The condition clearly implies that the induced unit form $q_b^{\tilde{\Delta}}$ is critical.

Theorem.[3] *If $n \geqslant 3$, each critical unit form on $\mathbb{Z}^n$ is equal to $q_b^{\tilde{\Delta}}$ for some extended Dynkin graph $\tilde{\Delta}$, some enumeration of the vertices and some clasping root basis b.*

The critical unit forms on $\mathbb{Z}^2$ are the functions $x \mapsto x_1^2 + x_2^2 - sx_1x_2$, $s \in \mathbb{N} \setminus \{0, 1\}$. On the other hand, if $\tilde{\Delta}$ has 2 vertices, it coincides with $\tilde{A}_1$; in this case $q_b^{\tilde{\Delta}}(x) = q^{\tilde{\Delta}}(x) = x_1^2 + x_2^2 - 2x_1x_2$ for all clasping root bases b.

The theorem opens the way for a complete classification of the critical unit forms.[4] For instance, the bigraphs of the critical forms $q_b^{\tilde{\Delta}}$ with $\tilde{\Delta} = \tilde{D}_5$ are depicted in Fig. 4. Be that as it may: For the practical purpose of detecting weakly positive unit forms, it suffices to know that the extension $\bar{q}$: $\mathbb{R}^n \to \mathbb{R}$ of a critical q is positive semi-definite and that its null space is generated by some vector in $\mathbb{N}^n \setminus \{0\}$.

6.4. Proof of Theorem 6.3.

a) If p is a quadratic form on $\mathbb{R}^n$, *the space* $N_p = \{x \in \mathbb{R}^n \setminus \{0\}, p(x) \leqslant 0\}$ *has at most 2 connected components*: Indeed, we may suppose that $p(x) = x_1^2 + \cdots + x_s^2 - x_{s+1}^2 - \cdots - x_{s+t}^2$, where $0 \leqslant s + t \leqslant n$. In case $t = 0$, the number of components of $N_p \xrightarrow{\sim} \mathbb{R}^{n-s} \setminus \{0\}$ is 0 if $s = n$, 2 if $s = n - 1$, 1 if $s \leqslant n - 2$. In case $t \geqslant 1$, we have a continuous surjection

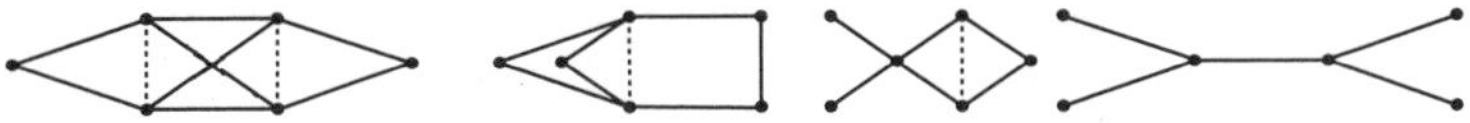

Fig. 4

$$f: \mathbb{B}^s \times \mathbb{S}^{t-1} \times (\mathbb{R}_{\geqslant 0} \times \mathbb{R}^{n-s-t} \setminus \{(0,0)\}) \to N_p, (y, z, u, v) \mapsto \begin{bmatrix} uy \\ uz \\ v \end{bmatrix},$$

where $\mathbb{B}^s = \{y \in \mathbb{R}^s: \sum_{i=1}^s y_i^2 \leqslant 1\}$, $\mathbb{S}^{t-1} = \{z \in \mathbb{R}^t: \sum_{j=1}^t z_j^2 = 1\}$ and $\mathbb{R}_{\geqslant 0} = \{x \in \mathbb{R}: x \geqslant 0\}$. Our statement follows from the fact that the domain of f is connected if $t \geqslant 2$ and has 2 components if $t = 1$.

b) *If a unit form p on $\mathbb{Z}^r$ is weakly positive, then $0 \neq x \in \mathbb{R}^r$ and $x_i \geqslant 0, \forall i$, imply $\bar{p}(x) > 0$*: The statement is clear indeed if $x \in \mathbb{Q}^r$. In the general case, we infer by continuity that $\bar{p}(x) \geqslant 0$. Now assume that $\bar{p}(x) = 0$, set $I = \{i: x_i \neq 0\} = \{i_1 < \cdots < i_s\}$ and $y = [x_{i_1} \ldots x_{i_s}]^T$. Then we have $y > 0$, $\bar{p}_I(y) = 0$ and $\bar{p}_I(z) > 0$ for all $z \in \mathbb{Q}^s$ near y. This is only possible if $\bar{p}_I$ is positive semidefinite. But then the null space of $\bar{p}_I$ is defined by linear equations with integral coefficients; it therefore contains some $z \in \mathbb{Q}^s$ near y which provides the required contradiction $\bar{p}_I(z) = 0$.

c) Set $\Pi = \{x \in \mathbb{R}^n: x_i > 0, \forall i\}$, $\partial\Pi = \bar{\Pi} \setminus \Pi = \mathbb{R}^n_{\geqslant 0} \setminus \Pi$ and $N = N_{\bar{q}}$, where q is a critical unit form on $\mathbb{Z}^n$. Since all the restrictions q^i are weakly positive, part b) above implies $N \cap \partial\Pi = \varnothing$, though $N \cap \Pi \neq \varnothing$. Since Π is a connected component of $\mathbb{R}^n \setminus \partial\Pi$, $N \cap \Pi$ is a union of connected components of N. Using part a), we infer that N has two connected components, namely $N \cap \Pi$ and $-N \cap \Pi$. *In particular, $\bar{q}(x) > 0$ if $x \notin \Pi \cup \{0\} \cup (-\Pi)$.*

d) Now suppose that $n \geqslant 3$, and consider two natural basis vectors e^i, e^j. Since the intersection of $\mathbb{R}e^i \oplus \mathbb{R}e^j$ with $\Pi \cup \{0\} \cup (-\Pi)$ is $\{0\}$, it follows from part c) that $q_{\{i,j\}}$ is positive (6.2). *In particular, $q_{ij} = q(e^i|e^j) = -1, 0$ or 1.*

e) Under the assumptions of c) and d), *there exists a root basis $c = (c^1, \ldots, c^n)$ such that q_c is critical and satisfies $q_{cin} \leqslant 0$ for all $i < n$*: Indeed, set $I = \{i \neq n: q_{in} = 1\}$, $c^i = e^i - e^n$ if $i \in I$ and $c^i = e^i$ if $i \notin I$. Then $q_{cin} = q_c(e^i|e^n) = q(e^i - e^n|e^n) = -1$ if $i \in I$ and $q_{cin} = q_{in} \leqslant 0$ if $i \notin I$. Moreover, q_c^j is positive if $j \neq n$ because $\bigoplus_{i \neq j} \mathbb{R}c^i = \bigoplus_{i \neq j} \mathbb{R}e^i$. Similarly, q_c^n is positive because $\bigoplus_{i \neq n} \mathbb{R}c^i$ intersects $\Pi \cup \{0\} \cup (-\Pi)$ in $\{0\}$ (part c). Finally, $x \in \mathbb{N}^n \setminus \{0\}$ and $q(x) \leqslant 0$ imply $q_c(x') = q(x) \leqslant 0$, where $x'_i = x_i$ for $i < n$ and $x'_n = x_n + \sum_{i \in I} x_i$.

f) Under the assumptions of c) and d), *there exists a root basis $c = (c^1, \ldots, c^n)$ such that q_c is critical and satisfies $q_{cij} \leqslant 0$ for all $i \neq j$*: We first construct a root basis d such that q_d is critical and satisfies $q_{din} \leqslant 0$ if $i < n$. We then consider the root bases c which satisfy $c^n = d^n$ and $\bigoplus_{i<n} \mathbb{R}c^i = \bigoplus_{i<n} \mathbb{R}d^i$ in addition to the requirements of part e). Among these bases we choose a basis c such that the number $\rho(c)$ of roots of the positive form q_d^n within $\sum_{i=1}^{n-1} \mathbb{N}c^i$ is maximal.

It remains to prove that $q_{cij} \leqslant 0$ if $i \neq j \neq n \neq i$: Otherwise, we could excerpt one of the 4 bigraphs of Fig. 5 from the bigraph of q_c. In the first three cases, we then could set $c'^i = c^i - c^j$ and $c'^l = c^l$ for $l \neq i$. In the fourth case, we could set

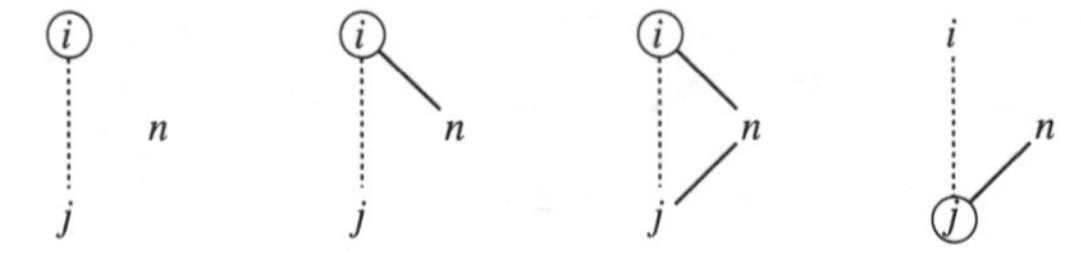

Fig. 5

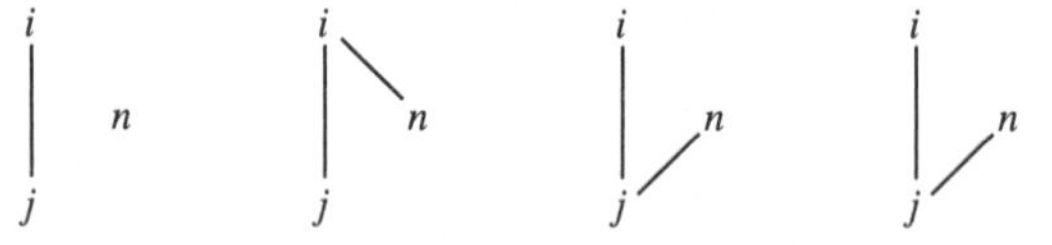

Fig. 6

$c'^j = c^j - c^i$ and $c'^l = c^l$ for $l \neq j$. In all cases, c' would be among the root bases considered above (the respective excerpts of the bigraphs of $q_{c'}$ are depicted in Fig. 6) and give rise to the contradiction $\rho(c') > \rho(c)$ (since $c'^i \notin \sum_{i=1}^{n-1} \mathbb{N}c^i$).

g) If the bigraph of a critical unit form contains no broken edge, the bigraph is an extended Dynkin graph (exercise! Compare with 6.2). Accordingly, the form q_c of part f) is equal to some $q^{\tilde{\Delta}}$. Now, if S is the matrix with columns $c^1, \ldots, c^n$ and b (resp. $-b$) the root basis formed by the columns of S^{-1} (resp. of $-S^{-1}$), we have $q = q_{cb} = q_b^{\tilde{\Delta}} = q_{-b}^{\tilde{\Delta}}$. Since q is critical, b or $-b$ is clasping. $\surd$

6.5. We now turn to the description of positive roots. To this effect, we remark that each root x of a unit form q on $\mathbb{Z}^n$ gives rise to an involutive *reflection*

$$\sigma_x\colon \mathbb{Z}^n \to \mathbb{Z}^n,\ y \mapsto y - q(y|x)x,$$

which maps x onto $-x$, induces the identity on the "hyperplane orthogonal to x" and preserves q ($q(\sigma_x(y)) = q(y)$, $\forall y \in \mathbb{Z}^n$). Accordingly, $\sigma_x(y)$ is a root if so is y.

In the particular case where x is a *simple root* (=natural basis vector) e^i, we simply write $\sigma_i := \sigma_{e^i}$. The subgroup W_q of the orthogonal group of q which is generated by $\sigma_1, \ldots, \sigma_n$ is called *Weyl group* of q. And the positive roots which lie in the orbit of a simple root under the action of W_q are called *real roots*.

Theorem.[5] *If the unit form q on $\mathbb{Z}^n$ is weakly positive or critical, each positive root of q is a real root.*

Proof. We call a positive root x corrigible if $0 < \sigma_i(x) < x$ for some i. Now suppose that the positive root x is not corrigible. Our problem then is to prove that x is simple: From $2 = 2q(x) = q(x|x) = \sum_i x_i q(x|e^i)$ we infer that $x_i > 0$ and $q(x|e^i) > 0$ for some i, hence that $\sigma_i(x) = x - q(x|e^i)e^i < x$. Since x is incorrigible, it follows that $\sigma_i(x)_i = x_i - q(x|e^i) < 0$ ($\sigma_i(x)_j = x_j$ if $j \neq i$).

On the other hand, $0 \leqslant q(x - x_i e^i) = q(x) + x_i^2 - x_i q(x|e^i) = 1 + x_i(x_i - q(x|e^i))$. It follows that $x_i = 1, x_i - q(x|e^i) = -1, q(x - x_i e^i) = 0$ and $x - x_i e^i = 0$, i.e. $x = e^i$ (since q^i is positive) $\surd$

6.6. A more precise statement holds if q is weakly positive.

Theorem. *Let $x < y$ be positive roots of a weakly positive unit form q. Then there is a sequence of integers $i_1, \ldots, i_r$ such that $\sigma_{i_s} \ldots \sigma_{i_1}(y) = y - e^{i_1} - \cdots - e^{i_s}$ for all s and $\sigma_{i_r} \ldots \sigma_{i_1}(y) = x$.*

Proof. From $q(x) = q(y) - q(y|y - x) + q(y - x)$ we deduce that $\sum_{i=1}^{n} (y_i - x_i) q(y|e^i) = q(y - x) > 0$, hence that $y_i > x_i$ and $q(y|e^i) > 0$ for some i. On the other hand, we have $2 - q(y|e^i) = q(y) - q(y|e^i) + q(e^i) = q(y - e^i) > 0$. It follows that $q(y|e^i) = 1$ and $x \leqslant \sigma_i(y) = y - e^i < y$. Thus we set $i_1 = i$ and repeat the argument. $\surd$

6.7. At least, Theorem 6.6 provides a theoretical means of computation of positive roots. The following theorem opens a more practical way.

Theorem.[6] *If the unit form q is weakly positive, the components x_i of its positive roots x are all $\leqslant 6$.*

Proof. Let s denote an integer $\geqslant 3$. Among the weakly positive unit forms p such that $s = \sup_{i,x} x_i$, where x runs through the positive roots of p, we choose a $q\colon \mathbb{Z}^n \to \mathbb{Z}$ having a minimal number of positive roots. If such p exist, we are free to further suppose that $s = x_n$ for some positive root x of q. We then choose a *maximal* positive root z of q satisfying $s = z_n$. Our aim is to collect enough information on q to exclude the case $s \geqslant 7$.

1) *$0 \leqslant q_{ij}$ if $i, j < n$*: Clearly, $q_{ij} \geqslant -1$ since $2 + q_{ij} = q(e^i + e^j) > 0$. In case $q_{ij} = -1$, we might further suppose that $z_i \leqslant z_j$ and set $b^i = e^i + e^j$, $b^l = e^l$ if $l \neq i$. The formula $q_b(x - x_i e^j) = q(x)$ then shows that $z' = z - z_i e^j$ is a positive root of q_b which satisfies $z'_n = s$. The map $y \mapsto y + y_i e^j$ also induces an injection from the positive roots of q_b to those of q. Since e^i is the image of the non-positive $e^i - e^j$, we infer that q_b has fewer positive roots than q: contradiction.

2) *$q_{ij} \leqslant 1$ if $i, j < n$*: To prove that $q_{ij} \leqslant 2$, we now may assume that $q(z|e^i) \leqslant q(z|e^j)$. The inequalities $1 \leqslant q(z + z_j e^i - z_j e^j) = q(z) + 2z_j^2 + z_j(q(z|e^i) - q(z|e^j)) - z_j^2 q(e^i|e^j) \leqslant 1 + (2 - q_{ij}) z_j^2$ then imply $2 \geqslant q_{ij}$. However, in case $2 = q_{ij}$, $z + z_j e^i - z_j e^j$ would give rise to a positive root y of $q^j\colon \mathbb{Z}^{n-1} \to \mathbb{Z}$ which satisfies $y_{n-1} = s$. The positive roots of q^j would correspond bijectively to the positive roots x of q such that $x_j = 0$. Therefore, q^j would have less positive roots than q: contradiction.

3) *$q_{in} = -1$ if $i < n$*, since z is omnipresent, $1 \leqslant q(z - e^i) = q(z) + q(e^i) - q(z|e^i) = 2 - 2z_i - \sum_{j \neq i,n} z_j q_{ij} - z_n q_{in} \leqslant -\sum_{j \neq i,n} z_j q_{ij} - z_n q_{in} \leqslant -z_n q_{in}$ implies $-1/z_n \geqslant q_{in}$, and $2 + q_{in} = q(e_i + e_n) > 0$ implies $q_{in} > -2$.

4) *There is an $a < n$ such that $z_a = 2$, $q(z|e^a) = 1$ and $q(z|e^j) = 0$ if $a \neq j$*: Indeed, $0 < q(z \pm e^j) = 2 \pm q(z|e^j)$ implies $|q(z|e^j)| < 2$. Accordingly, $\sigma_j(z) = z - q(z|e^j)e^j$ is a positive root of q. Since z is maximal, $q(z|e^j) = 0, 1$ for all j. Since $2 = q(z|z) = \sum_{j=1}^{n} z_j q(z|e^j)$, we have $q(z|e^n) = 0$ (for $z_n \geqslant 3$). If our claim was wrong, we should have $z_a = q(z|e^a) = z_b = q(z|e^b) = 1$ for some $a \neq b$. But then $z' = z - e^a$ would be a positive root with smaller support such that $z'_n = s$.

5) $q_{ij} = 1$ *for all distinct* $i, j \in I := \{l \neq a\colon q_{al} = 1\}$: Indeed, $z' = z - e^a + e^i + e^j + e^n$ is >0 and satisfies $z'_n = s + 1$, hence $2 \leqslant q(z') = 5 - q(z|e^a) - q_{ai} - q_{aj} - q_{an} + q_{ij} + q_{in} + q_{jn} = 1 + q_{ij}$.

6) $z_n = 3 + \sum_{i \in I} z_i = -1 + \sum_{j \in J} z_j$, *where* $J := \{j\colon q_{aj} = 0\}$: For, $1 = q(z|e^a) = 2z_a + \sum_{i \in I} z_i - z_n$ and $0 = q(z|e^n) = -z_a - \sum_{i \in I} z_i - \sum_{j \in J} z_j + 2z_n = (z_n - \sum_{i \in I} z_i) - z_a - \sum_{j \in J} z_j + z_n = 3 - 2 - \sum_{j \in J} z_j + z_n$.

7) $z_i = 1$ *and* $q_{ij} = 0$ *if* $i \in I$ *and* $j \in J$: For, $0 = q(z|e^i) = z_a + 2z_i + \sum_{i \neq l \in I} z_l + \sum_{j \in J} z_j q_{ij} - z_n = (\sum_{l \in I} z_l - z_n) + z_a + z_i + \sum_{j \in J} z_j q_{ij} = -1 + z_i + \sum_{j \in J} z_j q_{ij}$.

8) *If* $c \in J$ *is such that* $z_c \geqslant z_j$ *for all* $j \in J$, *there are two distinct* j, l *in* J *such that* $q_{cj} = q_{cl} = 0$: For, $0 = q(z|e^c) = 2z_c + \sum_{c \neq j \in J} z_j q_{cj} - z_n = z_c - \sum_{j \in J'} z_j + (\sum_{j \in J} z_j - z_n) = z_c + 1 - \sum_{j \in J'} z_j$, where $J' = \{j \in J\colon q_{cj} = 0\}$. Since $z_c \geqslant z_j$ for each $j \in J'$, J' has at least 2 elements.

9) *Conclusion*: Suppose that $s = z_n \geqslant 7$. By 6) and 7), I has at least 4 elements i_1, i_2, i_3, i_4. The "full" subbigraph of the bigraph of q which is formed by the vertices $a, i_1, i_2, i_3, i_4, c, j, l, n$ is therefore isomorphic to the fourth bigraph of Fig. 1 or to a bigraph having one broken edge less (between j and l). This leads us to the contradiction

$$q(e^a + e^{i_1} + e^{i_2} + e^{i_3} + e^{i_4} + 3e^c + 2e^j + 2e^l + 6e^n) = -4 + 4q_{jl} \leqslant 0. \checkmark$$

6.8. Remarks and References

1. In [135, 1977], A.V. Roiter examines functions on $\mathbb{Z}^n$ of the form $q(x) = \sum_i q_i x_i^2 + \sum_{i<j} q_{ij} x_i x_j$, where q_i, q_{ij}/q_i and q_{ij}/q_j are integers and $q_i > 0$.
2. The roots of a positive unit form provide a system of roots in the sense of [29, Bourbaki, 1968]. Dynkin graphs were introduced in [44, Dynkin, 1947].
3. [118, Ovsienko, 1978].
4. [75, von Höhne, 1988].
5. See [135, Roiter, 1977], where a more precise theorem is proved for integral quadratic forms.
6. [119, Ovsienko, 1979]. Ovsienko proves his theorem in the more general context of Roiter's integral forms. His key idea is to reduce the proof to the case of an L-bigraph (which satisfies the statements 1), 2) and 3) of 6.7). The faithful weakly positive L-bigraphs (=sincere weakly positive graphical forms) are classified in [132, Ringel, 1984]. We owe the details of our proof to K. Bongartz who leans on Ringel.

7. Representations of Quivers

In this section, we examine representations of a *finite* quiver Q over the algebraically closed field k. These representations are identified with left modules over the k-category of paths kQ. Unless otherwise stated, we assume that they are pointwise finite.

7.1. In our investigation, a central rôle is played by the quadratic form[1] $q_Q\colon \mathbb{Z}^{Q_v} \to \mathbb{Z}$ defined by

$$q_Q(d) = \sum_{a \in Q_v} d(a)^2 - \sum_{\alpha \in Q_a} d(t\alpha)d(h\alpha).$$

We are especially interested in the value $q_Q(\underline{\dim}\ V)$ of q_Q at the *dimension-function* $\underline{\dim}\ V: x \mapsto \dim V(x)$ of a representation V.

Theorem[2]. *The number of isoclasses of indecomposable representations of Q is finite if and only if q_Q is positive definite. If this is the case, the map $V \mapsto \underline{\dim}\ V$ provides a bijection between the set of these isoclasses and the set of positive roots of q_Q.*

As we know by 6.2, q_Q is positive definite if and only if Q is a disjoint union of Dynkin quivers. It follows that, if Q is Dynkin of type A_n, it has $\frac{1}{2}n(n+1)$ isoclasses of indecomposables. Among them, one only is *omnipresent*[3]. It may be delineated as follows:

$$k \overset{\mathbb{1}}{\text{———}} k \overset{\mathbb{1}}{\text{———}} k \text{ — } \cdots \text{ — } k \overset{\mathbb{1}}{\text{———}} k$$

If Q is Dynkin of type D_n, it has $(n-1)n$ isoclasses of indecomposables. Up to isomorphism, the omnipresent indecomposables are those of Fig. 1 (according to the orientations of the arrows, a is represented by the matrix $[0\ \ 1]$ or $[1\ \ 0]^T$, b by $[1\ \ 0]$ or $[0\ \ 1]^T$ and c by $[1\ \ 1]$ or $[1\ \ 1]^T$):

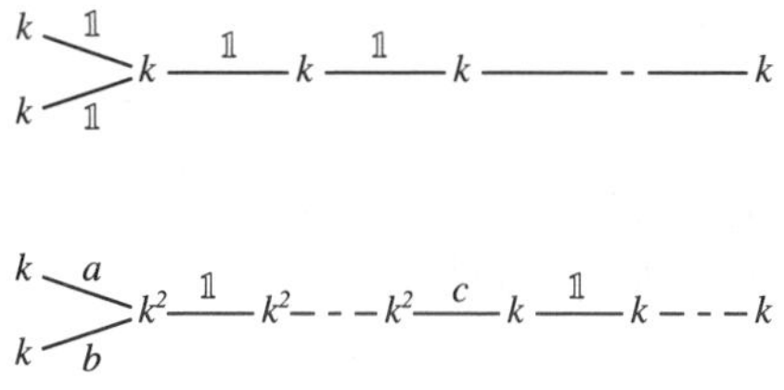

Fig. 1

If Q is Dynkin of type E_6, E_7 or E_8, it has 36, 63 or 120 isoclasses of indecomposables respectively. A concrete description will be produced in Sect. 10.

Example 1. A vector space V_0 together with 3 subspaces V_1, V_2, V_3 can be interpreted as a representation of the quiver $\begin{smallmatrix} x_1 \searrow \\ x_2 \nearrow \end{smallmatrix} x_0 \leftarrow x_3$. These representations can easily be classified "by hand". In particular, when V_0 has dimension 4 and V_1, V_2, V_3 are pairwise supplementary subspaces of dimension 2, the associated representation is a direct sum of 2 isomorphic indecomposables with dimension-function $\begin{smallmatrix} 1 \\ 1 \end{smallmatrix}\!\!>\!2\text{—}1$. In geometrical terms, this means that in the projective 3-space

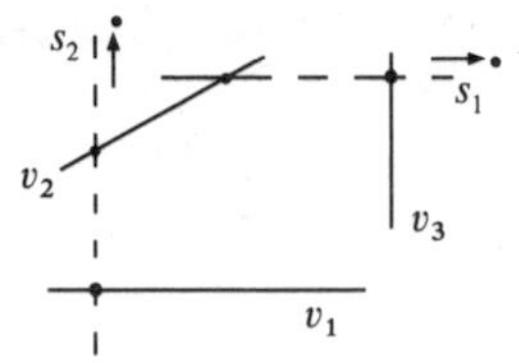

Fig. 2

three straight lines v_1, v_2, v_3 in skew position admit two common secants s_1, s_2 in skew position (Fig. 2).

Example 2. Let $V_n = k^n$ be the space of n-columns and V_i the i-dimensional subspace $\{x \in k^n: x_q = 0 \text{ if } i < q\}$. Each invertible $n \times n$-matrix g gives rise to the representation

$$V_1 \to V_2 - \cdots \to V_{n-1} \to V_n \leftarrow gV_{n-1} \leftarrow \cdots - gV_2 \leftarrow gV_1$$

of a Dynkin quiver of type A_{2n-1} (the maps are inclusions). Our classification of the indecomposables here means that V_n admits a basis $b^1, \ldots, b^n$ such that $b^i \in V_i \cap gV_{\sigma i}$, $V_i = \bigoplus_{h \leqslant i} kb^h$ and $gV_j = \bigoplus_{\sigma h \leqslant j} kb^h$ for some permutation σ (in terms of algebraic groups, two Borel subgroups of GL_n contain a common maximal torus[4]). Denoting by $e^1, \ldots, e^n$ the natural basis of k^n, by $\underline{\sigma}$ the permutation-matrix such that $\underline{\sigma}e^i = e^{\sigma i}$ and by $\underline{b}$ the upper triangular matrix with columns $b^1, \ldots, b^n$, we get $\underline{b}e^i = b^i$, $\underline{b}^{-1}gV_j = \bigoplus_{\sigma h \leqslant j} k\underline{b}^{-1}b^h = \bigoplus_{\sigma h \leqslant j} ke^h = \underline{\sigma}^{-1}V_j$ and $\underline{\sigma}\underline{b}^{-1}gV_i \subset V_i$. We infer that $g = \underline{b}\underline{\sigma}^{-1}\underline{c}$, where $\underline{\sigma}^{-1}$ is a permutation-matrix and $\underline{b}$, $\underline{c}$ are upper triangular.

7.2. Among the possible *proofs* of Theorem 7.1, we choose one which stresses the rôle of the quadratic form q_Q. It uses elementary notions of algebraic geometry and homological algebra.

We first notice that two representations V and W of Q give rise to an exact sequence

$$(*) \quad 0 \to \operatorname{Hom}(V, W) \xrightarrow{\gamma} \prod_{x \in Q_v} \operatorname{Hom}_k(V(x), W(x)) \xrightarrow{\delta} \prod_{\alpha \in Q_a} \operatorname{Hom}_k(V(t\alpha), W(h\alpha)) \to$$
$$\xrightarrow{\varepsilon} \operatorname{Ext}^1(V, W) \to 0,$$

where γ denotes the inclusion, δ maps a family $(f(x))_{x \in Q_v}$ onto $(f(h\alpha)V(\alpha) - W(\alpha)f(t\alpha))_{\alpha \in Q_a}$ and ε maps $(g(\alpha))_{\alpha \in Q_a}$ onto the equivalence class of the exact sequence $0 \to W \xrightarrow{i} E \xrightarrow{p} V \to 0$ such that $E(x) = W(x) \oplus V(x)$ for each $x \in Q_v$ and $E(\alpha) = \begin{bmatrix} W(\alpha) & g(\alpha) \\ 0 & V(\alpha) \end{bmatrix}$ for each $\alpha \in Q_a$ (the morphisms i and p are the obvious ones).

If d and e are the dimension-functions of V and W, (*) implies

$$\begin{aligned}\dim \operatorname{Hom}(V, W) - \dim \operatorname{Ext}^1(V, W) &= \sum_x \dim \operatorname{Hom}_k(V(x), W(x)) \\ &\quad - \sum_\alpha \dim \operatorname{Hom}_k(V(t\alpha), W(h\alpha)) \\ &= \sum_x d(x)e(x) - \sum_\alpha d(t\alpha)e(h\alpha)\end{aligned}$$

and in particular

$$\dim \operatorname{Hom}(V, V) - \dim \operatorname{Ext}^1(V, V) = \sum_x d(x)^2 - \sum_\alpha d(t\alpha)d(h\alpha) = q_Q(d).$$

Lemma. *If q_Q is positive definite, we have* $\operatorname{Hom}(V, V) = k\mathbb{1}_V$ *for each indecomposable representation V of Q.*

Proof[5]. Let V be a counterexample of minimal dimension and f a non-zero nilpotent endomorphism of V whose image I has minimal dimension. Set $K = \operatorname{Ker} f = K_1 \oplus \cdots \oplus K_s$, where each K_i is indecomposable.

Since $\dim I$ is minimal, I is indecomposable, we have $f^2 = 0$, hence $I \subset K$, and each non-zero projection $p_i: I \to K_i$ is injective. Since V is indecomposable, the equivalence class $\varepsilon = (\varepsilon_i) \in \operatorname{Ext}^1(I, K) \tilde{\to} \oplus_i \operatorname{Ext}^1(I, K_i)$ of the exact sequence $0 \to K \to V \to I \to 0$ is non-zero, and so is each ε_i.

Now, since $p_i: I \to K_i$ is injective, the exact sequences (*) applied to $V = K_i$, I and $W = K_i$ show that $\operatorname{Ext}^1(p_i, K_i): \operatorname{Ext}^1(K_i, K_i) \to \operatorname{Ext}^1(I, K_i)$ is surjective. It follows that $\operatorname{Ext}^1(K_i, K_i) \neq 0$. On the other hand, the minimality of $\dim V$ implies $\operatorname{Hom}(K_i, K_i) = k\mathbb{1}_{K_i}$, hence the required contradiction

$$\begin{aligned}0 < q_Q(\underline{\dim}\, K_i) &= \dim \operatorname{Hom}(K_i, K_i) - \dim \operatorname{Ext}^1(K_i, K_i) \\ &= 1 - \dim \operatorname{Ext}^1(K_i, K_i) \leqslant 0. \surd\end{aligned}$$

7.3. Proof of theorem 7.1. With the notations of 7.2, suppose that $V = W$ and that $V(x) = k^{d(x)}$ for each $x \in Q_v$. The representation V can then be identified with the family

$$(V(\alpha))_{\alpha \in Q_a} \in \prod_\alpha \operatorname{Hom}_k(V(t\alpha), V(h\alpha)) \tilde{\to} \prod_\alpha k^{d(h\alpha) \times d(t\alpha)}.$$

We denote this product by X_d and endow it with its natural structure of an algebraic variety of dimension $\sum_\alpha d(t\alpha)d(h\alpha)$.

On the other hand, the space $\prod_x \operatorname{Hom}_k(V(x), V(x))$ of (*) is identified with a product of matrix-algebras $\prod_x k^{d(x) \times d(x)}$. Its invertible elements form an algebraic group $G_d = \prod_x GL_{d(x)}$ of dimension $\sum_x d(x)^2$. The formula $(gV)(\alpha) = g(h\alpha)V(\alpha)g(t\alpha)^{-1}$ defines an action of G_d on X_d whose orbits correspond bijectively to the isoclasses of representations of Q with dimension-function d.

The *isotropy group* $G_{dV} = \{g \in G_d: gV = V\}$ is the group of automorphisms of V, i.e. of invertible elements of $\operatorname{Hom}(V, V)$. It is Zariski-open in $\operatorname{Hom}(V, V)$ and has the same dimension. It follows[6] that the orbit $G_d V = \{gV: g \in G_d\}$ has the dimension $\dim G_d V = \dim G_d - \dim G_{dV} = \dim G_d - \dim \operatorname{Hom}(V, V)$ and that

$$\dim \operatorname{Hom}(V, V) - \dim \operatorname{Ext}^1(V, V) = q_Q(d) = \dim G_d - \dim X_d$$
$$= \dim \operatorname{Hom}(V, V) - (\dim X_d - \dim G_d V).$$

These equalities imply $\dim X_d > \dim G_d V$ if $q_Q(d) \leqslant 0$. In this case, there are infinitely many orbits, hence infinitely many isoclasses of indecomposables with dimension-function $\leqslant d$. The case arises when q_Q is not positive definite, because then there is a $d > 0$ such that $q_Q(d) \leqslant 0$.

If q_Q is positive definite and V indecomposable, our Lemma 7.2 implies

$$0 < q_Q(d) = 1 - (\dim X_d - \dim G_d V)$$

hence
$$q_Q(d) = 1 \text{ and } \dim X_d = \dim G_d V.$$

It follows that $G_d V$ is Zariski-open[7] and dense in X_d. Therefore, it coincides with the orbit of any other indecomposable in X_d, and the map $V \mapsto \underline{\dim}\, V$ provides an injection from the set of isoclasses of indecomposables into the set of positive roots.

It remains to prove that each positive root d is the dimension-function of an indecomposable: We already know that the number of isoclasses of indecomposables is finite. It follows that X_d contains only finitely many orbits, and one of them, say $G_d V$, must have the same dimension as X_d. So we have $1 = q_Q(d) = \dim \operatorname{Hom}(V, V)$, $\operatorname{Hom}(V, V) = k\mathbb{1}_V$ is local, and V is indecomposable. $\surd$

7.4. Let us return to the *general* case of a *finite quiver* Q. The objective is to describe the subset of $\mathbb{Z}^{Q_v}$ formed by the dimension-functions of the indecomposable representations. For this we consider the bilinear form $q_Q(d|e) = q_Q(d + e) - q_Q(d) - q_Q(e)$ associated with q_Q. By 7.2, this form satisfies

$$q_Q(\underline{\dim}\, V | \underline{\dim}\, W) = \dim \operatorname{Hom}(V, W) + \dim \operatorname{Hom}(W, V) - \dim \operatorname{Ext}^1(V, W)$$
$$- \dim \operatorname{Ext}^1(W, V).$$

It is also determined by the following formulas, where $e^i(i) = 1$ and $e^i(j) = 0$ if $j \neq i \in Q_v$:

$$\tfrac{1}{2} q_Q(e^i|e^i) = 1 - \text{number of loops } \overset{i}{\cdot}\!\circlearrowleft$$
$$-q_Q(e^i|e^j) = \text{number of arrows between } i \text{ and } j \neq \text{i}.$$

In particular, we have $q_Q(e^i|e^i) = 2$ if e^i is a *simple root*, i.e. if there is no loop at i. The formula

$$\sigma_i(d) = d - q_Q(e^i|d)e^i, \qquad d \in \mathbb{Z}^{Q_v},$$

then defines the *reflection* in the direction e^i, i.e. the automorphism of $\mathbb{Z}^{Q_v}$ which maps e^i onto $-e^i$ and fixes the vectors orthogonal to e^i. The group generated by these reflections is the *Weyl group* W_Q. The positive functions belonging to the orbit $W_Q e^i$ of a simple root e^i are the *real roots* (6.5). We denote their set by R_Q^{re}.

The *fundamental cone* $K_Q \subset \mathbb{Z}^{Q_v}$ consists of the positive functions d which satisfy $q_Q(e^i|d) \leqslant 0$ for each simple root e^i and have a connected (non-empty) *support*. Under the action of W_Q it generates the set $R_Q^{im} = \bigcup_{w \in W_Q} wK_Q$ of *imaginary roots.*[8]

Theorem[9]. *If d is a real root, there is exactly one isoclass of indecomposables with dimension-function d. If d is an imaginary root, there are infinitely many such isoclasses. There is none if* $d \notin R_Q^{re} \cup R_Q^{im}$.

If q_Q is positive definite, there is no imaginary root, and the real roots coincide with the positive roots as follows from our theorems or from 6.5.

If Q is an extended Dynkin quiver, the quadratic form q_Q is positive semi-definite. The *isotropic* functions, on which q_Q vanishes, are then integral multiples of the *isotropic generator* δ^Q (6.3). In this case, we have $R_Q^{im} = K_Q = \{n\delta^Q: n \in \mathbb{N}\setminus\{0\}\}$, and it is easy[10] to exhibit an infinite family of non-isomorphic indecomposables with dimension-function $n\delta^Q$, $n > 0$. In the case of Example 1 below, the required family is

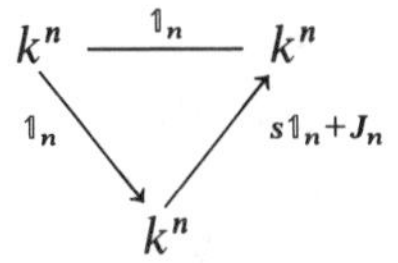

where $s\mathbb{1}_n + J_n$ denotes a "Jordan-block" with eigenvalue s (1.7).

If Q contains a component which *is neither Dynkin nor extended Dynkin*, there are functions $d \in K_Q$ such that $q_Q(d) < 0$, but *there is no*[11] *positive d with support Q such that* $q_Q(e^i|d) \geqslant 0$ *for all* $i \in Q_v$.

Example 1. 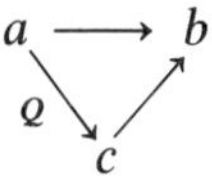

, $q_Q(xe^a + ye^b + ze^c) = x^2 + y^2 + z^2 - yz - xz - xy$

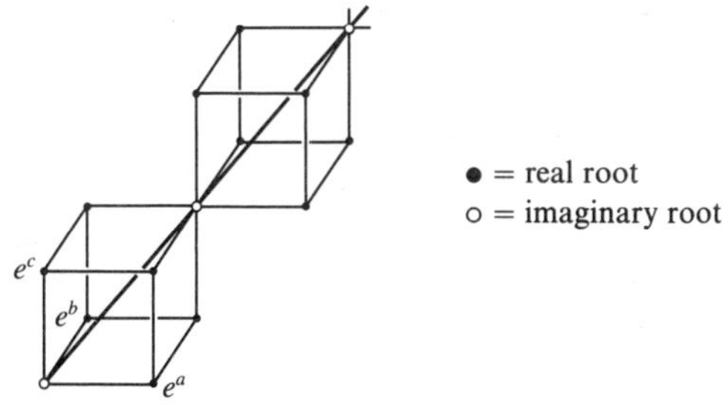

Fig. 3

Example 2. $a \xrightarrow[Q]{} b\circlearrowleft$, $q_Q(xe^a + ye^b) = x^2 - xy$

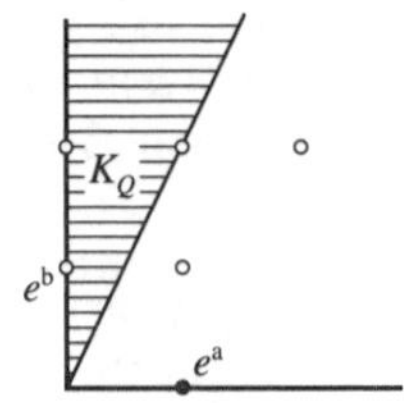

Fig. 4

Example 3. $a \underset{Q}{\overset{\longrightarrow}{\underset{\longrightarrow}{\longleftarrow}}} b$, $q_Q(xe^a + ye^b) = x^2 + y^2 - 3xy$

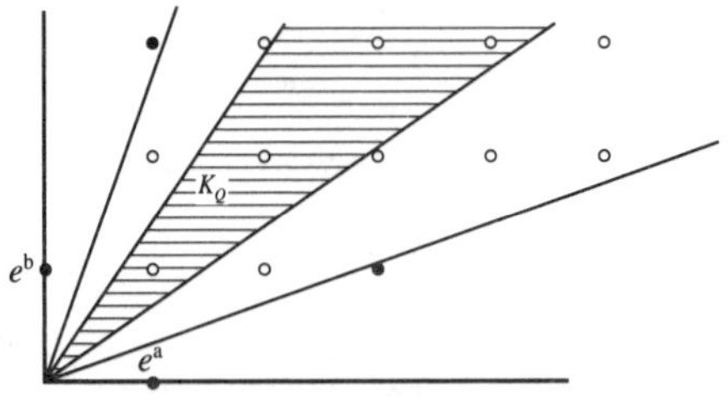

Fig. 5

7.5. Proof of theorem 7.4. Second part. The demonstration uses the following result: Denote by $\imath(d) \in \mathbb{N} \cup \{\infty\}$ the number of isoclasses of indecomposables with dimension-function d. *Then $\imath(d)$ remains invariant when an arrow of Q is replaced by an arrow in the reverse direction.* For instance, the quivers $a \overset{\longrightarrow}{\underset{\longrightarrow}{\longleftarrow}} b$ and $a \overset{\longrightarrow}{\underset{\longrightarrow}{\longrightarrow}} b$ give rise to the same numbers $\imath(xe^a + ye^b)$. Of course, this result must hold if the number $\imath(d)$ can be described in terms of the quadratic form q_Q. It is in fact the main constituent of the whole proof. Unfortunately, the proof of the invariance of $\imath(d)$ is so circuitous that the result remains mysterious. In 7.6–7.7 below, we present the parts of the proof which are not pure algebraic geometry.[12]

We first present the second part of the demonstration, which provides a new insight. In this part, R denotes the set of the dimension-functions of the indecomposables.

a) *If $d \in R$ is not a simple root e^i, we claim that*[13] $\imath(\sigma_i d) = \imath(d)$. Indeed, by the invariance of $\imath(d)$, we may suppose that i is a source of Q. If we denote the arrows starting at i by $i \xrightarrow{\alpha_p} j_p$ $(p = 1, \ldots, s)$, each representation V of Q then gives rise to an exact sequence

$$V(i) \xrightarrow{[V(\alpha_1) \ldots V(\alpha_s)]^T} \bigoplus_{p=1}^{s} V(j_p) \xrightarrow{[\beta_1 \ldots \beta_s]} C \longrightarrow 0,$$

where $C = \mathrm{Coker}[V(\alpha_1)\dots V(\alpha_s)]^T$. If we replace the α_p by arrows $i \overset{\alpha'_p}{\leftarrow} j_p$ in the reverse direction, we obtain a new quiver Q' and a representation V' of Q' such that $V'(i) = C$, $V'(j) = V(j)$ for each vertex $j \neq i$, $V'(\alpha'_p) = \beta_p$ and $V'(\gamma) = V(\gamma)$ for each arrow $\gamma \neq \alpha'_1, \dots, \alpha'_s$. The construction can be reversed if $[V(\alpha_1)\dots V(\alpha_s)]^T$ is an injection, hence if V is an indecomposable with dimension-function $\neq e^i$. In this case, the equalities $\dim V'(i) = \dim C = \sum_p \dim V(j_p) - \dim V(i)$ and $\dim V'(j) = \dim V(j)$ for $j \neq i$ are equivalent to $\underline{\dim}\, V' = \sigma_i(\underline{\dim}\, V)$ and prove our claim.

b) It follows from a) that the set $(R \cap R_Q^{re}) \cup -(R \cap R_Q^{re})$ is stable under W_Q. Since it contains the simple roots, it contains R_Q^{re}. We infer that $R \cap R_Q^{re} = R_Q^{re}$. Furthermore, since $\imath(e^i) = 1$ for each simple e^i, we have $\imath(d) = 1$ for each $d \in R_Q^{re}$.

c) It also follows from a) that $R \setminus R_Q^{re}$ is stable under W_Q. Since it contains K_Q by d) below, it contains R_Q^{im}. Conversely, if $c \in R \setminus R_Q^{re}$, let d be a minimal function in the orbit $W_Q c \subset R \setminus R_Q^{re}$. The formulas $\sigma_i(d) = d - q_Q(e^i|d)e^i$ then imply $q_Q(e^i|d) \leqslant 0$ for each simple e^i, hence $d \in K_Q$ and $c \in R_Q^{im} = R \setminus R_Q^{re}$. Since $\imath(d) = \infty$ if $d \in K_Q$ (see section d) below), we have $\imath(d) = \infty$ for each $d \in R_Q^{im}$.

d) Denote by S the support of a function $d \in K_Q$. The inequalities $q_Q(e^i|d) \leqslant 0$ then imply $2q_Q(d) = \sum_{i \in S} d(i) q_Q(e^i|d) \leqslant 0$. If S is extended Dynkin, we have $q_Q(d) = 0$ and there are infinitely many isoclasses of indecomposables with dimension-function d (7.4). If S is not extended Dynkin, X_d contains a subset Y as in the lemma e) below. On the other hand, we have

$$1 \leqslant \dim \mathrm{Hom}(U, U) = q_Q(d) + \dim X_d - \dim G_d U \leqslant \dim X_d - \dim G_d U$$

for each $U \in X_d$ (7.3). Therefore, the orbits of Y have dimension $< \dim Y = \dim X_d$, and their number is infinite.

e) **Lemma.** *If the support S of $d \in K_Q$ is not extended Dynkin, X_d contains a dense, open and G_d-stable subset Y which consists of indecomposable representations.*

Proof. Let $c \in \mathbb{Z}^{Q_v}$ be such that $0 < c < d$, set $c' = d - c$ and embed $X_c \times X_{c'}$ into X_d by mapping (U, U') onto the family V such that $V(\alpha) = \begin{bmatrix} U(\alpha) & 0 \\ 0 & U'(\alpha) \end{bmatrix}$ for each $\alpha \in Q_a$. If our lemma is false, $\bigcup_c G_d(X_c \times X_{c'})$ is dense in X_d. So there is a c for which $G_d(X_c \times X_{c'})$ is dense[14] in X_d. By Lemma f) below, this implies that $\mathrm{Ext}^1(U, U') = 0 = \mathrm{Ext}^1(U', U)$ for almost all $(U, U') \in X_c \times X_{c'}$; for these pairs, 7.4 then implies that

$$0 \leqslant \dim \mathrm{Hom}(U, U') + \dim \mathrm{Hom}(U', U) = q_Q(c|c') = \sum_{i,j \in S} c_i(d_j - c_j) q_Q(e^i|e^j)$$

$$= \sum_j \underbrace{\frac{c_j}{d_j}(d_j - c_j)}_{0 \leqslant} \underbrace{\left(\sum_i d_i q_Q(e^i|e^j)\right)}_{\leqslant 0} + \frac{1}{2} \sum_{i,j} \underbrace{\left(\frac{c_i}{d_i} - \frac{c_j}{d_j}\right)^2 d_i d_j}_{0 \leqslant} \underbrace{q_Q(e^i|e^j)}_{\leqslant 0} \leqslant 0$$

where $c_i = c(i)$, $d_i = d(i)$.

When infer that $\frac{c_i}{d_i} = \frac{c_j}{d_j}$ if $q_Q(e^i|e^j) \neq 0$, i.e. if i and j are the extremities of one arrow. Since S is connected, $\frac{c_i}{d_i} = \gamma$ is constant, and $0 = q_Q(c|d-c) = \gamma(1-\gamma)q_Q(d|d) = \sum_{i\in S} \underbrace{\gamma(1-\gamma)d_i}_{0<}\underbrace{q_Q(e^i|d)}_{\leqslant 0}$. We conclude that $q_Q(e^i|d) = 0$ for each $i \in S$, hence that S should be extended Dynkin (7.4).

f) **Lemma.** *With the notations of* e), *we have* $\mathrm{Ext}^1(U, U') = 0$ *for almost all* $(U, U') \in X_c \times X_{c'}$ *provided* $G_d(X_c \times X_{c'})$ *is dense in* X_d.

Proof.[15] We recall that the function $V \mapsto \dim \mathrm{Hom}(V, V) = \dim G_{dV}$ is upper semicontinuous on X_d: Indeed, consider the map $\pi\colon G_d \times X_d \to X_d \times X_d$, $(g, V) \mapsto (gV, V)$. Then we have an isomorphism $g \mapsto (g, V)$ of G_{dV} onto $\pi^{-1}\pi(1, V)$, and our claim follows from the upper semicontinuity of the fibre dimensions.[16]

Now denote by e the minimum of $V \mapsto \dim \mathrm{Hom}(V, V)$ on $X_c \times X_{c'} \subset X_d$, i.e. on $G_d(X_c \times X_{c'})$ and on its closure X_d. Then

$$Y = \{(U, U') \in X_c \times X_{c'}\colon \dim \mathrm{Hom}(U \oplus U', U \oplus U') = e\}$$

is open and dense in $X_c \times X_{c'}$. And the existence of a nonsplitting exact sequence $0 \to U' \to E \to U \to 0$, where $(U, U') \in Y$, would imply the contradiction $\dim \mathrm{Hom}(E, E) < e$, since Hom is left exact.

7.6. Proof of Theorem 7.4. First Part. We now turn to a proof-sketch for the invariance of $\imath(d)$. Kač proceeds in three steps: In the first step, he achieves an invariance result for finite fields. In the second, he proves the invariance of $\imath(d)$ for algebraic closures of finite fields. The extension to the general case uses arguments of algebraic geometry which are classical but lie beyond the scope of our monograph.[12]

First Step. Let p be a prime number and F_m the finite field with p^m elements. As in the case of algebraically closed fields, the isoclasses of representations of Q over F_m with dimension-function d correspond bijectively to the orbits of $X_d^m = \prod_{\alpha \in Q_a} F_m^{d(h\alpha)\times d(t\alpha)}$ under the action of the group G_d^m formed by the invertible elements of $\prod_{x \in Q_v} F_m^{d(x)\times d(x)}$. Let $\rho_m(d)$ be the number of these orbits.

a) $\rho_m(d)$ *is independent of the orientation of the arrows*: Indeed, denote by $\bar{Q}$ the quiver obtained from Q by replacing the arrow $x \xrightarrow{\beta} y$ by $x \xleftarrow{\bar\beta} y$. The action of G_d^m on X_d^m is induced by actions on the two factors $Y = \prod_{\alpha \neq \beta} F_m^{d(h\alpha)\times d(t\alpha)}$ and $H = F_m^{d(y)\times d(x)}$ of $X_d^m = Y \times H$. Now the representations of $\bar{Q}$ with dimension-function d correspond to orbits of the same group G_d in the space $\bar{X}_d^m = Y \times \bar{H}$, where $\bar{H} = F_m^{d(x)\times d(y)}$. So we must prove that $Y \times H$ and $Y \times \bar{H}$ have the same number of orbits. In fact, we shall show that the complex functions which are constant on these orbits form two complex vector spaces of the same dimension. For this it suffices to exhibit a G_d^m-equivariant isomorphism from $\mathbb{C}^{Y\times H} \xrightarrow{\sim}$

$\mathbb{C}^Y \otimes \mathbb{C}^H$ onto $\mathbb{C}^{Y\times \overline{H}} \xrightarrow{\sim} \mathbb{C}^Y \otimes \mathbb{C}^{\overline{H}}$, or one from $\mathbb{C}^H$ onto $\mathbb{C}^{\overline{H}}$. The required isomorphism $\mathfrak{F}: \mathbb{C}^H \xrightarrow{\sim} \mathbb{C}^{\overline{H}}$ and its inverse $\overline{\mathfrak{F}}$ are simply provided by Fourier transformation[17]:

$$(\mathfrak{F}f)(N) = \sum_{M\in H} f(M) \exp\left(\frac{2i\pi}{p}\tau(\mathrm{Tr}\,(MN))\right)$$

$$(\overline{\mathfrak{F}}\bar{f}\,)(M) = p^{-md(x)d(y)} \sum_{N\in \overline{H}} \bar{f}\,(N) \exp\left(-\frac{2i\pi}{p}\tau(\mathrm{Tr}\,(MN))\right).$$

In these formulas, $\tau: F_m \to F_1 = \mathbb{Z}/p$ denotes an arbitrary non-zero group homomorphism, and we set $\exp\left(\frac{2i\pi}{p}\bar{n}\right) = \exp\left(\frac{2i\pi}{p}n\right)$ if $\bar{n} \in \mathbb{Z}/p$ is the class of n modulo p.

b) *The number $\iota_m(d)$ of isoclasses of indecomposable representations of Q over F_m with dimension-function d is independent of the orientation of the arrows.* Indeed, let us call partition of d a function $\pi: \mathbb{N}^{Q_v} \to \mathbb{N}$ such that $d = \sum_e \pi(e)e$, where e ranges over $\mathbb{N}^{Q_v}$. Together with each partition π, we consider the isoclasses of representations of Q over F_m of the form $\bigoplus_e V_e$, where V_e is a direct sum of $\pi(e)$ indecomposables with dimension-function e. The unicity of the summands (3.3) then implies that the number of such isoclasses is equal to $\prod_e \binom{\iota_m(e) + \pi(e) - 1}{\pi(e)}$. This product reduces to $\iota_m(d)$ if π is the trivial partition of d ($\pi(d) = 1$ and $\pi(e) = 0$ if $e \neq d$). We infer that $\rho_m(d) = \iota_m(d) + \sum_\pi \prod_e \binom{\iota_m(e) + \pi(e) - 1}{\pi(e)}$, where π runs through the non-trivial partitions of d. The formula shows that $\iota_m(d)$ can be computed inductively in terms of the $\rho_m(e)$, $e \leqslant d$. Claim b) therefore follows from a).

7.7. Second Step. We denote by $F_\infty = \bigcup_{m\geqslant 1} F_m$ the algebraic closure of $F_1 = \mathbb{Z}/p$, by $\varphi: F_\infty \xrightarrow{\sim} F_\infty$ the automorphism $x \mapsto x^p$, by I_m and I_∞ sets of representatives of the indecomposable representations of Q over F_m and F_∞ (one representative for each isoclass!). The automorphism φ acts componentwise on $X_d^\infty = \prod_{\alpha\in Q_a} F_\infty^{d(h\alpha)\times d(t\alpha)}$. The induced action on I_∞ provides a bijection[18] $V \mapsto \overline{V}$ from I_m onto the space I_∞/φ^m of orbits of φ^m in I_∞. The bijection is determined by $F_\infty \otimes_{F_m} V \xrightarrow{\sim} \bigoplus_{W\in\overline{V}} W$, where $F_\infty \otimes_{F_m} V$ denotes the representation of Q over F_∞ which is defined by the same matrix-family $(V(\alpha))_{\alpha\in Q_a} \in X_d^m \subset X_d^\infty$ as V. In particular, the dimension-functions are subjected to the condition $\underline{\dim}\, V = |\overline{V}|\, \underline{\dim}\, W$, where $|\overline{V}|$ is the cardinality of $\overline{V}$.

Denote by $\iota(s, d)$ the number of orbits of φ in I_∞ which have cardinality s and consist of indecomposables with dimension-function d. This number is bounded by $\iota_1(sd)$, hence finite, and satisfies $\iota(d) = \sum_s s\iota(s, d)$ when $k = F_\infty$. In order to prove the invariance of $\iota(d)$ in the case $k = F_\infty$, it therefore suffices to show that $\iota(s, d)$ is independent of the orientation of the arrows. In fact, we will show that $\iota(s, d)$ can be calculated in terms of the invariant numbers $\iota_m(d)$:

Under the action of φ^m, each orbit B of cardinality s of φ in I_∞ decomposes into $m \vee s$ orbits of cardinality $\frac{s}{m \vee s}$, where $m \vee s$ is the greatest common divisor of m and s. Each B therefore gives rise to $m \vee s$ elements of I_m with dimension-function $\frac{s}{m \vee s} d$, where d is the dimension-function of the elements of B. It follows that

$$\iota_m(e) = \sum_{s,d} (m \vee s)\iota(s, d) = \sum_{t,n} n\iota\left(nt, \frac{1}{t} e\right),$$

where s, d are subjected to $\frac{s}{m \vee s} d = e$, and t, n to the following conditions: n divides m, t divides all the values $e(x)$ of e and $\frac{m}{n} \vee t = 1$. The resulting equality

$$m\iota(m, e) = \iota_m(e) - \sum_{n<m} n\iota(n, e) - \sum_{n} \sum_{1<t} n\iota\left(nt, \frac{1}{t} e\right)$$

then allows for a computation of the numbers $\iota(m, e)$ in terms of the $\iota_m(e)$ by induction on the lexicographically ordered pairs (e, m). $\surd$

7.8. Remarks and References

1. This quadratic form was introduced by Tits (see [52, Gabriel, 1970/72]). It is therefore also called *Tits form.*
2. The description of the finitely represented quivers as Dynkin quivers and the classification of their representations was presented in Oberwolfach (Spring 1970) and published in [52, Gabriel, 1972]. This article is indebted to the forerunners [145, Thrall, 1947], [78, Jans, 1954/57] and [148, Yoshii, 1956]. Among others, these authors examine modules over finite-dimensional algebras A satisfying $\mathscr{R}_A^2 = 0$. In our language, such modules can easily be reduced to representations of *alternating* quivers (see 4.12(2); alternating here means that each vertex is either a source or a sink). The remark that A is infinitely represented if the associated alternating quiver "contains" $\tilde{A}_n$ or $\tilde{D}_4$ is attributed to Brauer (1941). Thrall extends this observation to $\tilde{D}_n$; unfortunately, he also announces a converse, thus inaugurating an endless procession of blunders, of which representation theory seems to be fond. The results of Brauer and Thrall are made precise by Jans. Thrall's announcement suggests that he knew the representations of alternating quivers of type D_n, but it seems that there is no precise testimony for this. The result could therefore be credited to Yoshii, who also proves that A is infinitely represented if the associated alternating quiver contains $\tilde{E}_6$, $\tilde{E}_8$ or a quiver having one vertex more than $\tilde{E}_7$. Unfortunately, he also includes small mistakes (in the cases $\cdot\begin{smallmatrix}\nearrow\\ \rightarrow\\ \searrow\end{smallmatrix}$ and $\begin{smallmatrix}\nwarrow\\ \swarrow\end{smallmatrix}\cdot\rightarrow\cdot\begin{smallmatrix}\swarrow\\ \nwarrow\end{smallmatrix}$) and a big one (in the case $\tilde{E}_7$, for which he finds only 17 omnipresent indecomposables). And he simply refers to this unpropitious miss for the remaining cases E_6, E_7 and E_8. Yoshii's insight and his style – pellucid here, sibylline there – clearly impressed and discouraged the competitors in the year of publication of Cartan-Eilenberg's Homological algebra, which then attracted the interest of algebraists for more than one decade. Yoshii's paper is also corrected in [96, Krugliak, 1971/72].
3. V is *omnipresent* (faithful, *sincere*) if $V(x) \neq 0$ for each $x \in Q_v$.
4. [28, Borel, Cor. (14.12), 1969].
5. [131, Ringel, 1983].
6. [142, Springer, 4.3.3., 1980].
7. [142, Springer, 4.3.1., 1980].
8. The quadratic form attached to q_Q by enumeration of the vertices is a unit form if and only if Q has no loop. If this is the case, the real roots considered here coincide with the real roots of Sect. 6. But of course, imaginary roots are not roots in the sense of Sect. 6.

9. [81, Kač, 1979/80]. Our presentation, which is strongly indebted to Kač's paper, reproduces lectures in Ottawa, Trondheim, Mexico, Les Plans-sur-Bex (1979–85). For a similar, more complete version, see [94, Kraft/Riedtmann, 1985/86].
10. See Sect. 11.
11. Indeed, the considered component C strictly contains an extended Dynkin quiver $\tilde{\Delta}$. Let q be the quadratic form on $\mathbb{Z}^{Q_v}$ defined by $q(d) = q_{\tilde{\Delta}}(d')$ where $d' = d|\tilde{\Delta}_v$. Let $\delta \in \mathbb{N}^{Q_v}$ have support $\tilde{\Delta}_v$ and satisfy $q(\delta) = 0$, hence $q(\delta|d) = 0$ for all d. If $d \in (\mathbb{N}\setminus\{0\})^{Q_v}$ satisfies $q_Q(e^i|d) \geqslant 0$, $\forall i \in Q_v$, we infer that $0 = q(\delta|d) = \sum_a \delta_a q(e^a|d) \geqslant \sum_a \delta_a q_Q(e^a|d) \geqslant 0$, where a runs through $\tilde{\Delta}_v$. This contradicts the obvious inequality $q(e^b|d) > q_Q(e^b|d)$, where $b \in \tilde{\Delta}_v$ is an extremity of an arrow of C outside $\tilde{\Delta}$.
12. For a proof when $k \neq F_\infty$ (third step), see the 2 articles quoted in 9.
13. The argument originates in [19, Bernstein/Gelfand/Ponomarev, 1973].
14. [142, Springer, 1.9.6, 1980].
15. We owe this simplification of an earlier proof to D. Vossieck. Compare with [130, Ringel, Cor. 2.2, 1980].
16. If $f: X \to Y$ is a morphism of algebraic varieties, the function assigning to $x \in X$ the dimension of $f^{-1}(f(x))$ at x is upper semicontinuous.
17. The argument goes back to R. Brauer.
18. This seems to be hoary folklore: Let K be a perfect field with algebraic closure K' and Galois group Γ, A an algebra over K, I and I' sets of representatives of the indecomposable finite-dimensional modules over A and $K' \otimes_K A$. Let further E be the endomorphism algebra of some $M \in I$, Z the centre of the division algebra $\bar{E} = E/\mathscr{R}_E$, $z = [Z:K]$ and $r^2 = [\bar{E}:Z]$. The endomorphism algebra E' of $K' \otimes_K M$ over $K' \otimes_K A$ then satisfies $E'/\mathscr{R}_{E'} \tilde{\to} K' \otimes_K \bar{E} \tilde{\to} K' \otimes_K Z \otimes_Z \bar{E} \tilde{\to} K'^z \otimes_Z \bar{E} \tilde{\to} (K' \otimes_Z \bar{E})^z \tilde{\to} (K'^{r \times r})^z$. Associated with this decomposition of $E'/\mathscr{R}_{E'}$, we have a decomposition $K' \otimes_K M \tilde{\to} M_1^r \oplus \cdots \oplus M_z^r$, where $\overline{M} = \{M_1, \ldots, M_z\}$ is an orbit of cardinality z for the obvious action of Γ in $\bar{I}$. Thus we have a bijection $I \tilde{\to} I'/\Gamma$, $M \mapsto \overline{M}$.

8. Spectroids, Quivers, Coherence

Up to now, quivers appeared as examples among others. Their further-reaching "all-embracing" rôle is to visualize algebraic structures. In some sense, the relation quiver-algebra resembles the duality algebraic geometry-commutative algebra, though on a more elementary level. Before using quivers to "display the trophies" of further sections, we give a survey through examples.

As our examples show, the "range of our arrows" is limited by finiteness conditions. We devote the second part of this section to a discussion of various such limitations, which will later ensure the occurrence of almost split sequences.

Throughout 8.1–8.3, *k is an algebraically closed field.* In 8.4–8.6, the ground ring k is only supposed to be commutative.

8.1. The Quiver of a Spectroid. The *quiver*[1] $Q_{\mathscr{S}}$ *of a spectroid* $\mathscr{S}$ has the objects of $\mathscr{S}$ as vertices. The number of arrows from a vertex X to a vertex Y is $\dim \mathscr{R}_{\mathscr{S}}(X, Y)/\mathscr{R}^2_{\mathscr{S}}(X, Y)$. The quiver $Q_{\mathscr{S}}$ is a most evident combinatorial invariant of $\mathscr{S}$. Its importance is due to the following construction: Let us attach a radical morphism $\varphi(\alpha) \in \mathscr{R}_{\mathscr{S}}(X, Y)$ to each arrow $X \xrightarrow{\alpha} Y$. We only require that, for each pair (X, Y), the morphisms $\varphi(\alpha)$ with domain X and range Y form a basis of a supplement of $\mathscr{R}^2_{\mathscr{S}}(X, Y)$ in $\mathscr{R}_{\mathscr{S}}(X, Y)$. The map $\varphi: \alpha \mapsto \varphi(\alpha)$ then uniquely extends to a k-linear "*display-functor*" $\Phi: kQ_{\mathcal{S}} \to \mathcal{S}$ such that $\Phi(X) = X$ for each vertex X and $\Phi(\alpha) = \varphi(\alpha)$ for each arrow α. Since each $\bar{\mathcal{S}}(X, X) = \mathcal{S}(X, X)/\mathcal{R}_{\mathcal{S}}(X, X)$ is identified with k, the display-functor induces surjections $kQ_{\mathcal{S}}(X, Y) \to \mathcal{S}(X, Y)/\mathcal{R}^2_{\mathcal{S}}(X, Y)$, hence surjections $kQ_{\mathcal{S}}(X, Y) \to \mathcal{S}(X, Y)/\mathcal{R}^n_{\mathcal{S}}(X, Y)$ for each $n \in \mathbb{N}$, hence finally surjections

$$kQ_{\mathcal{S}}(X, Y) \to \mathcal{S}(X, Y)/\mathcal{R}^\infty_{\mathcal{S}}(X, Y),$$

where $\mathcal{R}^\infty_{\mathcal{S}}(X, Y) = \bigcap_{n \in \mathbb{N}} \mathcal{R}^n_{\mathcal{S}}(X, Y)$. The induced k-functor $\bar{\Phi}: kQ_{\mathcal{S}} \to \mathcal{S}/\mathcal{R}^\infty_{\mathcal{S}}$ induces an isomorphism $kQ_{\mathcal{S}}/\text{Ker}\,\bar{\Phi} \xrightarrow{\sim} \mathcal{S}/\mathcal{R}^\infty_{\mathcal{S}}$, where the ideal Ker $\bar{\Phi}$ of course highly depends on the choice of the map φ.

In the sequel, we also speak of *the quiver* $Q_{\mathcal{A}}$ *of an aggregate* $\mathcal{A}$. We then implicitly suppose that we have fixed a spectroid $\mathcal{I}\mathcal{A}$ formed by representatives of the indecomposables of $\mathcal{A}$ and that $Q_{\mathcal{A}} = Q_{\mathcal{I}\mathcal{A}}$. By abuse of language, we even speak of *the quiver*[1] $Q_{\mathcal{B}}$ *of a pointwise finite k-category* $\mathcal{B}$. We then mean the quiver of sp $\mathcal{B} = \mathcal{I}$ pro $\mathcal{B}$ (3.5).

Example 1. Let $\mathcal{P}$ be a poset and $\mathcal{S} = k\mathcal{P}$ (2.1). An arrow[2] $x \to y$ of $Q_{\mathcal{S}}$ then signifies that the interval $[x, y] = \{z \in \mathcal{P}: x \leqslant z \leqslant y\}$ is reduced to the two points $x, y \in \mathcal{P}$.

Example 2. Let A be the algebra of lower triangular $n \times n$-matrices, which we consider as a k-category with one object. As objects of the spectroid $\mathcal{S} = \text{sp}\, A$ of A we choose the modules e_iA, $1 \leqslant i \leqslant n$, and we identify $\text{Hom}_A(e_iA, e_jA)$ with e_jAe_i (3.4, Example 2). The quiver $Q_{\mathcal{S}}$ of $\mathcal{S}$ then is

$$e_1A \underset{\alpha_1}{\to} e_2A \underset{\alpha_2}{\to} e_3A - \cdots \to e_{n-1}A \underset{\alpha_{n-1}}{\to} e_nA.$$

We choose the morphisms $\varphi(\alpha_i)$ in such a way that $\varphi(\alpha_i)_{i+1,i} = 1$, all the other entries of the matrix $\varphi(\alpha_i) \in e_{i+1}Ae_i$ being 0. Then the display-functor $\Phi: kQ_{\mathcal{S}} \to \mathcal{S}$ is an isomorphism.

***Example 3.** Let $\mathfrak{S}_n$ be the group of permutations of $\{1, 2, \ldots, n\}$ and $\mathfrak{A}_n$ its subgroup of index 2. We intend to describe the quivers of the (spectroids of the) group-algebras $A_1 = k[\mathfrak{S}_3]$, $A_2 = k[\mathfrak{A}_4]$ and $A_3 = k[\mathfrak{S}_4]$ *when k has characteristic* 2. For this purpose, we consider the permutations $\sigma = (14)(23)$, $\tau = (13)(24)$, $\rho = (123)$, $\pi = (12)$ and the following elements of our group-algebras: $e = 1 + \rho + \rho^2$, $f = 1 + j\rho + j^2\rho^2$ where $1 \neq j \in k$ and $j^3 = 1$, $g = 1 + j^2\rho + j\rho^2$, $a = \sigma + j\tau + j^2\sigma\tau$, $b = \sigma + j^2\tau + j\sigma\tau$ and $c = \pi e - e$. The quivers then look as in Fig. 1.

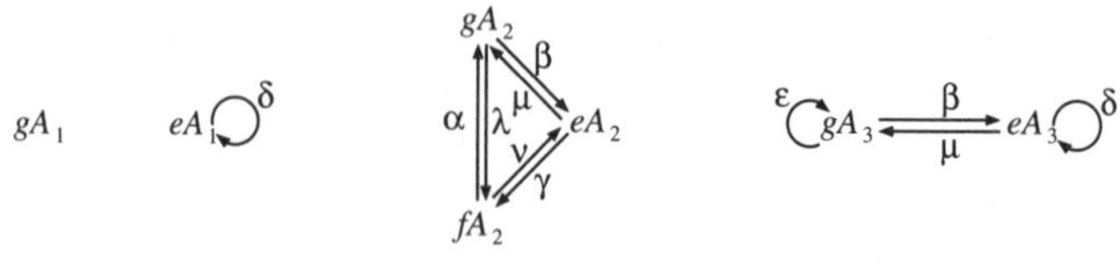

Fig. 1

In the case of A_1 and A_3, the spectroids have only two points because we then dispose of isomorphisms $fA_i \underset{\pi g}{\overset{\pi f}{\rightleftarrows}} gA_i$. In the 3 cases, we set $\varphi(\delta) = c$, $\varphi(\alpha) = ga$, $\varphi(\beta) = ea$, $\varphi(\gamma) = fa$, $\varphi(\lambda) = fb$, $\varphi(\mu) = gb$, $\varphi(\nu) = eb$ and $\varphi(\varepsilon) = jga\pi$. Then Ker Φ is generated by δ^2 in the first case, by $\beta\alpha$, $\gamma\beta$, $\alpha\gamma$, $\lambda\mu$, $\mu\nu$, $\nu\lambda$, $\nu\gamma - \beta\mu$, $\lambda\alpha - \gamma\nu$ and $\mu\beta - \alpha\lambda$ in the second, by $\beta\varepsilon$, $\varepsilon\mu$, δ^2, $\varepsilon^2 - \mu\beta$ and $\beta\mu\delta - \delta\beta\mu$ in the third $*$

***Example 4.** We suppose that *the characteristic p of k is* >0. We consider a finite group G which admits a normal cyclic Sylow p-subgroup S. We denote by K a supplement of S in G, i.e. a subgroup such that $S \cap K = \{1\}$ and $SK = G$. For instance, G might be the group of upper triangular 2×2-matrices $\begin{bmatrix} a & b \\ 0 & d \end{bmatrix}$ with entries in $\mathbb{Z}/p$ ($a \neq 0 \neq d$). In this case, S consists of the matrices $\begin{bmatrix} 1 & b \\ 0 & 1 \end{bmatrix}$ and K of the diagonal matrices.

If σ is a generator of $S = \{1, \sigma, \sigma^2, \ldots, \sigma^{p^a-1}\} = \sigma^{\mathbb{Z}/p^a}$, the action of K on S by conjugation is given by some formula

$$x^{-1}\sigma x = \sigma^{\chi(x)}, \qquad x \in K, \qquad \chi(x) \in \mathbb{Z}/p^a.$$

The induced action of K on the group-algebra $k[S]$ stabilizes the ideals

$$k[S] \supset (\sigma - 1)k[S] \supset (\sigma - 1)^2 k[S] \supset \cdots$$

Since the cardinality of K is prime to p, $k[S]$ is a semi-simple $k[K]$-module. So there is a $k[K]$-stable supplement of $(\sigma - 1)^2 k[S]$ in $(\sigma - 1)k[S]$ and an element $\pi \in (\sigma - 1)k[S]$ congruent to $\sigma - 1$ modulo $(\sigma - 1)^2$ and such that $x^{-1}\pi x \equiv x^{-1}(\sigma - 1)x = \sigma^{\chi(x)} - 1 = (\sigma - 1)(\sigma^{\eta-1} + \cdots + 1) \equiv (\sigma - 1)\chi(x) \equiv \pi\chi(x) \bmod (\sigma - 1)^2$, where $\eta \in \mathbb{Z}$ lies over $\chi(x)$. It follows that $x^{-1}\pi x = \pi\chi(x)$ for all $x \in K$.

Now consider the set $\mathscr{E}$ of isoclasses of simple $k[K]$-modules and choose representatives $E(e) \in e \in \mathscr{E}$. For each $k[K]$-module N, denote by N_χ the underlying vector space of N equipped with a new action $*$ of K such that $m{*}x = mx\chi(x)$ for all $m \in N$, $x \in K$. In this way, we obtain a permutation $e \mapsto e\chi$ of $\mathscr{E}$ such that $E(e)_\chi \xrightarrow{\sim} E(e\chi)$. The representatives $E(e)$ and each family of isomorphisms $u(e)\colon E(e)_\chi \xrightarrow{\sim} E(e\chi)$ then provide us with a description of Mod $k[G]$:

By restriction, each $M \in \operatorname{Mod} k[G]$ gives rise to a semisimple module $M|K$ over $k[K]$. Thus we have $M|K = \bigoplus_{e \in \mathscr{E}} M(e)$, where $M(e)$ is isotypic of type $E(e)$ and gives rise to a canonical isomorphism $\mu(e)\colon \operatorname{Hom}_{k[K]}(E(e), M) \otimes_k E(e) \xrightarrow{\sim} M(e)$. The formula $\pi x = x\pi\chi(x)$ then shows that the map $m \mapsto m\pi$ sends $M(e)$ into $M(e\chi)$ and induces a morphism $\pi(e)\colon M(e)_\chi \to M(e\chi)$. Thus we obtain the commutative square of Fig. 2,

$$\begin{array}{ccc}
\operatorname{Hom}_{k[K]}(E(e), M) \otimes_k E(e)_\chi & \xrightarrow[\sim]{\mu(e)_\chi} & M(e)_\chi \\
\big\downarrow & & \big\downarrow{\scriptstyle \pi(e)} \\
\operatorname{Hom}_{k[K]}(E(e\chi), M) \otimes_k E(e\chi) & \xrightarrow[\sim]{\mu(e\chi)} & M(e\chi)
\end{array}$$

Fig. 2

whose morphism on the left is $k[K]$-linear, hence of the form $f(e) \otimes_k u(e)$, where

$$f(e)\colon \operatorname{Hom}_{k[K]}(E(e), M) \to \operatorname{Hom}_{k[K]}(E(e\chi), M)$$

is k-linear.

Let now Q be the quiver which has $\mathscr{E}$ as set of vertices and one arrow $e \xrightarrow{\alpha_e} e\chi$ for each vertex e. This quiver is a disjoint union of "cycles". The maps $e \mapsto \operatorname{Hom}_{k[K]}(E(e), M)$ and $\alpha_e \mapsto f(e)$ provide a representation FM of Q which is annihilated by the paths of length p^a (because $\pi^{p^a} = 0$). Thus we finally obtain a functor

$$F\colon \operatorname{Mod} k[G] \to \operatorname{Mod}(kQ/\mathscr{J})^{op}, \qquad M \mapsto FM,$$

where $\mathscr{J}$ is the ideal generated by the paths of length p^a. In fact, the construction of a quasi-inverse functor is straightforward, so that F *is an equivalence*. It follows that *the spectroid of* $k[G]$ *is isomorphic to* $(kQ/\mathscr{J})^{op} \tilde{\to} kQ/\mathscr{J}$ $*$

Example 5. Let $\mathscr{A}$ be the aggregate formed by the finite-dimensional representations of the double arrow $\cdot \rightrightarrows \cdot$. For the description of the display-functor of Example 6 below, it seems convenient to choose slightly different representatives of the indecomposables than in 1.8. Thus we denote by $\mathscr{S}$ the spectroid formed by the representations associated with the following pairs of matrices:

$$\begin{array}{llll}
\text{a)} \begin{bmatrix} \mathbb{1}_n \\ 0 \end{bmatrix}, & \begin{bmatrix} 0 \\ \mathbb{1}_n \end{bmatrix} \in k^{(n+1)\times n}, & n \geqslant 0 \\
\text{b)}\ \mathbb{1}_n, & s\mathbb{1}_n + J_n^T, & n \geqslant 1,\ s \in k \\
\text{c)}\ J_n, & \mathbb{1}_n, & n \geqslant 1 \\
\text{d)}\ [\mathbb{1}_n\ 0], & [0\ \mathbb{1}_n] \in k^{n\times(n+1)}, & n \geqslant 0,
\end{array}$$

where J_n denotes the nilpotent upper triangular Jordan block (1.7).

a) The indecomposables of type a) yield a connected component of $Q_{\mathscr{S}}$ of the form

$$P_0 \underset{\rho_1}{\overset{\lambda_1}{\rightrightarrows}} P_1 \underset{\rho_2}{\overset{\lambda_2}{\rightrightarrows}} P_2 \underset{\rho_3}{\overset{\lambda_3}{\rightrightarrows}} P_3 \cdots$$

We choose the morphisms $\varphi(\lambda_n), \varphi(\rho_n)\colon P_{n-1} \rightrightarrows P_n$ in such a way that their tail and head components are represented by the following matrices:

$$\varphi(\lambda_n)\colon \begin{bmatrix} \mathbb{1}_{n-1} \\ 0 \end{bmatrix}, \begin{bmatrix} \mathbb{1}_n \\ 0 \end{bmatrix} \qquad \varphi(\rho_n)\colon \begin{bmatrix} 0 \\ \mathbb{1}_{n-1} \end{bmatrix}, \begin{bmatrix} 0 \\ \mathbb{1}_n \end{bmatrix}$$

b) For each $s \in k$, the indecomposables of type b) yield one connected component of $Q_{\mathscr{S}}$, whose arrows are associated with the morphisms produced below.

$$T_1^s \underset{\iota_1^s}{\overset{\pi_1^s}{\rightleftarrows}} T_2^s \underset{\iota_2^s}{\overset{\pi_2^s}{\rightleftarrows}} T_3^s \underset{\iota_3^s}{\overset{\pi_3^s}{\rightleftarrows}} T_4^s \cdots$$

$$\varphi(\pi_n^s)\colon [\mathbb{1}_n \ \ 0], [\mathbb{1}_n \ \ 0] \qquad \varphi(\iota_n^s)\colon \begin{bmatrix} 0 \\ \mathbb{1}_n \end{bmatrix}, \begin{bmatrix} 0 \\ \mathbb{1}_n \end{bmatrix}$$

c) The indecomposables of type c) give rise to a connected component of the same form as in b). The associated morphisms are chosen as follows:

$$T_1^\infty \underset{\iota_1^\infty}{\overset{\pi_1^\infty}{\rightleftarrows}} T_2^\infty \underset{\iota_2^\infty}{\overset{\pi_2^\infty}{\rightleftarrows}} T_3^\infty \underset{\iota_3^\infty}{\overset{\pi_3^\infty}{\rightleftarrows}} T_4^\infty \cdots$$

$$\varphi(\pi_n^\infty)\colon [0 \ \ \mathbb{1}_n], [0 \ \ \mathbb{1}_n] \qquad \varphi(\iota_n^\infty)\colon \begin{bmatrix} \mathbb{1}_n \\ 0 \end{bmatrix}, \begin{bmatrix} \mathbb{1}_n \\ 0 \end{bmatrix}$$

d) The indecomposables of type d) yield a last component whose associated morphisms are chosen as follows:

$$\cdots I_3 \underset{\bar\rho_3}{\overset{\bar\lambda_3}{\rightrightarrows}} I_2 \underset{\bar\rho_2}{\overset{\bar\lambda_2}{\rightrightarrows}} I_1 \underset{\bar\rho_1}{\overset{\bar\lambda_1}{\rightrightarrows}} I_0$$

$$\varphi(\bar\lambda_n)\colon [\mathbb{1}_n \ \ 0], [\mathbb{1}_{n-1} \ \ 0] \qquad \varphi(\bar\rho_n)\colon [0 \ \ \mathbb{1}_n], [0 \ \ \mathbb{1}_{n-1}]$$

We have $\mathscr{R}_{\mathscr{S}}^{\infty}(x, y) = 0$ if x and y belong to the same connected component of $Q_{\mathscr{S}}$. Otherwise, $\mathscr{R}_{\mathscr{S}}^{\infty}(x, y) = \mathscr{S}(x, y)$. For the given choice of the morphisms $\varphi(\alpha)$, the kernel of the display-functor is generated by $\rho_{i+1}\lambda_i - \lambda_{i+1}\rho_i$, $\pi_1^s\iota_1^s$, $\iota_i^s\pi_i^s - \pi_{i+1}^s\iota_{i+1}^s$ and $\bar\lambda_i\bar\rho_{i+1} - \bar\rho_i\bar\lambda_{i+1}$ ($i \geqslant 1$, $s \in k \cup \{\infty\}$).

***Example 6.** For each $n \in \mathbb{N}\backslash\{0\}$, let $\mathscr{S}_n$ be the full subspectroid of the spectroid $\mathscr{S}$ of Example 5 which is formed by the indecomposables P_i, I_i and T_j^s such that $i \leqslant n$ and $j \leqslant 2n$. By 3.2, Example 2, we then have $\mathscr{R}_{\mathscr{S}_n}^{\infty} = 0$, so that $\mathscr{S}_n$ is isomorphic to a suitable residue-category of $kQ_{\mathscr{S}_n}$. In fact, $Q_{\mathscr{S}_n}$ has the form described by Fig. 3. In particular, there are infinitely many arrows which start at P_n or stop at I_n.

We extend the definition of φ in Example 5 by setting

$$\begin{aligned}
&\varphi(\gamma^s)\colon && \begin{bmatrix} \exp sN_n \\ 0 \end{bmatrix} \in k^{2n\times n}, && \begin{bmatrix} \exp sN_{n+1} \\ 0 \end{bmatrix} \in k^{2n\times(n+1)} \\
&\varphi(\gamma^\infty)\colon && \begin{bmatrix} 0 \\ \mathbb{1}_n \end{bmatrix} \in k^{2n\times n}, && \begin{bmatrix} 0 \\ \mathbb{1}_{n+1} \end{bmatrix} \in k^{2n\times(n+1)} \\
&\varphi(\delta^s)\colon && [\exp sN_{n+1}^T \ \ 0]S \in k^{(n+1)\times 2n}, && [\exp sN_n^T \ \ 0]S \in k^{n\times 2n} \\
&\varphi(\delta^\infty)\colon && [0 \ \ \mathbb{1}_{n+1}]S \in k^{(n+1)\times 2n}, && [0 \ \ \mathbb{1}_n]S \in k^{n\times 2n},
\end{aligned}$$

where N_n and S are described in Fig. 4.

The display-functor Φ associated with φ induces an isomorphism $kQ_{\mathscr{S}_n}/\mathscr{I}_n \xrightarrow{\sim} \mathscr{S}_n$, where $\mathscr{I}_n = \operatorname{Ker}\Phi$ is the ideal generated by the following linear combinations of paths (ε^t denotes the endomorphism $\iota^t_{2n-1}\pi^t_{2n-1}$ of T^t_{2n}):

$$\begin{cases} \rho_{i+1}\lambda_i - \lambda_{i+1}\rho_i, \quad \bar{\lambda}_i\bar{\rho}_{i+1} - \bar{\rho}_i\bar{\lambda}_{i+1}, \quad 1 \leqslant i < n, \\ \pi^t_1\iota^t_1, \quad \pi^t_{j+1}\iota^t_{j+1} - \iota^t_j\pi^t_j, \quad 1 \leqslant j < 2n-1, \quad t \in k \cup \infty, \end{cases}$$

$$\begin{cases} \gamma^s\rho_n - \varepsilon^s\gamma^s\lambda_n - s\gamma^s\lambda_n, \quad \bar{\rho}_n\delta^s - \bar{\lambda}_n\delta^s\varepsilon^s - s\bar{\lambda}_n\delta^s, \quad s \in k, \\ \gamma^\infty\lambda_n - \varepsilon^\infty\gamma^\infty\rho_n, \quad \bar{\lambda}_n\delta^\infty - \bar{\rho}_n\delta^\infty\varepsilon^\infty, \end{cases}$$

$$\begin{cases} \delta^s(\varepsilon^s)^j\gamma^s - \sum_{h=0}^{j}\binom{2n-j-1+h}{h} s^h\delta^0(\varepsilon^0)^{j-h}\gamma^0, \quad s \in k, \quad 0 \leqslant j \leqslant 2n-1, \\ \delta^\infty(\varepsilon^\infty)^j\gamma^\infty - \delta^0(\varepsilon^0)^{2n-1-j}\gamma^0, \quad 0 \leqslant j \leqslant 2n-1, \end{cases}$$

$$\{\delta^0\gamma^0\lambda_n, \quad \bar{\lambda}_n\delta^0\gamma^0.$$

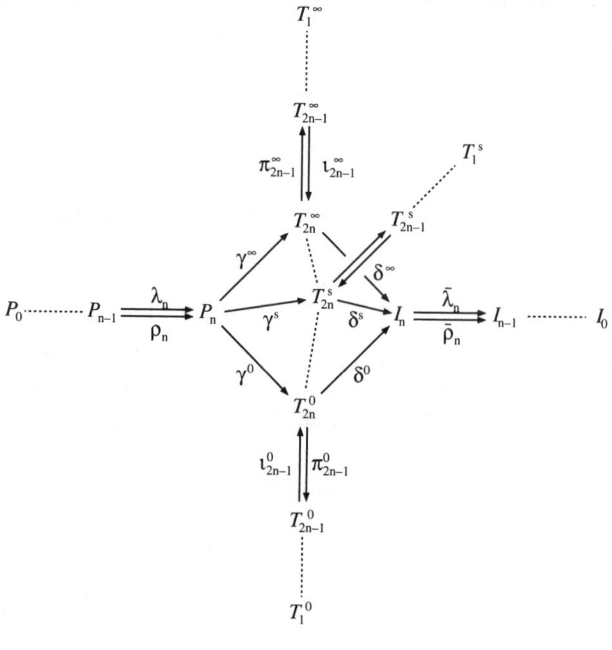

Fig. 3

$$N_n = \begin{bmatrix} 0 & 1 & 0 & 0 & 0 \\ 0 & 0 & 2 & 0 & 0 \\ 0 & 0 & 0 & 3 & 0 \\ 0 & 0 & 0 & 0 & 4 \\ 0 & 0 & 0 & 0 & 0 \end{bmatrix} \in k^{n \times n} \qquad S = \left[\begin{array}{cc|cc} 0 & 0 & 0 & 1 \\ 0 & 0 & 1 & 0 \\ \hline 0 & 1 & 0 & 0 \\ 1 & 0 & 0 & 0 \end{array}\right] \in k^{2n \times 2n}$$

Fig. 4

The isomorphism $kQ_{\mathcal{S}_n}/\mathcal{I}_n \xrightarrow{\sim} \mathcal{S}_n$ shows in particular that $\mathcal{S}_n(P_n, I_n) = \mathcal{S}_n(T_{2n}^t, I_n)\mathcal{S}_n(P_n, T_{2n}^t)$ for each $t \in k \cup \{\infty\}$. But the filtration of $\mathcal{S}_n(P_n, I_n)$ by the subspaces $\mathcal{S}_n(T_{2n}^t, I_n)\mathcal{R}^r_{\mathcal{S}_n}(T_{2n}^t, T_{2n}^t)\mathcal{S}_n(P_n, T_{2n}^t)$, $r \in \mathbb{N}$, varies with t ∗

8.2. If Q is an arbitrary quiver and $\mathcal{I}$ an ideal of kQ, the category $\mathrm{Mod}(kQ/\mathcal{I})$ is isomorphic to (and will be identified with) the full subcategory of Mod kQ formed by the representations V of Q^{op} such that $V(\mu) = 0$ for all elements μ of a set of generators of $\mathcal{I}$.

Thus, if the display-functor $\Phi\colon kQ_{\mathcal{S}} \to \mathcal{S}$ of 8.1 induces surjections of the morphism-spaces, the modules over $\mathcal{S}$ can be interpreted as representations of $Q_{\mathcal{S}}^{op}$ which annihilate Ker Φ. This ideal is not completely arbitrary, because the construction of Φ imposes the condition Ker $\Phi \subset k^2 Q_{\mathcal{S}}$ (=the ideal of $kQ_{\mathcal{S}}$ generated by the paths of length 2).

Proposition. *Let k be a field, Q a quiver and $\mathcal{I}$ a subideal of k^2Q. Then* $\mathrm{Mod}(kQ/\mathcal{I})$ *is hereditary* (*i.e.* Ext^1 *is right exact*) *if and only if* $\mathcal{I} = 0$.

Proof. a) To prove the right-exactness of the extension-functor if $\mathcal{I} = 0$, we provide a construction of $\mathrm{Ext}^1(V, W)$: For each exact sequence

$$0 \to W \xrightarrow{w} U \xrightarrow{v} V \to 0$$

of representations of Q^{op} and each vertex $x \in Q_v$, we choose a linear section $s(x)$ of $v(x)$ which furnishes the diagram

$$\begin{array}{ccccccccc}
0 & \longrightarrow & W(x) & \xrightarrow{w(x)} & U(x) & \xrightarrow{v(x)} & V(x) & \longrightarrow & 0 \\
 & & \| & & \uparrow{\scriptstyle [w(x)s(x)]} & & \| & & \\
0 & \longrightarrow & W(x) & \xrightarrow[{[1\,0]^T}]{} & W(x) \oplus V(x) & \xrightarrow[{[0\,1]}]{} & V(x) & \longrightarrow & 0
\end{array}$$

Each arrow $x \xrightarrow{\alpha} y$ of Q then yields a map $g(\alpha)\colon V(y) \to W(x)$ which makes the following square commutative

$$\begin{array}{ccc}
U(x) & \xleftarrow{U(\alpha)} & U(y) \\
{\scriptstyle [w(x)\,s(x)]}\uparrow & \cdot & \uparrow{\scriptstyle [w(y)\,s(y)]} \\
W(x) \oplus V(x) & \xleftarrow[{\begin{bmatrix} W(\alpha) & g(\alpha) \\ 0 & V(\alpha) \end{bmatrix}}]{} & W(y) \oplus V(y)
\end{array}$$

This means that the sequence (w, v) defines the same element of $\mathrm{Ext}^1(V, W)$ as $0 \to W \xrightarrow{i} E \xrightarrow{p} V \to 0$ where $E(x) = W(x) \oplus V(x)$, $i(x) = [1\ \ 0]^T$, $p(x) = [0\ \ 1]$ for each $x \in Q_v$ and $E(\alpha) = \begin{bmatrix} W(\alpha) & g(\alpha) \\ 0 & V(\alpha) \end{bmatrix}$ for each $\alpha \in Q_a$. In other words, we have a surjection $\varepsilon\colon \prod_{\alpha\in Q_a} \mathrm{Hom}_k(V(h\alpha), W(t\alpha)) \to \mathrm{Ext}^1(V, W)$ which maps the family $(g(\alpha))$ onto the class of the sequence (i, p). This ε is part of the exact sequence (*) below, where γ is the inclusion and δ maps $(f(x))_{x\in Q_v}$ onto

$$(f(t\alpha)V(\alpha) - W(\alpha)f(h\alpha))_{\alpha\in Q_a}.$$

The equality $\mathrm{Im}\,\gamma = \mathrm{Ker}\,\delta$ is just the definition of $\mathrm{Hom}(V, W)$, and $\mathrm{Im}\,\delta = \mathrm{Ker}\,\varepsilon$ paraphrases the definition of the equivalence relation on short exact sequences which gives rise to $\mathrm{Ext}^1(V, W)$ (compare with 7.2).

$$(*)\quad 0 \to \mathrm{Hom}(V, W) \xrightarrow{\gamma} \prod_{x\in Q_v} \mathrm{Hom}_k(V(x), W(x)) \xrightarrow{\delta} \prod_{\alpha\in Q_a} \mathrm{Hom}_k(V(h\alpha), W(t\alpha)) \xrightarrow{\varepsilon}$$
$$\xrightarrow{\varepsilon} \mathrm{Ext}^1(V, W) \to 0$$

The right exactness of Ext^1 now follows from the snake lemma and the exactness of the two middle terms of (*).

b) Now assume that $\mathscr{I} \neq 0$. Denoting by $\mathscr{J}$ the ideal of kQ generated by the arrows, we have $\mathscr{I}(z, x) \neq (\mathscr{I}\mathscr{J} + \mathscr{J}\mathscr{I})(z, x)$ if we choose the vertices x, z so that $\inf\{n\colon \mathscr{I}(z, x) \not\subset \mathscr{J}^n(z, x)\}$ is minimal. So it remains to verify that the dual of $\mathscr{I}(z, x)/(\mathscr{I}\mathscr{J} + \mathscr{J}\mathscr{I})(z, x)$ is identified with $\mathrm{Ext}^2_{\mathscr{T}}(x^-, z^-)$, where $x^- = kQ(?, x)/\mathscr{J}(?, x)$ is identified with a simple module over $\mathscr{T} = kQ/\mathscr{I}$ (compare with 3.7).

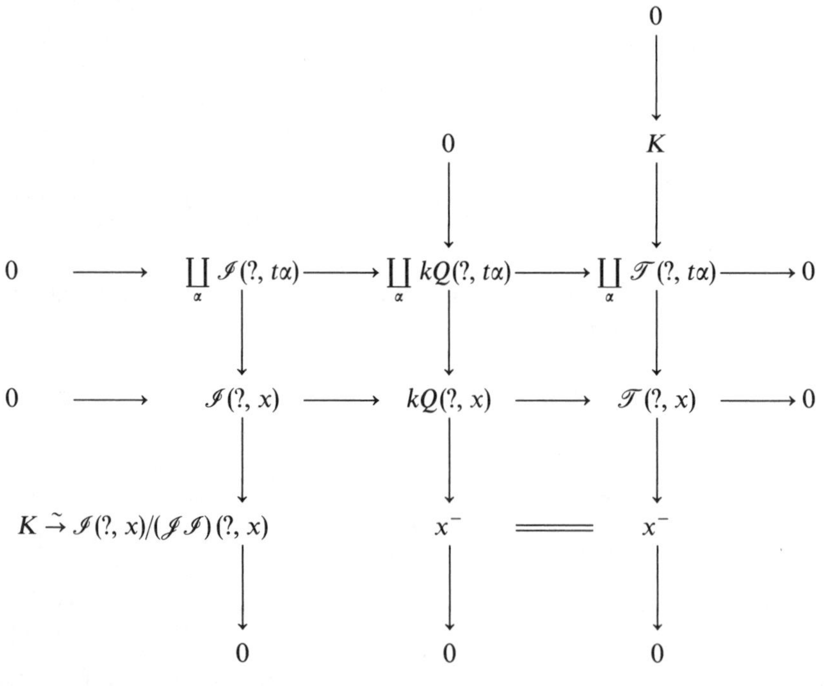

Fig. 5

For this we use the following beginning of a projective resolution of x^-:

$$0 \to \mathscr{I}(?, x)/(\mathscr{J}\mathscr{I})(?, x) \xrightarrow{\mu} \coprod_{h\alpha = x} \mathscr{T}(?, t\alpha) \xrightarrow{\nu} \mathscr{T}(?, x) \xrightarrow{\pi} x^- \to 0.$$

The morphism ν is induced by the arrows $t\alpha \xrightarrow{\alpha} x$, and μ maps the class $\bar{\varphi}$ of $\varphi = \sum_{h\alpha=x} \alpha\varphi_\alpha \in \mathscr{I}(y, x) \subset \mathscr{J}(y, x)$ onto $(\bar{\varphi}_\alpha)$. The exactness of the sequence follows from the snake lemma applied to the diagram of Fig. 5. The assumption $\mathscr{I} \subset \mathscr{J}^2$ then implies that $\varphi_\alpha \in \mathscr{J}(y, t\alpha)$ if $\varphi = \sum_{h\alpha=x} \alpha\varphi_\alpha \in \mathscr{I}(y, x)$, hence that $\mathrm{Hom}(\mu, z^-)$ vanishes and that $\mathrm{Ext}^2(x^-, z^-)$ is identified with $\mathrm{Hom}(\mathscr{I}(?, x)/(\mathscr{J}\mathscr{I})(?, x), z^-)$, i.e. with the dual of

$$\mathscr{I}(z, x)/(\mathscr{I}\mathscr{J} + \mathscr{J}\mathscr{I})(z, x). \surd$$

8.3. Locally Bounded Spectroids. We now turn to a particular class of spectroids of the form $kQ/\mathscr{I}$: A spectroid $\mathscr{S}$ is called *locally bounded* if, for each $X \in \mathscr{S}$, the number of $Y \in \mathscr{S}$ such that $\mathscr{S}(X, Y) \neq 0$ or $\mathscr{S}(Y, X) \neq 0$ is finite. It is equivalent to say that, for each $X \in \mathscr{S}$, the $\mathscr{S}$-modules $X^\wedge$ and $X^\vee$ have finite lengths.

Example 1. Finite spectroids are locally bounded.

Locally bounded spectroids are next to finite spectroids. They are met with in the investigation of finitely represented finite spectroids (Sect. 14).

Example 2. Let Q be a quiver. We call an ideal $\mathscr{I}$ of the k-category of paths kQ *admissible* if it satisfies the following two conditions: a) $\mathscr{I}$ is contained in the ideal k^2Q generated by the paths of length 2; b) For each $x \in Q_v$, there is an $l_x \in \mathbb{N}$ such that $\mathscr{I}$ contains all paths of length $\geqslant l_x$ which start or stop at x.

In case $\mathscr{I} \subset k^2Q$, it is easy to see that *$kQ/\mathscr{I}$ is a locally bounded spectroid if and only if $\mathscr{I}$ is admissible and Q locally finite* (i.e., for each $x \in Q_v$, there are only finitely many arrows which start or stop at x).

Example 3. With each spectroid $\mathscr{S}$ we associate a new spectroid, the *repetition*[3] $\mathscr{S}^{\mathbb{Z}}$ of $\mathscr{S}$. Its objects are the pairs $X^n = (X, n)$ where $X \in \mathscr{S}$ and $n \in \mathbb{Z}$. The morphisms $X^n \to Y^m$ are the pairs $f^n = (f, n)$ with $f \in \mathscr{S}(X, Y)$ if $m = n$, the pairs $\varphi^n = (\varphi, n)$ with $\varphi \in D\mathscr{S}(Y, X)$ if $m = n - 1$. They are zero if $m \notin \{n, n-1\}$. The composition is such that $f^n g^n = (fg)^n$, $\varphi^n g^n = \varphi(g?)^n$ and $h^{n-1}\varphi^n = \varphi(?h)^n$ if $g \in \mathscr{S}(W, X)$ and $h \in \mathscr{S}(Y, Z)$. For instance, if $\mathscr{S}$ equals $kQ/\mathscr{I}$, where Q is the quiver of Fig. 6 and $\mathscr{I}$ the ideal generated by cb, then $\mathscr{S}^{\mathbb{Z}}$ equals $kQ^{\mathbb{Z}}/\mathscr{I}^{\mathbb{Z}}$, where $\mathscr{I}^{\mathbb{Z}}$ is generated by $c^n b^n$, $b^{n-1}\varphi^n$, $a^{n-1}\psi^n$, $\psi^n\chi^{n+1}$, $\chi^n d^n$, $\xi^n c^n$, $\varphi^n a^n - \psi^n b^n$, $b^{n-1}\psi^n - \chi^n c^n$ and $c^{n-1}\chi^n - d^{n-1}\xi^n$, $n \in \mathbb{Z}$.

In the general case, the *translation* $X^n \mapsto X^{n-1}$ gives rise to an automorphism of $\mathscr{S}^{\mathbb{Z}}$. Moreover, the injective $\mathscr{S}^{\mathbb{Z}}$-module $(X^n)^\vee$ is canonically isomorphic to the projective $(X^{n-1})^\wedge$ for each $X \in \mathscr{S}$ and each $n \in \mathbb{Z}$.

If $\mathscr{S}$ is locally bounded, then so is $\mathscr{S}^{\mathbb{Z}}$.

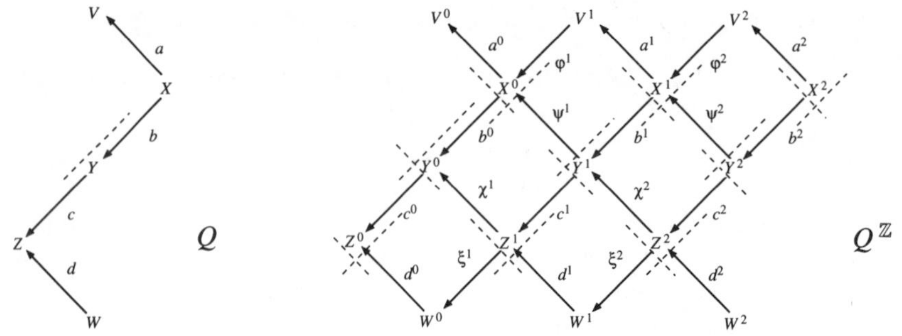

Fig. 6

Let us now return to an *arbitrary locally bounded* $\mathscr{S}$. Then the following statements are true:

a) $\bigcap_{n \in \mathbb{N}} \mathscr{R}^n_{\mathscr{S}} = \{0\}$: Indeed, if $X, Y \in \mathscr{S}$ are fixed, we have $\mathscr{R}^n_{\mathscr{S}}(X, Y) = \mathscr{R}^n_{\mathscr{T}}(X, Y)$ where $\mathscr{T}$ denotes the finite full subspectroid of $\mathscr{S}$ formed by the $Z \in \mathscr{S}$ such that $\mathscr{S}(X, Z) \neq 0 \neq \mathscr{S}(Z, Y)$. It therefore suffices to prove that $\mathscr{R}^n_{\mathscr{T}}(X, Y) = \mathscr{R}^n_{\mathscr{P}}(X^\wedge, Y^\wedge) = 0$ for large n, where $\mathscr{P} = \text{pro}\, \mathscr{T}$. This follows from 3.2, Example 2.

b) If $M \in \text{Mod}\, \mathscr{S}$ is not zero, the radical $\mathscr{R}M$ is $\neq M$: Otherwise, by 3.2, we would have $M(X) = \sum_{Y \in \mathscr{S}} M(Y)\mathscr{R}_{\mathscr{S}}(X, Y) = \sum_{Y, Z \in \mathscr{S}} M(Z)\mathscr{R}_{\mathscr{S}}(Y, Z)\mathscr{R}_{\mathscr{S}}(X, Y) = \sum_{Z \in \mathscr{S}} M(Z)\mathscr{R}^2_{\mathscr{S}}(X, Z) = \sum_{T \in \mathscr{S}} M(T)\mathscr{R}^n_{\mathscr{S}}(X, T) = 0$ since each $X \in \mathscr{S}$ satisfies $\mathscr{R}^n_{\mathscr{S}}(X, T) = 0$ for all T if n is large.

c) The finitely presented $\mathscr{S}$-modules coincide with the $\mathscr{S}$-modules of finite length: $\text{mod}\, \mathscr{S} = \text{mod}_f\, \mathscr{S}$. The projectives of $\text{mod}\, \mathscr{S}$ are the finite sums of modules $X^\wedge$, the injectives the finite sums of modules $X^\vee$, $X \in \mathscr{S}$.

It follows in particular that the projectives of $\text{mod}\, \mathscr{S}^{\mathbb{Z}}$ coincide with the injectives. We express this property by saying that the spectroid $\mathscr{S}^{\mathbb{Z}}$ is *selfinjective*.

d) Define the *transfer* $\mathscr{T}: \text{mod}\, \mathscr{S} \to \text{mod}\, \mathscr{S}$ by $(\mathscr{T}M)(Y) = D\,\text{Hom}(M, Y^\wedge)$. This functor maps $X^\wedge$ onto $X^\vee$ because $(\mathscr{T}X^\wedge)(Y) = D\,\text{Hom}(X^\wedge, Y^\wedge) \tilde{\to} D\mathscr{S}(X, Y) = X^\vee(Y)$. It therefore maps $\text{pro}\, \mathscr{S}$ into the full subcategory $\text{inj}\, \mathscr{S}$ of $\text{mod}\, \mathscr{S}$ formed by the injectives. In fact, $\mathscr{T}$ induces an *equivalence* $\text{pro}\, \mathscr{S} \tilde{\to} \text{inj}\, \mathscr{S}$, as follows from $\text{Hom}(X^\wedge, Y^\wedge) \tilde{\leftarrow} \mathscr{S}(X, Y) \tilde{\to} DD\mathscr{S}(X, Y) = DX^\vee(Y) \tilde{\to} \text{Hom}(X^\vee, Y^\vee)$ (3.7b).

8.4. Induced and Finitely Presented Modules. Let $\mathscr{S}$ be a spectroid. As we have seen above, $\mathscr{S}$-modules can be interpreted as representations of the quiver $Q^{op}_{\mathscr{S}}$ if $\mathscr{R}^\infty_{\mathscr{S}} = 0$, hence if $\mathscr{S}$ is finite or locally bounded. If $\mathscr{S}$ is arbitrary, a similar interpretation can be given for $\mathscr{S}$-modules which are determined by their restrictions to a finite full subspectroid of $\mathscr{S}$. We devote the end of Sect. 8 to $\mathscr{S}$-modules having this property.

Let $F\colon \mathscr{F} \to \mathscr{S}$ be a k-functor between two spectroids and N a module over $\mathscr{F}$. We call *induction*[4] of N to $\mathscr{S}$ (along F) and denote by $N_{\mathscr{S}}$ the cokernel of the morphism

$$v_N\colon \coprod_{Y,Z \in \mathscr{F}} N(Z) \otimes_k \mathscr{F}(Y, Z) \otimes_k (FY)^\wedge \to \coprod_{Y \in \mathscr{F}} N(Y) \otimes_k (FY)^\wedge$$

defined by $v_N(n \otimes \varphi \otimes \sigma) = n\varphi \otimes \sigma - n \otimes (F\varphi)\sigma$. (For each $V \in \operatorname{Mod} k$ and each $M \in \operatorname{Mod} \mathscr{S}$, $V \otimes_k M$ denotes the $\mathscr{S}$-module $X \mapsto V \otimes_k M(X)$.) The induction gives rise to natural bijections

$$\operatorname{Hom}_{\mathscr{F}}(N, M|\mathscr{F}) \xrightarrow{\sim} \operatorname{Hom}_{\mathscr{S}}(N_{\mathscr{S}}, M),$$

where $M|\mathscr{F}$ is the *restriction* $Y \mapsto M(FY)$ of an $\mathscr{S}$-module M to $\mathscr{F}$: The morphism $\mu' \in \operatorname{Hom}_{\mathscr{S}}(N_{\mathscr{S}}, M)$ associated with some $\mu \in \operatorname{Hom}_{\mathscr{F}}(N, M|\mathscr{F})$ maps the residue class of $n \otimes \sigma \in N(Y) \otimes_k \mathscr{S}(X, FY)$ onto $\mu(n)\sigma \in M(X)$.

We are especially interested in the case $N = M|\mathscr{F}$ and $\mu = \mathbb{1}_{M|\mathscr{F}}$. We then call $\mu'\colon (M|\mathscr{F})_{\mathscr{S}} \to M$ the *canonical morphism*. This is of course an isomorphism when $\mathscr{F} = \mathscr{S}$ and $F = \mathbb{1}_{\mathscr{S}}$.

Example 1. In case $N = Y^\wedge$, the isomorphisms $\operatorname{Hom}_{\mathscr{S}}((FY)^\wedge, M) \xrightarrow{\sim} M(FY) \xrightarrow{\sim} \operatorname{Hom}_{\mathscr{F}}(Y^\wedge, M|\mathscr{F}) \xrightarrow{\sim} \operatorname{Hom}_{\mathscr{S}}(N_{\mathscr{S}}, M)$ show that $N_{\mathscr{S}} \xrightarrow{\sim} (FY)^\wedge$.

Example 2. Suppose that $\mathscr{S} = kS$ where S is a poset, that T is a full subposet of S and $F\colon kT \to kS$ the inclusion, and that $N = k_T$ (3.6, Example 2). For each $s \in S$, $N_{\mathscr{S}}(s)$ is a free k-module with a natural basis whose elements correspond bijectively to the connected components of the poset $T \cap [s$ (3.7).

A similar description of $N_{\mathscr{S}}$ is possible when $N = k_J$ is produced by an interval J of T. For instance, if S is linearly ordered, we have $N_{\mathscr{S}} \xrightarrow{\sim} k_I$ where I is the interval of S formed by the s such that $s \leqslant t \leqslant j$ with $t \in T$ and $j \in J$ implies $t \in J$ and that $s \leqslant j$ for some $j \in J$.

Proposition. *Let M be a pointwise finite module over the spectroid $\mathscr{S}$. Then M is finitely presented if and only if there exists a finite full subspectroid $\mathscr{F}$ of $\mathscr{S}$ for which the canonical morphism $(M|\mathscr{F})_{\mathscr{S}} \to M$ is invertible.*

Proof. If M is finitely presented, there is by definition an exact sequence

$$\bigoplus_{\beta \in B} Y_\beta^\wedge \to \bigoplus_{\alpha \in A} X_\alpha^\wedge \to M \to 0,$$

where A and B are finite. Let $\mathscr{F}$ be the full subcategory of $\mathscr{S}$ formed by the X_α and Y_β: By Example 1 above, the canonical morphisms $(X_\alpha^\wedge|\mathscr{F})_{\mathscr{S}} \to X_\alpha^\wedge$ and $(Y_\beta^\wedge|\mathscr{F})_{\mathscr{S}} \to Y_\beta^\wedge$ are invertible. Hence, so is $(M|\mathscr{F})_{\mathscr{S}} \xrightarrow{\sim} M$.

Conversely, suppose that $M \xrightarrow{\sim} N_{\mathscr{S}}$ where N is a pointwise finite module over a finite full subspectroid $\mathscr{F}$ of $\mathscr{S}$. Then we obviously have an exact sequence

$$\bigoplus_{\beta \in B} V_\beta^\wedge \to \bigoplus_{\alpha \in A} U_\alpha^\wedge \to N \to 0$$

in $\operatorname{mod}_f \mathscr{F}$, where A and B are finite. The induced sequence of $\operatorname{mod}_{pf} \mathscr{S}$ formed by the inductions to $\mathscr{S}$ remains exact and provides a finite presentation of $M \xrightarrow{\sim} N_{\mathscr{S}}$.

Corollary. *If $\mathscr{S}$ is a spectroid, the category* mod $\mathscr{S}$ *formed by the finitely presented $\mathscr{S}$-modules is an aggregate.*

8.5. The Indicators of Generation and Relations. A module M over the spectroid $\mathscr{S}$ is *finitely generated* if there are finitely many elements $m_\alpha \in M(X_\alpha)$, $\alpha \in A$, such that each $m \in M(X)$ can be written as $m = \sum_{\alpha \in A} m_\alpha \varphi_\alpha$ where $\varphi_\alpha \in \mathscr{S}(X, X_\alpha)$. It is equivalent to say that M is a quotient of a finitely free module $\bigoplus_{\alpha \in A} X_\alpha^\wedge$, or to say that M is pointwise finite and that the canonical morphism $(M|\mathscr{F})_\mathscr{S} \to M$ is an epimorphism for some finite full subspectroid $\mathscr{F}$ of $\mathscr{S}$.

The *generation indicator* of an $\mathscr{S}$-module M is by definition the set of all $X \in \mathscr{S}$ such that $M(X) \neq \mathscr{R}M(X)$ or, equivalently, such that $\mathrm{Hom}_\mathscr{S}(M, X^-) \neq 0$. If M is a quotient of $\coprod_{\alpha \in A} X_\alpha^\wedge$, the generation indicator of M is of course contained in $\{X_\alpha: \alpha \in A\}$; accordingly it is finite if M is finitely generated.

Suppose that M is finitely generated, and let G be its generation indicator. For each $X \in G$, $\overline{M}(X) = M(X)/(\mathscr{R}M)(X)$ is a finite-dimensional vector space over the division algebra $\overline{\mathscr{S}}(X, X) = \mathscr{S}(X, X)/\mathscr{R}_\mathscr{S}(X, X)$. We choose a projective covering[5] $p^X: P^X \to \overline{M}(X)$ of the $\mathscr{S}(X, X)$-module underlying $\overline{M}(X)$. (If d is the dimension of $\overline{M}(X)$ over $\overline{\mathscr{S}}(X, X)$, P^X is isomorphic to $\mathscr{S}(X, X)^d$, and the induction $P^X_\mathscr{S}$ of P^X to $\mathscr{S}$ is isomorphic to $(X^\wedge)^d$.) We further choose a factorization $P^X \xrightarrow{q^X} M(X) \xrightarrow{can.} \overline{M}(X)$ of p^X. By 8.4, the maps q^X provide morphisms $r^X: P^X_\mathscr{S} \to M$ and $r = (r^X)_{X \in G}: \bigoplus_{X \in G} P^X_\mathscr{S} \to M$ such that the composition $\bigoplus_{X \in G} P^X_\mathscr{S} \to M \to M/\mathscr{R}M$ is an epimorphism. It easily follows that $\mathrm{Im}\, r$ is contained in no maximal submodule of M, hence that r is an epimorphism. In fact, r is a *projective covering*.

The *module of relations* $\Omega M = \mathrm{Ker}\, r$ is uniquely determined up to isomorphism by M. Its generation indicator R is by definition the *indicator of relations* of M. It coincides with the set of $X \in \mathscr{S}$ such that $\mathrm{Ext}^1_\mathscr{S}(M, X^-) \neq 0$. If M is finitely presented, ΩM is finitely generated and admits a projective covering which, together with r, provides a *minimal projective presentation*

$$(*) \quad \bigoplus_{\beta \in B} Y_\beta^\wedge \to \bigoplus_{\alpha \in A} X_\alpha^\wedge \to M \to 0$$

where A, B are finite, $X_\alpha \in G$ for each α and $Y_\beta \in R$ for each β; moreover, if $\mathscr{F}$ is a finite full subcategory of $\mathscr{S}$ which contains $G \cup R$, the restrictions $Y_\beta^\wedge|\mathscr{F}$ and $X_\alpha^\wedge|\mathscr{F}$ are projective in $\mathrm{mod}_f\, \mathscr{F}$; restricting (*) to $\mathscr{F}$, we therefore obtain a minimal projective presentation of $M|\mathscr{F}$. It follows that G is also the generation indicator and R the indicator of relations of $M|\mathscr{F}$. This implies the necessity of the following equivalence, whose sufficiency is not difficult to prove: *A pointwise finite module N over the spectroid $\mathscr{S}$ is finitely presented if and only if there exists a finite full subcategory $\mathscr{F}_0 \subset \mathscr{S}$ such that the generation indicator and the indicator of relations of $N|\mathscr{F}$ are contained in $\mathscr{F}_0$ for each finite full subcategory $\mathscr{F}$ of $\mathscr{S}$ containing $\mathscr{F}_0$.*

Example 1. Let I be an interval of a finite poset S and $\mathscr{S} = kS$. The generation indicator G of k_I (3.6) is the set of maximal elements of I. The indicator of relations of k_I is formed by the maximal elements of

$$\{s \in S \backslash I: s \leqslant j \text{ for some } j \in I\}$$

and by the elements $i \in I \backslash G$ such that the poset $\{j \in I: i < j\}$ is not connected.

Example 2. *Let $\mathscr{S}$ be locally bounded.* It then follows from 8.3b) that the construction of projective coverings extends to arbitrary $\mathscr{S}$-modules M. If M is *pointwise finite*, then so is its projective cover. Furthermore, the following statements are then equivalent: (i) M has finite length; (ii) M is finitely presented; (iii) the generation indicator of M is finite; (iv) the *top* $M/\mathscr{R}M$ of M has finite length; (v) the projective cover of M has finite length.

By duality, if $\mathscr{S}$ is locally bounded, each $\mathscr{S}$-module M admits an *injective envelopment*[5]; if M is pointwise finite, then so is its injective envelope; furthermore, M is then of finite length if and only if so is its *socle* (=maximal semisimple submodule)...

8.6. Coherent Spectroids. For an exact sequence of modules $0 \to L \to M \to N \to 0$, the following assertions are well known: If N is finitely presented and M finitely generated, then L is finitely generated; if L and N are finitely presented, then so is M; if L is finitely generated and M finitely presented, then N is finitely presented.

These assertions have the following implications: Call a module *coherent*[6] if it is finitely generated and if all its finitely generated submodules are finitely presented. Then, for each morphism between coherent modules, the image, the kernel and the cokernel are coherent. Moreover, if in the exact sequence above L and N are coherent, then so is M.

The coherent modules over a spectroid $\mathscr{S}$ therefore form an exactly embedded subcategory of Mod $\mathscr{S}$. This subcategory may contain only zero modules as our first example will show. Nonetheless, we shall encounter spectroids $\mathscr{S}$ such that all the representable modules $X^\wedge$ are coherent. In this case, the coherent modules coincide with the finitely presented modules, and we say that $\mathscr{S}$ is *right coherent*.

We are especially interested *in the case where $\mathscr{S}$ is right coherent and all the injectives $X^\vee$, $X \in \mathscr{S}$, are coherent.* Since each simple module X^- is identified with the image of some morphism $X^\wedge \to X^\vee$ (3.7d), it then follows that *the simple $\mathscr{S}$-modules are coherent.*

Besides right coherent spectroids, we will also encounter left coherent spectroids: By definition, $\mathscr{S}$ is *left coherent* if $\mathscr{S}^{op}$ is right coherent. It is equivalent to say that the $\mathscr{S}$-modules N which admit a *finite copresentation*

$$0 \to N \to \bigoplus_{\alpha \in A} X_\alpha^\vee \to \bigoplus_{\beta \in B} Y_\beta^\vee, \quad A \text{ and } B \text{ finite},$$

form an exactly embedded full subcategory of Mod $\mathscr{S}$.

In the sequel, $\mathrm{mod}^\vee \mathscr{S}$ denotes the category formed by such *finitely copresented* $\mathscr{S}$-modules N.

Example 1. Let $S = \mathbb{R}^2$ be equipped with the partial order such that $(x, y) < (x', y')$ is equivalent to $x < x'$. Then the coherent modules over $\mathscr{S} = kS$ are zero.

Example 2. Let S be a poset, $\mathscr{S} = kS$ and $t \in S$. Then $t^\wedge$ is coherent if and only if, for all $r,s \in t] = \{x \in S: x \leqslant t\}$, the subset $r] \cap s]$ is a finite union $m_1] \cup \cdots \cup m_n]$ of subsets $m_i]$.

Example 3. If the poset S is linearly ordered, $\mathscr{S} = kS$ is right and left coherent. Accordingly, mod $\mathscr{S}$ and $\text{mod}^\vee \mathscr{S}$ are exactly embedded in Mod $\mathscr{S}$. If $S \neq \varnothing$, the condition $\text{mod}^\vee \mathscr{S} \subset \text{mod}\, \mathscr{S}$ means that S and the intervals $t[= \{x \in S: x < t\}$, where $t \in S$ is not minimal, have maximal elements. Of course, this does not imply $\text{mod}^\vee \mathscr{S} = \text{mod}\, \mathscr{S}$.

Example 4. A locally bounded spectroid $\mathscr{S}$ is right and left coherent and satisfies $\text{mod}\, \mathscr{S} = \text{mod}^\vee \mathscr{S}$.

Example 5. Let $\mathscr{S}$ be a spectroid and ind $\mathscr{S} = \mathscr{I}$ mod $\mathscr{S}$ the full subcategory of mod $\mathscr{S}$ formed by representatives of the indecomposables. Since mod $\mathscr{S}$ admits cokernels, the spectroid ind $\mathscr{S}$ is *left coherent*.

If $\mathscr{S}$ is right coherent, ind $\mathscr{S}$ is left and right coherent. In this case, $\text{mod}^\vee \mathscr{S} \subset \text{mod}\, \mathscr{S}$ implies $\text{mod}^\vee(\text{ind}\, \mathscr{S}) \subset \text{mod}(\text{ind}\, \mathscr{S})$: Indeed, each $N \in \text{ind}\, \mathscr{S}$ admits a finite presentation $\bigoplus_\beta Y_\beta^\wedge \to \bigoplus_\alpha X_\alpha^\wedge \to N \to 0$ in mod $\mathscr{S}$, and this induces an exact sequence $\bigoplus_\beta Y_\beta^{\wedge\vee} \to \bigoplus_\alpha X_\alpha^{\wedge\vee} \to N^\vee \to 0$. Now, the isomorphisms Hom $(M, X^\vee) \tilde{\to} D\, \text{Hom}(X^\wedge, M)$ of 3.7c mean that $X^{\vee\wedge} \tilde{\to} X^{\wedge\vee}$. It follows that $Y_\beta^{\wedge\vee}$, $X_\alpha^{\wedge\vee} \in \text{mod}(\text{ind}\, \mathscr{S})$, hence $N^\vee \in \text{mod}(\text{ind}\, \mathscr{S})$. This implies our claim because mod(ind $\mathscr{S}$) is "closed under kernels".

Example 6. For each spectroid $\mathscr{S}$, $\text{ind}^\vee \mathscr{S} = \mathscr{I}\, \text{mod}^\vee \mathscr{S}$ (2.5) is right coherent. If $\mathscr{S}$ is left coherent, then so is $\text{ind}^\vee \mathscr{S}$, and $\text{mod}\, \mathscr{S} \subset \text{mod}^\vee \mathscr{S}$ implies $\text{mod}(\text{ind}^\vee \mathscr{S}) \subset \text{mod}^\vee(\text{ind}^\vee \mathscr{S})$. If $\mathscr{S}$ satisfies $\text{mod}\, \mathscr{S} = \text{mod}^\vee \mathscr{S}$, then $\mathscr{S}$ is left and right coherent and we have $\text{mod}(\text{ind}\, \mathscr{S}) = \text{mod}^\vee(\text{ind}\, \mathscr{S})$.

8.7. Remarks and References

1. This terminology was introduced in [52, 53, 54, Gabriel 1970–1973]. The first occurrence of the concept seems to be found in the few lines of [144, Thrall, 1947 = the year of Dynkin's work [44]!]. To each finite-dimensional algebra A with vanishing radical square Thrall attaches a quiver Q'_A which is obtained from Q_A by identifying parallel arrows and subdividing the vertices v of Q_A into tail-vertices v_1 and head-vertices v_2. From his point of view, Q'_A is the "graph" of the square Cartan-matrix of the bimodule $\mathscr{R}_A$. The quiver Q'_A is replaced by Q_A (without parallel arrows) in [78, Jans, 1954/1957], though Q'_A seems more appropriate to the formulation of the criterions proved by Brauer, Thrall and Jans.

The concept of a path-algebra is extraneous to the work of these authors (for some weak substitute in the form of "Cartan-bases" see [116, Nesbitt, 1938]). Accordingly, they make no attempt to interpret representations of A in terms of Q_A. As we were informed by W.H. Gustafson, the essence of quiver-algebras is to be found in [74, Hochschild, 1947]. There Hochschild observes that each finite-dimensional algebra A is a quotient of the tensor algebra of the bimodule $\mathscr{R}_A/\mathscr{R}_A^2$ over $A/\mathscr{R}_A$. But graphs and modular equivalence seem extraneous to his work, which Brauer reviews in [30, 1947]. Tensor-algebra, graphs and Brauer-Morita thus had to wait for convergence.

2. The quiver $Q_{k\mathscr{P}}$ is also called *Hasse-diagram.*
3. [107, Müller, 1973/74], [143, Tachikawa, 1979/80], [76, Hughes/Waschbüsch, 1983]. The main application 4.2 aimed at in the third article is based on wrong premises.
4. The induction is also called *Kan extension* [100, MacLane, 1971].

5. A morphism of modules $p: P \to M$ is a *projective covering* if P is projective and p surjective and if the surjectivity of a composition $p \circ q$ implies the surjectivity of q. The condition determines the "*projective cover*" P up to isomorphism. Similarly, $j: M \to I$ is an *injective envelopment* if I is injective and j an injection and if the injectivity of some $h \circ j$ implies the injectivity of h. The condition determines the "*injective envelope*" I up to isomorphism.
6. This notion, used in Complex Analysis by H. Cartan and Oka, was adapted to algebra by Serre [139, 1955].

9. Almost Split Sequences

Used in combination with the geometrical device of Sect. 8 and with numerical implements due to I.M. Gelfand and V.A. Ponomarev, almost split sequences appear as a deus ex machina disentangling the jungle of indecomposables.[1] Roughly speaking, they are "minimal" non-split short exact sequences. Their discovery is the achievement of M. Auslander and I. Reiten. They reflect the additional structure imposed upon aggregates of finite-dimensional modules by the existence of kernels and cokernels. We present them in the needed general context of exact categories.

The first paragraph is classical preliminary spadework. There the ground ring k is only supposed to be commutative. Elsewhere it is also local artinian. If V is a k-module, we then set $DV = \mathrm{Hom}_k(V, \check{k})$, where $\check{k}$ is an injective envelope of the residue field of k (3.7). We further recall that, whenever we write "category", we really mean a k-category ...

9.1. Exact Categories. Let $\mathscr{A}$ be an additive category. A pair (i, d) of composable morphisms $X \xrightarrow{i} Y \xrightarrow{d} Z$ of $\mathscr{A}$ is called *exact* if X is identified by i with the *kernel*[2] of d and Z by d with the *cokernel*[2] of i.

Let $\mathscr{E}$ be a class of exact pairs which is *closed under isomorphisms*[3]. We call $\mathscr{E}$ an *exact structure* on $\mathscr{A}$ and $(\mathscr{A}, \mathscr{E})$ an *exact category* if the following axioms $E1$, $E2$, $E3$ and $E3^{op}$ are satisfied. The *deflations* considered in these axioms are by definition the second components d of the pairs $(i, d) \in \mathscr{E}$. The first components i will be called *inflations*.

$E1$: The composition of two deflations is a deflation.
$E2$: For each $f \in \mathscr{A}(Z', Z)$ and each deflation $d \in \mathscr{A}(Y, Z)$, there is a $g \in \mathscr{A}(Y', Y)$ and a deflation $d': Y' \to Z'$ such that $dg = fd'$.
$E3$: Identities are deflations; if de is a deflation, then so is d.
$E3^{op}$: Identities are inflations; if ji is an inflation, then so is i.

Proposition.[4] *Let $\mathscr{A}$ be a svelte additive category and $\mathscr{E}$ a class of exact pairs which is closed under isomorphisms. Then the following three assertions are equivalent:*

(i) *$\mathscr{E}$ is an exact structure on $\mathscr{A}$.*
(ii) *The pairs (d, i), where $(i, d) \in \mathscr{E}$, form an exact structure on $\mathscr{A}^{op}$.*

(iii) *There is an algebra A, a full additive subcategory $\mathcal{M}$ of* Mod *A and an equivalence* $F\colon \mathcal{A} \xrightarrow{\sim} \mathcal{M}$ *such that $\mathcal{M}$ is closed*[5] *under extensions and kernels of retractions and that $\mathcal{E}$ is formed by the composable pairs* (i, d) *inducing exact sequences of A-modules*

$$0 \to FX \xrightarrow{Fi} FY \xrightarrow{Fd} FZ \to 0.$$

Example 1. Suppose that $\mathcal{A}$ is additive and admits kernels of retractions. Then the retractions are the deflations of an exact structure, the *split* structure, whose inflations are the sections.

Example 2. An additive category $\mathcal{A}$ is called *abelian* if each morphism has a kernel and a cokernel and is the composition of a *normal*[6] epimorphism with a normal monomorphism. These conditions imply that the class of all exact pairs is an exact structure on $\mathcal{A}$. Unless otherwise stated, abelian categories will implicitly be equipped with this structure.

For each svelte $\mathcal{B}$, the categories Mod $\mathcal{B}$ (2.2), $\text{mod}_{pf}\,\mathcal{B}$ (3.5) and $\text{mod}_f\,\mathcal{B}$ (3.1) are abelian. The coherent modules over $\mathcal{B}$ also form an abelian category. If $\mathcal{B}$ carries an exact structure $\mathcal{F}$, the left exact $\mathcal{B}$-modules M (transforming $(i, d) \in \mathcal{F}$ into exact sequences $0 \to MZ \xrightarrow{Md} MY \xrightarrow{Mi} MX$) form an abelian category.[7]

Example 3. Let $(\mathcal{A}, \mathcal{E})$ be an exact category and $\mathcal{B}$ a full subcategory which is *closed under extensions* (if a pair $X \xrightarrow{i} Y \xrightarrow{d} Z$ belongs to $\mathcal{E}$ and X, Z to $\mathcal{B}$, then $Y \in \mathcal{B}$) and kernels of retractions. The pairs of $\mathcal{E}$ with components in $\mathcal{B}$ then form an exact structure on $\mathcal{B}$.

Our assumptions are satisfied in particular when $\mathcal{P}$ is a poset, $\mathcal{A}$ the category of left modules over $k\mathcal{P}^*$ (2.3) and $\mathcal{B}$ the subcategory of modules isomorphic to $\mathcal{P}$-spaces. In this case, $\mathcal{B}$ is even closed in $\mathcal{A}$ under subobjects. As a consequence, the exact structure induced by $\mathcal{A}$ on $\mathcal{B}$ consists of all the exact pairs of $\mathcal{B}$.

Example 4. Let k be local artinian and $\mathcal{S}$ a spectroid whose objects are assorted into two disjoint full subspectroids $\mathcal{I}$ and $\mathcal{P}$ such that $\mathcal{S}(y, x) = 0$ whenever $x \in \mathcal{I}$ and $y \in \mathcal{P}$. We call an $\mathcal{S}$-module M *prinjective* if the restrictions $M|\mathcal{P}$ and $M|\mathcal{I}$ are finitely projective and *finitely injective*[8] respectively. The full subcategory prin $\mathcal{S}$ of Mod $\mathcal{S}$ formed by the prinjective modules is closed under extensions and kernels of retractions. It therefore carries an exact structure inherited from Mod $\mathcal{S}$.

***Example 5.** Let k be local artinian, $\mathcal{J}$ and $\mathcal{Q}$ two aggregates over k and B a *bimodule* over $\mathcal{J}$ and $\mathcal{Q}$. By definition, B consists of abelian groups $B(Q, J)$ (one for each pair of objects $Q \in \mathcal{Q}, J \in \mathcal{J}$) and of maps $\mathcal{J}(J, J') \times B(Q, J) \to B(Q, J')$, $(f, b) \mapsto fb$ and $B(Q, J) \times \mathcal{Q}(Q', Q) \to B(Q', J), (b, g) \mapsto bg$ which satisfy the usual axioms: For all $J \in \mathcal{J}$ and $Q \in \mathcal{Q}$, $B(?, J)$ and $B(Q, ?)$ are right and left modules over $\mathcal{Q}$ and $\mathcal{J}$ respectively; moreover, $(fb)g = f(bg)$ and $(\lambda \mathbb{1}_J)b = b(\lambda \mathbb{1}_Q)$ for all $\lambda \in k$. Moreover, we suppose here that $\dim_k B(Q, J) < \infty$ for all Q, J.

The bimodule B gives rise to an exact category $(\mathcal{A}, \mathcal{E})$ whose objects are the triples (Q, b, J) where $J \in \mathcal{J}$, $Q \in \mathcal{Q}$, and $b \in B(Q, J)$. A morphism $(Q, b, J) \to (Q', b', J')$ consists of an $f \in \mathcal{J}(J, J')$ and a $g \in \mathcal{Q}(Q, Q')$ such that $fb = b'g$. By definition, it is an inflation or deflation of $\mathscr{E}$ according as f, g are sections or retractions.

A typical case is the following: Let A be a finite-dimensional algebra, E an ideal of A, D the right annihilator of E, $\mathscr{J} = \operatorname{mod} A/D$, $\mathscr{Q} = \operatorname{mod} A/E$ and $B(Q, J) = \operatorname{Ext}^1_A(Q, J)$ (compare with 4.1–2).

In the general case, Example 5 is closely related to Example 4: To B we can attach a spectroid $\mathscr{S}$ whose set of objects is the disjoint union of the sets of objects of spectroids $\mathscr{I}$ and $\mathscr{P}$ of $\mathscr{J}$ and $\mathscr{Q}$ (3.5). We identify $\mathscr{I}$ and $\mathscr{P}$ with full subspectroids of $\mathscr{S}$ and set $\mathscr{S}(x, y) = DB(y, x)$ (3.7), $\mathscr{S}(y, x) = 0$ if $x \in \mathscr{I}$, $y \in \mathscr{P}$. We finally obtain an equivalence[9] $\mathscr{A} \xrightarrow{\sim} \operatorname{prin} \mathscr{S}$, $(Q, b, J) \mapsto M$ such that $M(x) = D\mathscr{J}(J, x)$ if $x \in \mathscr{I}$ and $M(y) = \mathscr{Q}(y, Q)$ if $y \in \mathscr{P}$. In case $g \in M(y)$ and $\sigma \in \mathscr{S}(x, y)$, $g\sigma \in M(x)$ is the map $f \mapsto \sigma(fbg) \in \check{k}$ $*$

9.2. *From now on, $\mathscr{E}$ denotes an exact structure on an aggregate $\mathscr{A}$ over the local artinian ground ring k.* Proposition 9.1 allows us to extend classical arguments of Module Theory to $(\mathscr{A}, \mathscr{E})$. In particular, we say that two *conflations* (=elements of $\mathscr{E}$) $X \xrightarrow{i} E \xrightarrow{d} Z$ and $X \xrightarrow{j} F \xrightarrow{e} Z$ are equivalent (in the sense of Baer) if $j = fi$ and $d = ef$ for some $f \in \mathscr{A}(E, F)$. For fixed extremities X and Z, the equivalence classes are called *extensions* of X by Z. They form an abelian group $\operatorname{Ext}^1_{\mathscr{E}}(Z, X)$. If Z varies, the groups $\operatorname{Ext}^1_{\mathscr{E}}(Z, X)$ give rise to an $\mathscr{A}$-module $\operatorname{Ext}^1_{\mathscr{E}}(?, X)$ which is a filtered union of finitely presented submodules: Indeed, each conflation $X \xrightarrow{i} Y \xrightarrow{d} Z$ induces an exact sequence of $\mathscr{A}$-modules

$$0 \to X^\wedge \xrightarrow{i^\wedge} Y^\wedge \xrightarrow{d^\wedge} Z^\wedge \to \operatorname{Ext}^1_{\mathscr{E}}(?, X)$$

and identifies the finitely presented $\mathscr{A}$-module $\operatorname{Coker} d^\wedge$ with the submodule of $\operatorname{Ext}^1_{\mathscr{E}}(?, X)$ which is generated by the extension $\varepsilon \in \operatorname{Ext}^1_{\mathscr{E}}(Z, X)$ containing (i, d).

Lemma. *Let X and X_1 be two objects of $\mathscr{A}$. Then each morphism* $\operatorname{Ext}^1_{\mathscr{E}}(?, X) \xrightarrow{\varphi} \operatorname{Ext}^1_{\mathscr{E}}(?, X_1)$ *is induced by some $f \in \mathscr{A}(X, X_1)$.*

Proof. Suppose that $\varphi(\varepsilon) = \varepsilon_1$, where ε and ε_1 denote the extensions containing the conflations $X \xrightarrow{i} Y \xrightarrow{d} Z$ and $X_1 \xrightarrow{i_1} Y_1 \xrightarrow{d_1} Z$ respectively. Then φ can be inserted as follows into a diagram with commutative squares

$$\begin{array}{ccccccccc}
0 & \longrightarrow & X^\wedge & \xrightarrow{i^\wedge} & Y^\wedge & \xrightarrow{d^\wedge} & Z^\wedge & \longrightarrow & \operatorname{Ext}^1_{\mathscr{E}}(?, X) \\
 & & \downarrow{\scriptstyle f^\wedge} & & \downarrow{\scriptstyle g^\wedge} & & \| & & \downarrow{\scriptstyle \varphi} \\
0 & \longrightarrow & X_1^\wedge & \xrightarrow{i_1^\wedge} & Y_1^\wedge & \xrightarrow{d_1^\wedge} & Z^\wedge & \longrightarrow & \operatorname{Ext}^1_{\mathscr{E}}(?, X_1)
\end{array}$$

By construction, f then satisfies $f\varepsilon = \varepsilon_1$. Therefore it suffices to choose (i, d) so "large" that ε has the same annihilator in $\mathscr{A}(X, X_1)$ as $\operatorname{Ext}^1_{\mathscr{E}}(?, X)$. $\surd$

In the sequel we shall denote by $\underline{\mathscr{A}}$ the residue-category of $\mathscr{A}$ modulo the kernel of the functor $X \mapsto \operatorname{Ext}^1_{\mathscr{E}}(?, X)$. If each $X \in \mathscr{A}$ is the tail of some conflation $X \to E \to Z$ with *$\mathscr{E}$-injective* center E, i.e. such that $\operatorname{Ext}^1_{\mathscr{E}}(?, E) = 0$, the considered kernel consists of the morphisms of $\mathscr{A}$ which can be factorized through an $\mathscr{E}$-injective.

Dually, we write $\overline{\mathscr{A}}$ for the residue-category of $\mathscr{A}$ modulo the kernel of $Z \mapsto \mathrm{Ext}^1_{\mathscr{E}}(Z, ?)$.

***Example 1.** Let us return to the bimodule B over the aggregates $\mathscr{J}$ and $\mathscr{Q}$ considered in 9.1, Example 5. By $(\mathscr{A}, \mathscr{E})$ we again denote the exact category formed by the triples (Q, b, J) where $J \in \mathscr{J}$, $Q \in \mathscr{Q}$ and $b \in B(Q, J)$. *We suppose that the left module* $B(Q, ?)$ *over* $\mathscr{J}$ *is finitely generated for each* $Q \in \mathscr{Q}$.

Our objective is to describe the *$\mathscr{E}$-projectives*, i.e. the $P \in \mathscr{A}$ such that $\mathrm{Ext}^1_{\mathscr{E}}(P, ?) = 0$: Each indecomposable $J \in \mathscr{J}$ gives rise to an indecomposable $\mathscr{E}$-projective $(0, 0, J)$. Furthermore, for each indecomposable $Q \in \mathscr{Q}$, we choose a projective covering $^\wedge b \colon {}^\wedge I \to B(Q, ?)$, i.e. an element $b \in B(Q, I)$ such that each $c \in B(Q, K)$ has the form $c = fb$ for some $f \in \mathscr{J}(I, K)$ and that $rb \neq 0$ for each retraction $r \neq 0$; then (Q, b, I) is an indecomposable $\mathscr{E}$-projective. Each $\mathscr{E}$-projective is a sum of indecomposables of the described forms $(0, 0, J)$ or (Q, b, I), and each object of $\mathscr{A}$ is the range of a deflation with $\mathscr{E}$-projective domain *

***Example 2.** We aim at a concrete description of the residue-category $\overline{\mathscr{A}}$ in the particular case where k is a field, $\mathscr{J} = \oplus k\mathscr{P}$ the aggregate associated with a finite poset $\mathscr{P}$, $\mathscr{Q} = \mathrm{mod}\, k$, $B(V, X) = \mathrm{Hom}_k(V, M^{\mathscr{P}}(X))$ and $\mathscr{A} = M^{\mathscr{P}k}$ (see Example 1 and 4.1).

For each $s \in \mathscr{P}$, let M_s denote the submodule of $M := M^{\mathscr{P}}$ such that $M_s(t) = 0$ if $t \leqslant s$ and $M_s(t) = M(t) = k$ if $t \in \mathscr{P}$ and $t \nleqslant s$. To each $(V, f, X) \in \mathscr{A}$ we then attach the $\mathscr{P}^{op}$-space formed by V and the subspaces $V(s) = f^{-1}(M_s(X))$. Denoting by $\mathscr{B}$ the aggregate formed by the finite-dimensional $\mathscr{P}^{op}$-spaces, we claim that *the functor*

$$F\colon \mathscr{A} \to \mathscr{B}, (V, f, X) \mapsto V$$

is quasi-surjective.[10]

The kernel of the quasi-surjective functor $F\colon \mathscr{A} \to \mathscr{B}$ *consists of the morphisms which can be factored through an $\mathscr{E}$-projective of the form* $(0, 0, X)$. Therefore, the "unique" indecomposable $\mathscr{E}$-projective not annihilated by F is $(k, \mathrm{diag}, \oplus m)$, where m runs through the minimal points of $\mathscr{P}$. Its image $F(k, \mathrm{diag}, \oplus m)$ is the $\mathscr{P}^{op}$-space k_0 supported by k and such that $k_0(s) = 0$ for all $s \in \mathscr{P}$. Now, the $\mathscr{P}^{op}$-spaces V which admit no summand of the form k_0 are precisely those satisfying $V = \sum_{s \in \mathscr{P}} V(s)$. They form a full subcategory $\mathscr{B}^0$ of $\mathscr{B}$ which is equivalent to the residue-category of $\mathscr{B}$ modulo $k_0 = F(k, \mathrm{diag}, \oplus m)$. *As a consequence, $\mathscr{B}^0$ is equivalent to the residue-category $\overline{\mathscr{A}}$ of $\mathscr{A} = M^{\mathscr{P}k}$ modulo the $\mathscr{E}$-projectives* *

9.3. We now enter the work of M. Auslander and I. Reiten, to which the end of Sect. 9 is devoted.

Theorem. *Let* $\varepsilon \in \mathrm{Ext}^1_{\mathscr{E}}(W, U)$ *be the extension attached to a conflation* $U \xrightarrow{i} E \xrightarrow{d} W$ *with indecomposable ends* U, W. *Then the following statements are equivalent:*

(i) $0 \to U^\wedge \xrightarrow{i^\wedge} E^\wedge \xrightarrow{d^\wedge} W^\wedge \xrightarrow{can} W^- \to 0$ *is a minimal projective resolution of the simple $\mathscr{A}$-module* $W^- = W^\wedge / \mathscr{R}_{\mathscr{A}}(?, W)$.

(ii) *ε is non-zero and belongs to the socle of the $\mathscr{A}$-module* $\operatorname{Ext}^1_{\mathscr{E}}(?, U)$.

(iii) *d is not a retraction, and each radical morphism $g \in \mathscr{R}_{\mathscr{A}}(W', W)$ factors through d.*

(iv) $0 \to {}^\wedge W \xrightarrow{{}^\wedge d} {}^\wedge E \xrightarrow{{}^\wedge i} {}^\wedge U \xrightarrow{can} {}^- U \to 0$ *is a minimal projective resolution of the simple left $\mathscr{A}$-module* ${}^- U = {}^\wedge U / \mathscr{R}_{\mathscr{A}}(U, ?)$.

(v) *ε is non-zero and belongs to the socle of the left $\mathscr{A}$-module* $\operatorname{Ext}^1_{\mathscr{E}}(W, ?)$.

(vi) *i is not a section, and each $f \in \mathscr{R}_{\mathscr{A}}(U, U')$ factors through i.*

Moreover, if the socle of $\operatorname{Ext}^1_{\mathscr{E}}(?, U)$ *or of* $\operatorname{Ext}^1_{\mathscr{E}}(W, ?)$ *is non-zero, then both socles are simple, and sequences $U \xrightarrow{i} E \xrightarrow{d} W$ satisfying* (i)–(vi) *are uniquely determined up to isomorphism by U and uniquely by W.*

Proof. (iii) is just a reformulation of (ii) and is equivalent to the exactness of $E^\wedge \xrightarrow{d^\wedge} W^\wedge \xrightarrow{can} W^-$. As a consequence, (i) implies (iii), and (iii) obviously implies that the sequence of (i) is a projective resolution. It follows that the sequence is a (direct) sum of a minimal projective resolution and of a resolution of zero. If the latter was non-trivial, $U^\wedge$ would be a summand of $E^\wedge$, hence U a summand of E. Therefore, we have (i)$\Leftrightarrow$(ii)$\Leftrightarrow$(iii) and (iv)$\Leftrightarrow$(v)$\Leftrightarrow$(vi) by duality.

Now suppose that (iii) holds. Then each "non-split" conflation (j, e) with head W can be embedded into a diagram with commutative squares.

$$\begin{array}{ccccc} X & \xrightarrow{\;j\;} & F & \xrightarrow{\;e\;} & W \\ {\scriptstyle f}\downarrow & & \downarrow{\scriptstyle g} & & \| \\ U & \xrightarrow{\;i\;} & E & \xrightarrow{\;d\;} & W \end{array}$$

In other words, for each non-zero $\eta \in \operatorname{Ext}^1_{\mathscr{E}}(W, X)$, there is an $f \in \mathscr{A}(X, U)$ such that $f\eta = \varepsilon$. Equivalently: Each non-zero submodule of $\operatorname{Ext}^1_{\mathscr{E}}(W, ?)$ contains ε. This means that the submodule of $\operatorname{Ext}^1_{\mathscr{E}}(W, ?)$ generated by ε is simple and admits $\operatorname{Ext}^1_{\mathscr{E}}(W, ?)$ as an essential extension[11]. As a consequence, the socle of $\operatorname{Ext}^1_{\mathscr{E}}(W, ?)$ is simple, and each non-zero extension η in the socle satisfies $f\eta = \varepsilon$ for some non-radical $f \in \mathscr{A}(U, U)$. $\surd$

In the terminology of M. Auslander and I. Reiten, a conflation $U \xrightarrow{i} E \xrightarrow{d} W$ satisfying the equivalent conditions of the preceding theorem is an *almost split sequence* of $(\mathscr{A}, \mathscr{E})$. We shall further call W a *translate*[12] of U and U a *cotranslate* of W.

Example 1. Let k be a field, S a linearly ordered poset and $\mathscr{A}$ the full subcategory of Mod kS formed by the finite sums of indecomposables k_I (3.6, Example 2). It is easy to prove that $\mathscr{A}$ is abelian and exactly embedded in Mod kS. Furthermore, an indecomposable k_I is the tail of an almost split sequence of $\mathscr{A}$ if and only if $I = [a, b[$[13] for some $a, b \in S$. In this case,

$$k_{[a,b[} \xrightarrow{\begin{bmatrix} \kappa \\ -\kappa \end{bmatrix}} k_{[a,b]} \oplus k_{]a,b[} \xrightarrow{[\kappa\ \kappa]} k_{]a,b]}$$

is an almost split sequence (by κ we denote the canonical morphisms).

Example 2. Let k be a field, S the direct sum of two linearly ordered posets R, T and $\mathscr{A}$ the category of finite-dimensional S-spaces equipped with the exact structure of 9.1, Example 3. With each *final*[14] subset F of S we associate an indecomposable S-space k_F with underlying vector space k such that $k_F(s) = k$ if $s \in F$ or else $k_F(s) = 0$. It is easy to see that each indecomposable of $\mathscr{A}$ is isomorphic to some k_F. Furthermore, k_F is the tail of an almost split sequence of $\mathscr{A}$ if and only if $F =]r \cup]t = \{s \in S\colon r < s \text{ or } t < s\}$ for some $r \in R$, $t \in T$. In this case,

$$k_{]r \cup]t} \xrightarrow{\begin{bmatrix} \kappa \\ -\kappa \end{bmatrix}} k_{[r \cup]t} \oplus k_{]r \cup [t} \xrightarrow{[\kappa\ \kappa]} k_{[r \cup [t}$$

is an almost split sequence (κ = "canonical" morphism).

Example 3. Let $\mathscr{C}$ be the category of conflations of $\mathscr{A}$ (a full subcategory of the category of composable pairs of morphisms; 2.1, Example 10). We equip $\mathscr{C}$ with the exact structure whose inflations are the morphisms of $\mathscr{C}$ consisting of 3 inflations of $\mathscr{A}$. If $X \xrightarrow{i} Y \xrightarrow{d} Z$ is an almost split sequence of $\mathscr{A}$, the diagram of Fig. 1 provides an almost split sequence of $\mathscr{C}$.

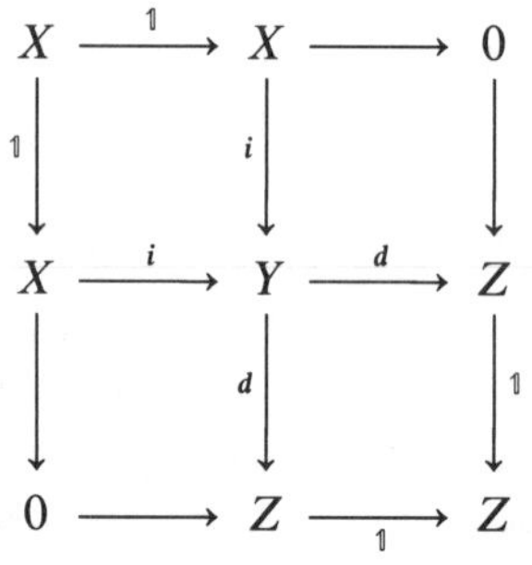

Fig. 1

9.4. Corollary. *Under the assumptions of Theorem* 9.3, *let*

$$\varphi\colon \operatorname{Ext}^1_{\mathscr{E}}(W, U) \to \check{k}$$

be k-linear and such that $\varphi(\varepsilon) \neq 0$ (3.7). *Then the equivalent statements* (i)–(vi) *imply that the maps*

$$\operatorname{Ext}^1_{\mathscr{E}}(Z, U) \to D\overline{\mathscr{A}}(W, Z),\ \eta \mapsto (h \mapsto \varphi(\eta h))$$

and $$\operatorname{Ext}^1_{\mathscr{E}}(W, X) \to D\underline{\mathscr{A}}(X, U),\ \eta \mapsto (f \mapsto \varphi(f\eta))$$

are bijective for all $X, Z \in \mathscr{A}$.

Proof. In order to prove the injectivity of the first map, we start from a non-split conflation $U \xrightarrow{j} F \xrightarrow{e} Z$ which we integrate into a diagram of the following form using 9.3(vi). We infer that $\varphi(\eta h) = \varphi(\varepsilon) \neq 0$, where η is the extension containing (j, e) and $h \in \overline{\mathcal{A}}(W, Z)$ the residue-class of g.

$$\begin{array}{ccccc} U & \xrightarrow{i} & E & \xrightarrow{d} & W \\ \| & & \downarrow & & \downarrow{\scriptstyle g} \\ U & \xrightarrow{j} & F & \xrightarrow{e} & Z \end{array}$$

We now turn to the surjectivity, which is equivalent to the injectivity of the associated map $\overline{\mathcal{A}}(W, Z) \to D\,\mathrm{Ext}^1_{\mathcal{E}}(Z, U)$, $h \mapsto (\eta \mapsto \varphi(\eta h))$: Let $g \in \mathcal{A}(W, Z)$ have non-zero image h in $\overline{\mathcal{A}}(W, Z)$, hence satisfy $\theta g \neq 0$ for some $\theta \in \mathrm{Ext}^1_{\mathcal{E}}(Z, X)$. Using 9.3(iii), we can now choose a conflation $(j, e) \in \theta$ and integrate it into a diagram of the form of Fig. 2. We infer that $0 \neq \varphi(\varepsilon) = \varphi(f\theta g) = \varphi(\eta g)$, where $\eta = f\theta$. $\surd$

$$\begin{array}{ccccc} X & \xrightarrow{j} & Y & \xrightarrow{e} & Z \\ \| & & \uparrow & & \uparrow{\scriptstyle g} \\ X & \longrightarrow & T & \longrightarrow & W \\ {\scriptstyle f}\downarrow & & \downarrow & & \| \\ U & \xrightarrow{i} & E & \xrightarrow{d} & W \end{array}$$

Fig. 2

9.5. Corollary. *If $W \in \mathcal{A}$ is indecomposable and not $\mathcal{E}$-projective, the following statements are equivalent:*

(i) *There is an almost split sequence $U \xrightarrow{i} E \xrightarrow{d} W$ with head W.*

(ii) *The simple $\mathcal{A}$-module W^- is finitely presented.*

(iii) *The $\mathcal{A}$-module $D\,\mathrm{Ext}^1_{\mathcal{E}}(W, ?)$ is finitely generated.*

Proof. (i) $\Rightarrow$ (ii) is clear because (ii) is part of statement (i) of 9.3.

(ii) $\Rightarrow$ (iii): Let $E'^\wedge \xrightarrow{d'^\wedge} W^\wedge \to W^- \to 0$ be a presentation of W^-. Since W is not $\mathcal{E}$-projective, there exists a non-split conflation (j, e) heading for W. Then e factors through d', and d' is a deflation with kernel, say $U' \xrightarrow{i'} E'$. Denoting by $\varepsilon' \in \mathrm{Ext}^1_{\mathcal{E}}(W, U')$ the class of (i', d') and by φ': $\mathrm{Ext}^1_{\mathcal{E}}(W, U') \to \check{k}$ a k-linear map such that $\varphi'(\varepsilon') \neq 0$, we show as in 9.4 that $\mathrm{Ext}^1_{\mathcal{E}}(W, ?) \to D\underline{\mathcal{A}}(?, U')$, $\eta \mapsto (f \mapsto \varphi'(f\eta))$ is a monomorphism, hence that the associated $\mathcal{A}(?, U') \to D\,\mathrm{Ext}^1_{\mathcal{E}}(W, ?)$, $g \mapsto (\eta \mapsto \varphi'(g\eta))$ is an epimorphism.

(iii) $\Rightarrow$ (i): If $D\,\mathrm{Ext}^1_{\mathcal{E}}(W, ?)$ is finitely generated and non-zero, it has a simple quotient. By duality, $\mathrm{Ext}^1_{\mathcal{E}}(W, ?)$ has a simple submodule. $\surd$

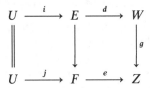

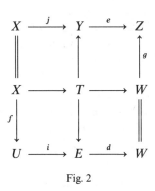

9.6. Corollary. *The full subcategory* $\overline{\mathscr{A}}'$ *of* $\overline{\mathscr{A}}$ *formed by the* C *such that* $D\,\mathrm{Ext}^1_{\mathscr{E}}(C, ?)$ *is finitely generated is equivalent to the full subcategory* $\underline{\mathscr{A}}'$ *of* $\underline{\mathscr{A}}$ *formed by the* A *such that* $D\,\mathrm{Ext}^1_{\mathscr{E}}(?, A)$ *is finitely generated.*

Proof. The indecomposable summands of each $C \in \overline{\mathscr{A}}'$ in $\mathscr{A}$ are $\mathscr{E}$-projective or the heads of almost split sequences. Using 9.4, we can therefore find objects $\tau C \in \underline{\mathscr{A}}'$ and isomorphisms $\varphi_C: \underline{\mathscr{A}}'(?, \tau C) \xrightarrow{\sim} D\,\mathrm{Ext}^1_{\mathscr{E}}(C, ?)$. By duality, we can find a $\bar{\tau} A \in \overline{\mathscr{A}}'$ and an isomorphism $\psi^A: \overline{\mathscr{A}}'(\bar{\tau} A, ?) \xrightarrow{\sim} D\,\mathrm{Ext}^1_{\mathscr{E}}(?, A)$ for each $A \in \underline{\mathscr{A}}'$. The isomorphisms ψ^A and φ_C provide functors $\bar{\tau}: \underline{\mathscr{A}}' \to \overline{\mathscr{A}}'$, $\tau: \overline{\mathscr{A}}' \to \underline{\mathscr{A}}'$ and bijections $\overline{\mathscr{A}}'(\bar{\tau} A, C) \xrightarrow{\sim} \underline{\mathscr{A}}'(A, \tau C)$. In fact, Lemma 9.2 and its dual show that τ and $\bar{\tau}$ are fully faithful. We infer that the right adjoint τ of $\bar{\tau}$ really is a quasi-inverse of $\bar{\tau}$ $\surd$

In the sequel, $\bar{\tau}: \underline{\mathscr{A}}' \to \overline{\mathscr{A}}'$ will be called the *translation-functor.*

Example. Suppose that $\mathscr{A}$ has only finitely many isoclasses of indecomposables. In this case, the condition 9.5(ii) is satisfied for each indecomposable $W \in \mathscr{A}$. Therefore, each non-$\mathscr{E}$-projective indecomposable is the head of an almost split sequence, and each non-$\mathscr{E}$-injective indecomposable is the tail of one. The translation-functor is an equivalence $\underline{\mathscr{A}} \xrightarrow{\sim} \overline{\mathscr{A}}$, and $\mathscr{A}$ has the same number of isoclasses of $\mathscr{E}$-injective indecomposables and of $\mathscr{E}$-projective indecomposables.

9.7. Definition. A spectroid $\mathscr{S}$ over k is called *dualizing* if $\mathrm{mod}\,\mathscr{S} = \mathrm{mod}^{\vee}\,\mathscr{S}$ (8.6), i.e. if the classes of finitely presented and of finitely copresented $\mathscr{S}$-modules coincide. A pointwise finite k-category $\mathscr{A}$ is called *dualizing* if so is its spectroid $\mathrm{sp}\,\mathscr{A} = \mathscr{I}\,\mathrm{pro}\,\mathscr{A}$.

Of course, the condition $\mathrm{mod}\,\mathscr{S} = \mathrm{mod}^{\vee}\,\mathscr{S}$ is equivalent to saying that $\mathscr{S}$ is left and right coherent and that, for each $x \in \mathscr{S}$, the left and right $\mathscr{S}$-modules $D\mathscr{S}(?, x)$ and $D\mathscr{S}(x, ?)$ are finitely presented.

Theorem. *Let* $\mathscr{A}$ *be a dualizing aggregate. Then, for each indecomposable* $W \in \mathscr{A}$, *the simple modules* W^- *and* ${}^-W$ *admit minimal projective resolutions in* $\mathrm{mod}\,\mathscr{A}$ *and* $\mathrm{mod}\,\mathscr{A}^{\mathrm{op}}$ *respectively.*

If $\mathscr{E}$ *is an exact structure on* $\mathscr{A}$, *each non-*$\mathscr{E}$*-projective indecomposable is the head of an almost split sequence, each non-*$\mathscr{E}$*-injective indecomposable is the tail of one.*

Proof. The simple modules and the projective modules are coherent by 8.6. This implies the first statement. The second then follows from 9.5. $\surd$

If $U \xrightarrow{i} E \xrightarrow{d} W$ is an almost split sequence, the minimal projective resolution of W^- has the form given in 9.3(i). It follows that W^- has *projective dimension*[15] 2 if the indecomposable $W \in \mathscr{A}$ is not $\mathscr{E}$-projective. In general, the projective dimension of W^- is $\neq 2$.

Example 1. Let S be a non-empty linearly ordered poset. Then the spectroid kS is dualizing if and only if the following three conditions are satisfied: a) S has a minimum and a maximum; b) if $s \in S$ is not minimum, $s[= \{x \in S: x < s\}$ has a maximum; c) if $s \in S$ is not maximum, $]s = \{x \in S: s < x\}$ has a minimum.

If $s \in S$ is minimal, the simple kS-module s^- coincides with $s^\wedge$ and is projective. Otherwise it has projective dimension 1.

Example 2. Each locally bounded spectroid is dualizing.

Example 3. If the spectroid $\mathscr{S}$ is dualizing, then so is mod $\mathscr{S}$ (8.6, Example 6). Furthermore, if $W = x^\wedge \in \text{mod}\ \mathscr{S}$ is a projective indecomposable, the minimal projective resolution of W^- is

$$0 \to \mathscr{R}_{\mathscr{S}}(?, x)^\wedge \underset{i^\wedge}{\to} \mathscr{S}(?, x)^\wedge \to W^- \to 0,$$

where i denotes the inclusion.

Example 4. *Let $\mathscr{S}$ be a right coherent spectroid and $\mathscr{T}$ a full subspectroid such that, for each $x \in \mathscr{S}$, the restriction $\mathscr{S}(?, x)|\mathscr{T}$ is finitely generated. Then $\mathscr{T}$ is right coherent, and the condition* $\text{mod}^\vee \mathscr{S} \subset \text{mod}\ \mathscr{S}$ *implies* $\text{mod}^\vee \mathscr{T} \subset \text{mod}\ \mathscr{T}$.

Indeed, since $\mathscr{S}$ is right coherent, each finite family of morphisms $f_i \in \mathscr{T}(y_i, y)$ is associated with some exact sequence

$$\bigoplus_j \mathscr{S}(?, z_j) \xrightarrow{\ g\ } \bigoplus_i \mathscr{S}(?, y_i) \xrightarrow{[\mathscr{S}(?, f_i)]} \mathscr{S}(?, y)$$

of mod $\mathscr{S}$. By assumption, each z_j gives rise to some epimorphism $\bigoplus_l \mathscr{T}(?, t_{jl}) \to \mathscr{S}(?, z_j)|\mathscr{T}$ of mod $\mathscr{T}$, hence to the exact sequence

$$\bigoplus_{j,l} \mathscr{T}(?, t_{jl}) \longrightarrow \bigoplus_i \mathscr{T}(?, y_i) \xrightarrow{[\mathscr{T}(?, f_i)]} \mathscr{T}(?, y).$$

We infer that $\mathscr{T}$ is right coherent.

As for the second assertion, it suffices to prove that $D\mathscr{T}(t, ?)$ is finitely presented for each $t \in \mathscr{T}$, or, more generally, that $E|\mathscr{T}$ is finitely presented for each $E \in \text{mod}\ \mathscr{S}$. To this effect, we may restrict the proof to the case $E = \mathscr{S}(?, s)$, where $s \in \mathscr{S}$: Our assumptions then imply the existence of an epimorphism

$$[\mathscr{T}(?, p_i)]: \bigoplus_i \mathscr{T}(?, t_i) \to \mathscr{S}(?, s)|\mathscr{T},$$

of an exact sequence

$$\bigoplus_j \mathscr{S}(?, s_j) \to \bigoplus_i \mathscr{S}(?, t_i) \to \mathscr{S}(?, s),$$

and of epimorphisms

$$\bigoplus_j \mathscr{T}(?, t_{jl}) \to \mathscr{S}(?, s_j)|\mathscr{T},$$

which provide the required exact sequence

$$\bigoplus_{j,l} \mathscr{T}(?, t_{jl}) \to \bigoplus_i \mathscr{T}(?, t_i) \to \mathscr{S}(?, s)|\mathscr{T} \to 0$$

of mod $\mathscr{T}$ (all the considered sums are finite) $\surd$

As an example of application we obtain that, *for each finite poset $\mathscr{P}$, the finite-dimensional $\mathscr{P}$-spaces form a dualizing aggregate*: Indeed, this aggregate is identified with a full subcategory $\mathscr{B}$ of the dualizing aggregate $\mathscr{A}$ formed by the left modules over $k\mathscr{P}^*$, where $\mathscr{P}^*$ is the poset obtained by adding a maximum ω to $\mathscr{P}$ (2.3, Example 2). Furthermore, for each $A \in \mathscr{A}$, $\mathscr{A}(A, ?)|\mathscr{B}$ is identified with $\mathscr{B}(A(\omega), ?)$, where the submodules equipping $A(\omega)$ are the images of the $A(s)$, $s \in \mathscr{P}$. In particular, $\mathscr{A}(A, ?)|\mathscr{B}$ is finitely generated, and the dual of the above assertion implies that mod $\mathscr{B} \subset \text{mod}^{\vee} \mathscr{B}$. In order to obtain the reverse inclusion $\text{mod}^{\vee} \mathscr{B} \subset \text{mod}\, \mathscr{B}$, we finally observe that $\mathscr{B}^{op}$ is equivalent to the category of all finite-dimensional $\mathscr{P}^{op}$-spaces (to each $B \in \mathscr{B}$ we attach the dual space DB equipped with the subspaces $B(s)^{\perp}$). $\surd$

Example 5. Example 4 implies that *a full subspectroid $\mathscr{T}$ of a dualizing spectroid $\mathscr{S}$ is dualizing if $\mathscr{S}(?, x)|\mathscr{T}$ and $\mathscr{S}(x, ?)|\mathscr{T}$ are finitely generated for each $x \in \mathscr{S}$.*

The assumptions imposed upon $\mathscr{T}$ *are satisfied* for instance *if $\mathscr{T}$ contains almost all objects of $\mathscr{S}$*: Indeed, let us examine the case of $\mathscr{S}(?, x)|\mathscr{T}$. Since $\mathscr{S}$ is dualizing, the iterated radicals $\mathscr{R}^n_{\mathscr{S}}(?, x)$ are coherent for all n. We choose n in such a way that

$$\dim \mathscr{R}^n_{\mathscr{S}}(s, x) = \dim \mathscr{R}^{n+1}_{\mathscr{S}}(s, x)$$

for all $s \in \mathscr{S} \setminus \mathscr{T}$. Equivalently, we may suppose that the semisimple $\mathscr{S}$-module $\mathscr{R}^n_{\mathscr{S}}(?, x)/\mathscr{R}^{n+1}_{\mathscr{S}}(?, x)$ has no summand isomorphic to s^-, $s \in \mathscr{S} \setminus \mathscr{T}$. We then have an epimorphism $\bigoplus_i \mathscr{S}(?, t_i) \to \mathscr{R}^n_{\mathscr{S}}(?, x)$, where $t_i \in \mathscr{T}$, and infer that $\mathscr{R}^n_{\mathscr{S}}(?, x)|\mathscr{T}$ is finitely generated. On the other hand,

$$R = \mathscr{S}(?, x)/\mathscr{R}^n_{\mathscr{S}}(?, x)$$

has finite length, and so does $R|\mathscr{T}$ because the restriction of a simple $\mathscr{S}$-module s^- is simple or null according as $s \in \mathscr{T}$ or $s \in \mathscr{S} \setminus \mathscr{T}$. It follows that $R|\mathscr{T}$ is finitely generated, and so is $\mathscr{S}(?, x)|\mathscr{T}$ $\surd$

Example 6. Let us now resume 9.1, Example 4. If $\mathscr{S}(?, y)|\mathscr{I}$ is finitely presented for each $y \in \mathscr{P}$, the finitely presented $\mathscr{S}$-modules coincide with the $\mathscr{S}$-modules M such that $M|\mathscr{P}$ and $M|\mathscr{I}$ are finitely presented. Therefore, *$\mathscr{S}$ is dualizing if $\mathscr{I}$ and $\mathscr{P}$ are dualizing and if $\mathscr{S}(?, y)|\mathscr{I}$ and $\mathscr{S}(x, ?)|\mathscr{P}$ are finitely presented for all $x \in \mathscr{I}$ and $y \in \mathscr{P}$.*

Moreover, *under these conditions*, prin $\mathscr{S}$ *is dualizing*: Indeed, according to the preceding example, it suffices to prove that $M^{\wedge}|\text{prin}\, \mathscr{S}$ and ${}^{\wedge}M|\text{prin}\, \mathscr{S}$ are finitely generated for each $M \in \text{mod}\, \mathscr{S}$. We detail the second proof: Our problem is to construct a morphism $M \xrightarrow{u} N$ with prinjective range N such that each morphism with domain M and prinjective range factors through u. Since $\mathscr{P}$ is left coherent, there is an exact sequence

$$\bigoplus_m \mathscr{P}(?, a_m) \xrightarrow{h} \bigoplus_l \mathscr{P}(?, z_l) \xrightarrow{v} M|\mathscr{P} \to 0$$

and a morphism $\bigoplus_l \mathscr{P}(?, z_l) \xrightarrow{g} \bigoplus_j \mathscr{P}(?, y_j)$ such that $gh = 0$ and that each $\bigoplus_l$

$\mathscr{P}(?, z_l) \xrightarrow{f} \mathscr{P}(?, y)$ satisfying $fh = 0$ factors through g. And there is a commutative diagram

$$\begin{array}{ccccc} \bigoplus_l \mathscr{S}(?, z_l)|\mathscr{I} & \xrightarrow{\bar{g}} & \bigoplus_j \mathscr{S}(?, y_j)|\mathscr{I} & & \\ \downarrow{\scriptstyle \bar{v}} & & \downarrow{\scriptstyle p} & & \\ M|\mathscr{I} & \xrightarrow{q} & S & \xrightarrow{e} & \bigoplus_i D\mathscr{I}(x_i, ?), \end{array}$$

where $\bar{g}$ and $\bar{v}$ are induced by g and v, S is the *amalgamated sum*[16] of $(\bar{v}, \bar{g})$ and e an injective envelopment. We set $N|\mathscr{P} = \bigoplus_j \mathscr{P}(?, y_j)$ and $N|\mathscr{I} = \bigoplus_i D\mathscr{I}(x_i, ?)$; in addition, for all $y \in \mathscr{P}$, $x \in \mathscr{I}$ and $\mu \in \mathscr{S}(x, y)$, we define $N(\mu)$ to be the composition

$$\bigoplus_j \mathscr{S}(y, y_j) \xrightarrow{\oplus \mathscr{S}(\mu, y_j)} \bigoplus_j \mathscr{S}(x, y_j) \xrightarrow{p(x)} S(x) \xrightarrow{e(x)} \bigoplus_i D\mathscr{I}(x_i, x)$$

The required morphism $M \xrightarrow{u} N$ is induced by g, q and e. $\surd$

Example 7. We now return to 9.1, Example 5. The category $\mathscr{A}$ formed by the triples (Q, b, J) with $b \in B(Q, J)$ is equivalent to some prin $\mathscr{S}$. It follows that $\mathscr{A}$ is dualizing if $\mathscr{J}$, $\mathscr{Q}$ are dualizing and if $B(Q, ?)$, $B(?, J)$ are finitely presented for all $J \in \mathscr{J}$, $Q \in \mathscr{Q}$. This holds in particular in the case $\mathscr{J} = \operatorname{mod} A/D$, $\mathscr{Q} = \operatorname{mod} A/E$, $B(Q, J) = \operatorname{Ext}^1_A(Q, J)$ specified in 9.1.

***Example 8.** *For each object X of a dualizing aggregate $\mathscr{A}$, the residue-category*[17] *$\mathscr{A}/X$ of $\mathscr{A}$ modulo X is dualizing*: This follows from the fact that a left or right module over $\mathscr{A}/X$ is finitely presented if and only if it is finitely presented over $\mathscr{A}$.

Our assumptions are satisfied in particular if $\mathscr{P}$ is a finite poset, $\mathscr{A} = M^{\mathscr{P}k}$ and $X = (0, 0, \bigoplus_{s \in \mathscr{P}} s)$. If k is a field, $\mathscr{A}/X$ is then equivalent to the aggregate of all finite-dimensional $\mathscr{P}^{op}$-spaces (9.2, Example 2; compare with Example 4 above)$_*$

***Example 9.** Let $\mathscr{A}$ be a dualizing aggregate, $\mathscr{C}^b\mathscr{A}$ the category of bounded differential complexes X (X^p is zero if $|p|$ is large enough) and $\mathscr{H}^b\mathscr{A}$ the residue-category of $\mathscr{C}^b\mathscr{A}$ modulo the null-homotopic morphisms. It is easy to prove that $\mathscr{C}^b\mathscr{A}$ is dualizing. (First prove that $\mathscr{C}^b\mathscr{A}$ is left and right coherent. Then prove that $D\mathscr{C}^b\mathscr{A}(K, ?)$ and $D\mathscr{C}^b\mathscr{A}(?, K)$ are finitely presented if the derivation $K^n \to K^{n+1}$ is invertible for some n and $K^p = 0$ for $p \neq n, n+1$.) As for $\mathscr{H}^b\mathscr{A}$, it is easily seen that a module M over $\mathscr{H}^b\mathscr{A}$ is finitely presented if and only if it is finitely presented over $\mathscr{C}^b\mathscr{A}$. It follows that M is finitely presented if and only if so is DM, hence that $\mathscr{H}^b\mathscr{A}$ is dualizing$_*$

9.8. Let us now return to the almost split sequence $U \xrightarrow{i} E \xrightarrow{d} W$ of $(\mathscr{A}, \mathscr{E})$ examined in Theorem 9.3. If $\mathscr{S} = \mathscr{I}\mathscr{A}$ is a spectroid of $\mathscr{A}$, E is isomorphic to $\bigoplus_{V \in \mathscr{S}} V^{\mu(V)}$ for some well determined "multiplicities" $\mu(V) \in \mathbb{N}$. On the other hand, since $d^\wedge$ induces a projective covering $E^\wedge \to \mathscr{R}_{\mathscr{A}}(?, W)$, we have

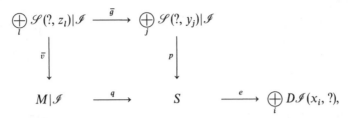

$$\bigoplus_{V \in \mathscr{S}} (V^-)^{\mu(V)} \xrightarrow{\sim} E^\wedge/\mathscr{R}_{\mathscr{A}}(?, E) \xrightarrow{\sim} \mathscr{R}_{\mathscr{A}}(?, W)/\mathscr{R}^2_{\mathscr{A}}(?, W),$$

hence

$$\mu(V) = \dim_{\bar{E}_V} \mathscr{R}_{\mathscr{A}}(V, W)/\mathscr{R}^2_{\mathscr{A}}(V, W), \quad \text{where} \quad \bar{E}_V = \mathscr{A}(V, V)/\mathscr{R}_{\mathscr{A}}(V, V),$$

and by duality

$$\mu(V) = \dim_{\bar{E}_V} \mathscr{R}_{\mathscr{A}}(U, V)/\mathscr{R}^2_{\mathscr{A}}(U, V).$$

Therefore, if we know how to construct the almost split sequences with tail U or head W, we can determine the spaces $\mathscr{R}_{\mathscr{A}}(U, V)/\mathscr{R}^2_{\mathscr{A}}(U, V)$ and $\mathscr{R}_{\mathscr{A}}(V, W)/\mathscr{R}^2_{\mathscr{A}}(V, W)$, thus getting a chance to use the following simple proposition.

Proposition.[18] *Let $\mathscr{S}$ be a spectroid and $\mathscr{T}$ a finite set of objects of $\mathscr{S}$ having the following properties:*

a) *For each $V \in \mathscr{T}$, the simple modules ${}^-V$ and V^- are finitely presented.*

b) *The assumption $U \in \mathscr{T}$, $V \in \mathscr{S}$ and $\mathscr{R}_{\mathscr{A}}(U, V) \neq \mathscr{R}^2_{\mathscr{A}}(U, V)$ implies $V \in \mathscr{T}$.*

c) *The assumption $W \in \mathscr{T}$, $V \in \mathscr{S}$ and $\mathscr{R}_{\mathscr{A}}(V, W) \neq \mathscr{R}^2_{\mathscr{A}}(V, W)$ implies $V \in \mathscr{T}$.*

Then we have $\mathscr{S}(S, T) = 0 = \mathscr{S}(T, S)$ for all $T \in \mathscr{T}$ and all $S \in \mathscr{S} \setminus \mathscr{T}$.

Proof. Let $[g_i^\wedge]\colon \bigoplus_i T_i^\wedge \to \mathscr{R}_{\mathscr{S}}(?, T)$ be a projective covering. Then each T_i belongs to $\mathscr{T}$ and each morphism $f \in \mathscr{S}(S, T) = \mathscr{R}_{\mathscr{S}}(S, T)$ admits a decomposition $f = \sum_i g_i f_i$ where $f_i \in \mathscr{S}(S, T_i) = \mathscr{R}_{\mathscr{S}}(S, T_i)$. Similarly, each f_i admits a decomposition $f_i = \sum_j g_{ij} f_{ij}$, where $f_{ij} \in \mathscr{R}_{\mathscr{S}}(S, T_{ij})$ and $g_{ij} \in \mathscr{R}_{\mathscr{S}}(T_{ij}, T_i)$ for some $T_{ij} \in \mathscr{T}$. Again, each f_{ij} can be written as $f_{ij} = \sum_h g_{ijh} f_{ijh}$, where $f_{ijh} \in \mathscr{R}_{\mathscr{S}}(S, T_{ijh})$ and $g_{ijh} \in \mathscr{R}_{\mathscr{S}}(T_{ijh}, T_{ij})$ for some $T_{ijh} \in \mathscr{T}$. This yields $f = \sum_i g_i f_i = \sum_{i,j} g_i g_{ij} f_{ij} = \sum_{i,j,h} g_i g_{ij} g_{ijh} f_{ijh}$. By further iteration, we obtain decompositions $f = \sum_\alpha g_1^\alpha g_2^\alpha \ldots g_n^\alpha f_\alpha$ where $n \geqslant 1$ is arbitrary, $f_\alpha \in \mathscr{R}_{\mathscr{S}}(S, T_n^\alpha)$, $g_p^\alpha \in \mathscr{R}_{\mathscr{S}}(T_p^\alpha, T_{p-1}^\alpha)$, $T_p^\alpha \in \mathscr{T}$ and $T_0^\alpha = T$. Since we have $g_1^\alpha g_2^\alpha \ldots g_n^\alpha \in \mathscr{R}^n_{\mathscr{T}}(T_n^\alpha, T) = 0$ for large values of n (3.2), we infer that $f = 0$ and $\mathscr{S}(S, T) = 0$. By duality, $\mathscr{S}(T, S) = 0$. $\surd$

An *example* initially borne in mind was the following: Let $\mathscr{P}$ be a finite linearly ordered poset and $\mathscr{S} = \operatorname{ind} k\mathscr{P}$ a spectroid of mod $k\mathscr{P}$ containing $\mathscr{T} = \{k_I\colon I = \text{interval}\}$ (3.6). The conditions above then hold by 9.3, Example 1, and 9.7, Example 2. Hence we recover the known fact that $\mathscr{S} = \mathscr{T}$. Substantial applications will be given in Sect. 10 and 14.

9.9. The Construction of Almost Split Sequences. Let $\mathscr{M}$ be a dualizing aggregate and $W \in \mathscr{M}$ an indecomposable. Our objective is to construct a minimal projective presentation of the simple $\mathscr{M}$-module W^-. For this we need a minimal projective presentation

$$\mathscr{M}(?, T_1) \xrightarrow{m^\wedge} \mathscr{M}(?, T_0) \to D\mathscr{M}(W, ?) \to 0$$

and a monomorphism $\mu\colon W^- \to D\mathscr{M}(W, ?) = W^\vee$ induced by some non-zero linear form on $\bar{\mathscr{M}}(W, W) = \mathscr{M}(W, W)/\mathscr{R}_{\mathscr{M}}(W, W)$ (3.7d). Since $\mathscr{M}(?, W) = W^\wedge$ is projective, the composition $W^\wedge \xrightarrow{\text{can}} W^- \xrightarrow{\mu} W^\vee$ factors through some $W^\wedge \xrightarrow{n^\wedge} T_0^\wedge$. Since $\mathscr{M}$ is coherent, the kernel of $[-m^\wedge \;\; n^\wedge]\colon T_1^\wedge \oplus W^\wedge \to T_0^\wedge$ has a projective cover $E^\wedge$ which gives rise to the following commutative diagram with exact rows

$$(*)\qquad \begin{array}{ccccccc} \mathcal{M}(?, T_1) & \xrightarrow{m^\wedge} & \mathcal{M}(?, T_0) & \longrightarrow & D\mathcal{M}(W, ?) & \longrightarrow & 0 \\ \uparrow & & \uparrow{\scriptstyle n^\wedge} & & \uparrow{\scriptstyle \mu} & & \\ \mathcal{M}(?, E) & \longrightarrow & \mathcal{M}(?, W) & \xrightarrow{\text{can}} & W^- & \longrightarrow & 0 \end{array}$$

This construction can be given a more explicit form in case $\mathcal{M} = \text{mod}\, \mathcal{S}$, where $\mathcal{S}$ is a dualizing spectroid. The definition of the *transfer*[19] $\mathcal{T}: \text{mod}\, \mathcal{S} \to \text{mod}\, \mathcal{S}$ given in 8.3d) formally extends to this case and gives rise to a natural transformation

$$D\,\text{Hom}(N, M) \xrightarrow{\nu} \text{Hom}(M, \mathcal{T}N)$$

such that the image of $\varphi \in D\,\text{Hom}(N, M)$ maps $m \in M(Y)$ onto the linear form

$$\text{Hom}(N, Y^\wedge) \to \check{k}, f \mapsto \varphi(m^\wedge f).$$

It is easy to verify that ν is bijective if N is projective (see 3.7c and 8.3d).

The transfer provides a precise construction of the objects T_0 and T_1 considered above: Indeed, each minimal projective presentation

$$P_1 \xrightarrow{p} P_0 \to W \to 0$$

induces an exact sequence of $\mathcal{M}$-modules

$$D\,\text{Hom}(P_1, ?) \to D\,\text{Hom}(P_0, ?) \to D\,\text{Hom}(W, ?) \to 0$$

i.e. a projective presentation

$$\text{Hom}(?, \mathcal{T}P_1) \xrightarrow{(\mathcal{T}p)^\wedge} \text{Hom}(?, \mathcal{T}P_0) \longrightarrow D\,\text{Hom}(W, ?) \longrightarrow 0$$

Like p the morphism $\mathcal{T}p$ admits no invertible proper summand (8.3d). If W is not projective, P_1 is $\neq 0$. Accordingly, $U = \text{Ker}\, \mathcal{T}p$ is $\neq 0$ and

$$0 \longrightarrow U \longrightarrow \mathcal{T}P_1 \xrightarrow{\mathcal{T}p} \mathcal{T}P_0$$

is a minimal injective copresentation. Moreover, E is identified with the fibre product of $\mathcal{T}P_1 \xrightarrow{\mathcal{T}p} \mathcal{T}P_0 \xleftarrow{n} W$, and (*) can be embedded into the following commutative diagram with exact rows

$$\begin{array}{ccccccccccc} 0 & \longrightarrow & U^\wedge & \longrightarrow & (\mathcal{T}P_1)^\wedge & \xrightarrow{(\mathcal{T}p)^\wedge} & (\mathcal{T}P_0)^\wedge & \longrightarrow & D\mathcal{M}(W, ?) & \longrightarrow & 0 \\ & & \| & & \uparrow & & \uparrow & & \uparrow & & \\ 0 & \longrightarrow & U^\wedge & \longrightarrow & E^\wedge & \longrightarrow & W^\wedge & \longrightarrow & W^- & \longrightarrow & 0 \end{array}$$

In particular, we obtain a minimal projective resolution of W^- and the following statement.

Theorem.[20] *Let $\mathcal{S}$ be a dualizing spectroid and $P_1 \xrightarrow{p} P_0 \to W \to 0$ a minimal projective presentation of a non-projective indecomposable $W \in \text{mod}\, \mathcal{S}$. Then* Ker $\mathcal{T}p$ *is a cotranslate of W.*

***9.10.** The preceding theorem allows us to determine the almost split sequences of mod $\mathcal{S}$ when $\mathcal{S}$ is dualizing. In the case of prin $\mathcal{S}$ (9.7, Example 5), we can apply the following proposition.

Proposition. *Let $(\mathcal{A}, \mathcal{E})$ be an exact category, $\mathcal{B}$ a full subcategory of $\mathcal{A}$ closed under extensions and kernels of retractions and $\mathcal{F}$ the exact structure on $\mathcal{B}$ induced by $\mathcal{E}$. Let further $W \in \mathcal{B}$ be indecomposable and not $\mathcal{F}$-projective, $U \to E \to W$ an almost split sequence of $(\mathcal{A}, \mathcal{E})$ and $\mathcal{B}(?, \tilde{U}) \xrightarrow{p^\wedge} \mathcal{A}(?, U)|\mathcal{B}$ a projective covering.*

Then $\tilde{U}$ is the sum of an $\mathcal{F}$-injective and of the tail V of an almost split sequence $V \to F \to W$ of $(\mathcal{B}, \mathcal{F})$.

Proof. $\mathrm{Ext}^1_{\mathcal{F}}(W, ?)$ is isomorphic to $D\underline{\mathcal{A}}(?, U)|\mathcal{B}$ (9.4), hence to a submodule of $D\mathcal{B}(?, \tilde{U})$, which is an injective envelope of its socle (3.7d). Therefore, $\mathrm{Ext}^1_{\mathcal{F}}(W, ?)$ contains a simple submodule and $(\mathcal{B}, \mathcal{F})$ an almost split sequence $V \to F \to W$. Besides the epimorphism

$$\mathcal{B}(?, \tilde{U}) \to \underline{\mathcal{A}}(?, U)|\mathcal{B} \xrightarrow{\sim} D\,\mathrm{Ext}^1_{\mathcal{F}}(W, ?)$$

induced by p, we thus obtain a projective covering

$$\mathcal{B}(?, V) \to \underline{\mathcal{B}}(?, V) \xrightarrow{\sim} D\,\mathrm{Ext}^1_{\mathcal{F}}(W, ?),$$

hence an isomorphism $V \oplus S \xrightarrow{\sim} \tilde{U}$ such that the composition

$$\mathcal{B}(?, S) \to \mathcal{B}(?, \tilde{U}) \to \underline{\mathcal{A}}(?, U)|\mathcal{B}$$

is zero, i.e. such that $S \to \tilde{U} \xrightarrow{p} U$ annihilates $\mathrm{Ext}^1_{\mathcal{E}}(?, S)$. This implies that, for each inflation $S \xrightarrow{i} T$ of $\mathcal{F}$, p admits a factorization $p = qj$, where j denotes the composition $\tilde{U} \xrightarrow{\sim} V \oplus S \xrightarrow{1 \oplus i} V \oplus T$. On the other hand, since $\mathcal{A}(V \oplus T, p)$ is surjective, we have $q = pr$ for some r, hence $p(\mathbb{1} - rj) = 0$. It follows that $\mathbb{1} - rj$ lies in the radical of $\mathcal{B}(?, \tilde{U})$, hence that rj is invertible, that j and i are sections and that $\mathrm{Ext}^1_{\mathcal{F}}(?, S) = 0$ *

9.11. Remarks and References

1. [63, Gelfand/Ponomarev, 1970/72], [19, Bernstein/Gelfand/Ponomarev, 1973], [8, Auslander/Reiten, 1975–1978], [7, Auslander/Reiten, 1974], [15, Bautista/Martinez, 1978/79], [9, Auslander/Smalø, 1981]. According to I. Reiten, almost split sequences were discovered in 1973 independently of reflection functors. The link between the two points of view is examined in [6, Auslander/Platzek/Reiten, 1979] and [60, Gabriel, Sect. 5, 1979–80]. In our monograph, the two methods will converge in Sect. 10 and Sect. 11. The fact that Auslander-Reiten insisted on working so systematically without diagrams seems to have been important for linking reflection and tilting functors (Sect. 12).
2. [99, MacLane, chap. 9].
3. This means that $\mathcal{E}$ contains all pairs isomorphic to (i, d) (in the category of composable pairs, 2.1, Example 10) if $\mathcal{E}$ contains (i, d).
4. [105, Mitchell, VI, § 7], [49, Freyd, chap. 7, G].
5. $\mathcal{M}$ is *closed under extensions* if, for each exact sequence $0 \to L \to M \to N \to 0$ of Mod A, the assertion $L, N \in \mathcal{M}$ implies $M \in \mathcal{M}$. It is *closed under kernels of retractions* if, for each diagram $M \underset{s}{\overset{r}{\rightleftarrows}} N$ of mod A, the assertions $M, N \in \mathcal{M}$ and $rs = \mathbb{1}_N$ imply Ker $r \in \mathcal{M}$.
6. A monomorphism $i: K \to L$ is *normal* if there is an $f: L \to M$ such that $fi = 0$ and that each $g: N \to L$ satisfying $fg = 0$ factors through i. For epimorphisms, use duality.
7. [51, Gabriel, II, § 2, 1961/62].

8. A module over $\mathcal{I}$ is *finitely injective* if it is isomorphic to a finite sum of modules of the form $s^\vee$, where $s \in \mathcal{I}$.
9. For this point of view, originally due to D. Simson, see [122, Peña/Simson, 1988] and [146, Vossieck, 1988].
10. Indeed, if $\mathcal{F}$ is the exact structure on $\mathcal{B}$ defined in 9.1, Example 3, each $s \in \mathcal{P}$ gives rise to an indecomposable $\mathcal{F}$-injective k_s supported by k and such that $k_s(t)$ is 0 whenever the inequality $s \leqslant t$ holds in $\mathcal{P}$, and that $k_s(t) = k$ otherwise. Besides these k_s there is a further $\mathcal{F}$-injective k_1 supported by k and such that $k_1(t) = k, \forall t \in \mathcal{P}$. Now each $\mathcal{F}$-injective is a sum of the described indecomposables and each $V \in \mathcal{B}$ is the tail of an inflation $V \xrightarrow{i} I$ with $\mathcal{F}$-injective head I. If $\tilde{\mathcal{A}}$ is the aggregate formed by these inflations, it obviously follows that

$$\tilde{F}: \tilde{\mathcal{A}} \to \mathcal{B}, \qquad (V \xrightarrow{i} I) \mapsto V$$

is quasi-surjective. Thus it remains to verify that $\tilde{F}$ factorizes through F (If $\mathcal{I}$ is the aggregate of all $\mathcal{F}$-injectives, we have a quasi-surjective functor $S: \mathcal{I} \to \oplus k\mathcal{P}$ which maps k_s onto s, annihilates k_1 and identifies $M^{\mathcal{P}}(SI)$ with $I/\bigcap_{s \in \mathcal{P}} I(s)$; the wanted functor $\tilde{\mathcal{A}} \to \mathcal{A}$ maps $V \xrightarrow{i} I$ onto $(V, can \circ i, SI)$.) $\surd$ Compare with [55, Gabriel, 1972/73].
11. A module M is an *essential extension* of a submodule N if each submodule $L \neq 0$ of M satisfies $L \cap N \neq 0$.
12. What we call *translate* is generally called *cotranslate*, and reversely.
13. $[a, b[= \{x \in S: a \leqslant x < b\},]a, b] = \cdots$
14. A subset F of a poset S is *final* if $s \leqslant t$ and $s \in F$ imply $t \in F$.
15. The *projective dimension* of a module M is (if it exists) the smallest number $d \in \mathbb{N}$ for which there is an exact sequence $0 \to P_d \to \cdots \to P_0 \to M \to 0$ with projective terms P_n.
16. This means that the square formed by $\bar{v}, \bar{g}, p$ and q is universal in the sense of [99, MacLane, Chap. 12, § 1]. Amalgamated sums are also called *push-outs.*
17. $\mathcal{A}/X = \mathcal{A}/\mathcal{I}$, where $\mathcal{I}$ is the ideal of $\mathcal{A}$ generated by $\mathbb{1}_X$.
18. [4, Auslander, 1976/78].
19. $\mathcal{T}$ is also called *Nakayama-functor.*
20. [8, Auslander/Reiten, 1975–1978], [35, Butler, 1979/80], [60, Gabriel, 1979/80].

10. Postprojective Components

Throughout Sect. 10 we suppose that k *is an algebraically closed field and* $\mathcal{S}$ *a locally bounded spectroid over* k. The finitely presented $\mathcal{S}$-modules then have finite dimension over k and form a pointwise finite category mod $\mathcal{S}$. Our purpose is to examine the quiver $\Gamma_{\mathcal{S}}$ of the spectroid ind $\mathcal{S} = \mathcal{I}$ mod $\mathcal{S}$ formed by chosen representatives of the isoclasses of indecomposables of mod $\mathcal{S}$. The almost split sequences of Auslander and Reiten provide precise information on this quiver and sometimes enable us to compute components of $\Gamma_{\mathcal{S}}$ by purely combinatorial means.

10.1. In order to be distinguished from the (ordinary) quiver $Q_{\mathcal{S}}$ of $\mathcal{S}$, $\Gamma_{\mathcal{S}} = Q_{\text{ind}\,\mathcal{S}}$ is called *representation-quiver*[1] of $\mathcal{S}$. The arrows of $\Gamma_{\mathcal{S}}$ which start at a non-injective vertex $U \in \Gamma_{\mathcal{S}}$ are furnished by the almost split sequences of mod $\mathcal{S}$ which start at U: In fact, if

$$U \xrightarrow{i} \bigoplus_{V \in \Gamma_{\mathcal{S}}} V^{\mu(V)} \xrightarrow{d} W$$

is such a sequence, $\Gamma_{\mathscr{S}}$ admits $\mu(V)$ arrows from U to V (9.8). This number coincides with the number of arrows from V to W, if W is the chosen (non-projective) representative of its isoclass. In this case, we set $W = \bar{\tau}(U)$ and call W the *translate* of U. The thus defined *translation* $\bar{\tau}$ imposes upon $\Gamma_{\mathscr{S}}$ the structure of a *translation-quiver*: By definition, this means that $\Gamma_{\mathscr{S}}$ is equipped with a bijection $\bar{\tau}$ between two sets of vertices of $\Gamma_{\mathscr{S}}$ (the "*non-injectives*" and the "*non-projectives*"); furthermore, for each vertex V and each non-injective U, a bijection $\alpha \mapsto \sigma\alpha$ is given between the arrows $U \xrightarrow{\alpha} V$ with tail U and the arrows $V \xrightarrow{\sigma\alpha} \bar{\tau}U$ with head $\bar{\tau}U$.

Now consider a *projective vertex* $s^\wedge = \mathscr{S}(?, s)$ of $\Gamma_{\mathscr{S}}$, and set $\mathscr{A} = \operatorname{mod} \mathscr{S}$. If $X \in \mathscr{A}$ is indecomposable, the non-invertible morphisms $X \to s^\wedge$ are factorized through the radical $\mathscr{R}s^\wedge = \mathscr{R}_{\mathscr{S}}(?, s)$ of $s^\wedge$. We infer that

$$\mathscr{R}_{\mathscr{A}}(X, s^\wedge) \tilde{\to} \mathscr{A}(X, \mathscr{R}_{\mathscr{S}}(?, s)) \qquad \text{and} \qquad \mathscr{R}^2_{\mathscr{A}}(X, s^\wedge) \tilde{\to} \mathscr{R}_{\mathscr{A}}(X, \mathscr{R}_{\mathscr{S}}(?, s)),$$

hence that $\mathscr{R}_{\mathscr{A}}(X, s^\wedge)/\mathscr{R}^2_{\mathscr{A}}(X, s^\wedge) \tilde{\to} \bigoplus_V (\mathscr{A}(X, V)/\mathscr{R}_{\mathscr{A}}(X, V))^{\pi(V)}$, if $\mathscr{R}_{\mathscr{S}}(?, s) \tilde{\to} \bigoplus_{V \in \Gamma_{\mathscr{S}}} V^{\pi(V)}$. As a consequence, $\pi(V)$ is the number of arrows of $\Gamma_{\mathscr{S}}$ from V to $s^\wedge$.

By dual arguments, we can prove that the number of arrows of $\Gamma_{\mathscr{S}}$ from an *injective* $s^\vee = D\mathscr{S}(s, ?)$ to V is equal to $\iota(V)$, if the quotient of $s^\vee$ modulo its socle is isomorphic to $\bigoplus_{V \in \Gamma_{\mathscr{S}}} V^{\iota(V)}$.

Finally, we observe that *the representation-quiver of a locally bounded spectroid admits no loop*: Indeed, if the modules V and W are indecomposable, a morphism $f \in \mathscr{R}_{\mathscr{A}}(V, W) \backslash \mathscr{R}^2_{\mathscr{A}}(V, W)$ cannot be properly decomposed into an epimorphism and a monomorphism; therefore, f must be a proper epimorphism or a proper monomorphism, and $V \not\to W$.

10.2. Example: Representations of Quivers. Let Q be a quiver which is finite, connected and *directed* (it contains no *closed* path, i.e. no path of length >0 whose origin and terminus coincide). Then the k-category of paths $\mathscr{S} = kQ^{op}$ is locally bounded, and right modules over $\mathscr{S}$ are identified with representations of Q.

If s is a vertex of Q and $^\wedge s = kQ(s, ?)$ the associated projective representation, the arrows $s \xrightarrow{\alpha} t_\alpha$ of Q with tail s induce monomorphisms $^\wedge\alpha\colon {}^\wedge t_\alpha \to {}^\wedge s$ and an isomorphism of $\bigoplus_\alpha {}^\wedge t_\alpha$ onto the radical $\mathscr{R}^\wedge s$ of $^\wedge s$. By 10.1, it follows that the maps $s \mapsto {}^\wedge s$, $\alpha \mapsto {}^\wedge\alpha$ induce an isomorphism of Q^{op} onto a *full* subquiver $^\wedge Q$ of the representation-quiver $\Gamma_{\mathscr{S}} = \Gamma_Q$. Our purpose is to describe the connected component of Γ_Q which contains $^\wedge Q$.

For this purpose, we associate a *translation-quiver* $\mathbb{N}R$ with each quiver R: The vertices of $\mathbb{N}R$ are the pairs (n, x) where $n \in \mathbb{N}$ and $x \in R$; furthermore, each arrow $x \xrightarrow{\beta} y$ of R gives rise to two series of arrows $(n, x) \xrightarrow{(n, \beta)} (n, y)$ and $(n, y) \xrightarrow{\sigma(n, \beta)} (n + 1, x)$ of $\mathbb{N}R$. Finally, we set $\bar{\tau}(n, x) = (n + 1, x)$. In particular, $\mathbb{N}R$ has no injective vertex, and the projectives have the form $(0, x)$.

Proposition. *If the finite, connected, directed quiver Q is not a Dynkin quiver, the translates $\bar{\tau}^n({}^\wedge s)$ are defined for all $n \in \mathbb{N}$ and all vertices s of Q. Furthermore, the map $(n, s) \mapsto \bar{\tau}^n({}^\wedge s)$ extends to an isomorphism of $\mathbb{N}Q^{op}$ onto the connected component of Γ_Q which contains the projectives.*

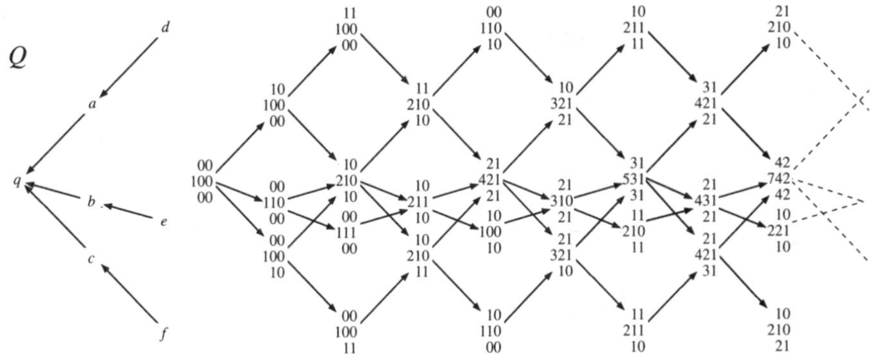

Fig. 1

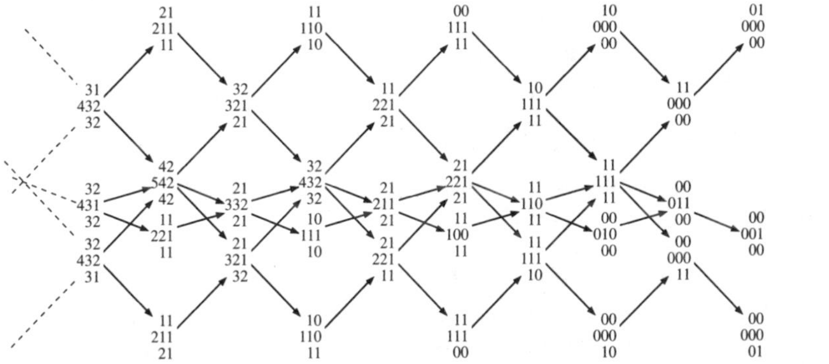

Fig. 2

The proof of the proposition is easy.[2] It essentially uses the arguments of 10.1 and the fact that the quotient of an injective is injective. In the example illustrated by Fig. 1, indecomposables are represented by their *dimension-functions*. For instance, $\bar{\tau}^2(\hat{}a)$ has the dimensions 3, 1, 2, 2, 0, 1 and 1 at the points q, a, b, c, d, e and f respectively. We reckon these dimensions as follows: If we fix some vertex of Q, say b, the function d_b defined on the vertices of Γ_Q by $d_b(V) = \dim V(b)$ is *additive*. This means that each non-injective $U \in \Gamma_Q$ subjects d_b to one equation $d_b(\bar{\tau}U) + d_b(U) = \sum_\alpha d_b(h\alpha)$, where $t\alpha \xrightarrow{\alpha} h\alpha$ ranges over the arrows of Γ_Q with tail $t\alpha = U$. This equation follows from the exactness of the almost split sequence $U \to \bigoplus_\alpha h\alpha \to \bar{\tau}U$. Altogther these equations allow us to compute the dimensions $d_b(\bar{\tau}^n(\hat{}s))$ in terms of the values $d_b(\hat{}t)$ of d_b on $\hat{}Q$.

Using dual arguments, we can also construct the connected component of Γ_Q which contains the indecomposable injectives (Fig. 2). In the particular case of our example, and more generally for extended Dynkin quivers, the other components will be constructed in Sect. 11.

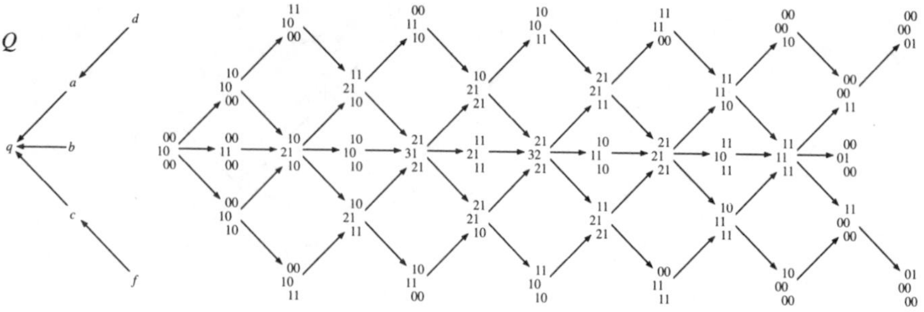

Fig. 3

On the contrary, if Q is a Dynkin quiver, Γ_Q is finite (7.1) and has only one component (see 9.8). We give one example in Fig. 3. There again we have reckoned the dimension functions from the left to the right. But we stopped the construction at the vertices where the algorithm would have produced negative dimensions. For instance, if $\bar{\tau}^6(^\wedge q)$ was defined, its dimension at the vertex q would be determined by

$$d_q(\bar{\tau}^6(^\wedge q)) = -d_q(\bar{\tau}^5(^\wedge q)) + d_q(\bar{\tau}^5(^\wedge a)) + d_q(\bar{\tau}^5(^\wedge b)) + d_q(\bar{\tau}^5(^\wedge c)) = -1.$$

In fact, the appearance of -1 means that $\bar{\tau}^5(^\wedge q)$ is the injective $^\vee q$ with socle $^- q$ and that $^\vee q/^- q \tilde{\to} \bar{\tau}^5(^\wedge a) \oplus \bar{\tau}^5(^\wedge b) \oplus \bar{\tau}^5(^\wedge c) = {}^\vee c \oplus {}^\vee b \oplus {}^\vee a$.

10.3. A connected component Π of $\Gamma_{\mathscr{S}}$ is called *postprojective* if, for each vertex V, the *supremum* $p(V) \in \mathbb{N} \cup \{\infty\}$ of the lengths of the paths with terminus V is $< \infty$. The condition implies that each vertex of Π is an iterated translate $\bar{\tau}^n(P)$ of a projective $P \in \Pi$. The converse is true if $p(P) < \infty$ for each projective $P \in \Pi$. In the case $\mathscr{S} = kQ^{op}$ examined in 10.2, the postprojective component is *complete* in the sense that it contains all the projective vertices of $\Gamma_{\mathscr{S}}$. On the other hand, $k[T]/T^r$ has no postprojective component at all if $r \geqslant 2$ (Fig. 4).

We also use the dual notion of a *preinjective* component. For each vertex V, such a component admits only finitely many paths with origin V. In the case of

$\mathscr{S} = k[T]/T^4$
$M_i = k[T]/T^i$
$\bar{\tau}(M_i) = M_i$ if $i \leqslant 3$

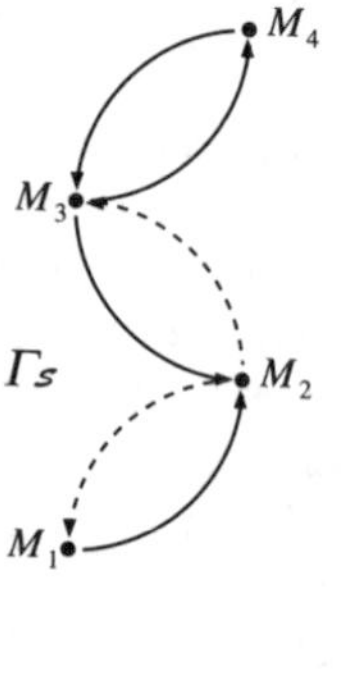

Fig. 4

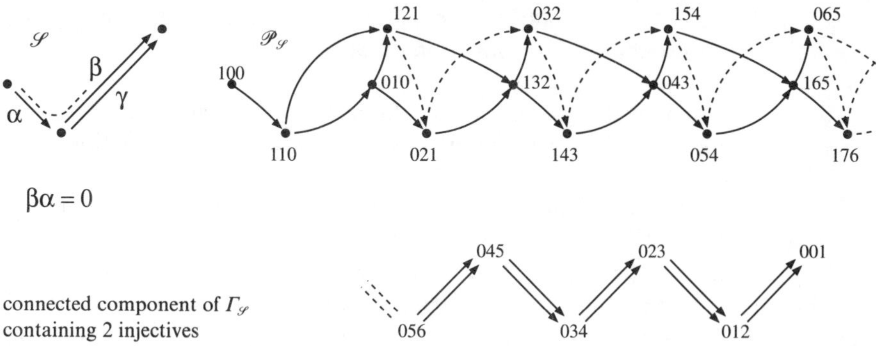

Fig. 5

Fig. 5, $\Gamma_{\mathcal{S}}$ has a complete projective component and a non-complete preinjective component.

In the sequel, we shall call a module over $\mathcal{S}$ *postprojective*[3] (resp. *preinjective*) if its indecomposable summands are isomorphic to vertices of postprojective (resp. preinjective) components of $\Gamma_{\mathcal{S}}$.

The important point about postprojective components is that the full subcategory of mod $\mathcal{S}$ formed by the postprojective vertices $V \in \Gamma_{\mathcal{S}}$ admits a purely combinatorial description. For this sake, we first consider a locally finite translation-quiver Γ. With each non-injective vertex $V \in \Gamma$ we then associate the so-called *mesh-sum* $\sigma_V = \sum_\alpha (\sigma\alpha)\alpha \in k\Gamma(V, \bar{\tau}V)$, where α ranges over the arrows of Γ with tail V. When V varies, the mesh-sums generate an ideal $(\sigma_V)_{V \in \Gamma}$ of the k-category of paths. The associated residue-category $k_\tau\Gamma = k\Gamma/(\sigma_V)_{V \in \Gamma}$ is the so-called *mesh-category* of Γ.

Theorem. *Let $\mathcal{S}$ be a locally bounded spectroid and Π a postprojective component of its representation-quiver $\Gamma_{\mathcal{S}}$. Then*:

a) $\mathrm{Hom}(W, V) = 0$ *for each $V \in \Pi$ and each $W \in \Gamma_{\mathcal{S}} \setminus \Pi$.*

b) *Suppose that V is a vertex of Π, W a vertex of $\Gamma_{\mathcal{S}}$ and that* $\dim V(x) = \dim W(x)$ *for each $x \in \mathcal{S}$. Then $V = W$.*

c) *There exists a display-functor $\Phi\colon k\Gamma_{\mathcal{S}} \to$* ind $\mathcal{S}$ (8.1) *which induces an isomorphism of the mesh-category $k_\tau\Pi$ onto the full subcategory of* ind $\mathcal{S}$ *formed by the vertices of Π.*

The *proof* of the theorem is elementary. Statement a) is proved by induction on $p(V)$. It is clear if $p(V) = 0$, i.e. if V is projective and simple. If $p(V) > 0$, suppose that $0 \neq f \in \mathrm{Hom}(W, V)$. Then f factors through the middle term of the almost split sequence stopping at V or through the radical $\mathcal{R}V$ (if V is projective). In both cases there exists an arrow $V' \to V$ of Π and a non-zero $g \in \mathrm{Hom}(W, V')$. This contradicts our induction hypothesis, since $p(V') < p(V)$.

For statement b), the idea is to compare the dimensions of $\mathrm{Hom}(U, V)$ and $\mathrm{Hom}(U, W)$ by induction on $p(U)$.[4] Finally, for statement c), the problem is to choose basis vectors $\varphi(\alpha)$ of supplements of the subspaces $\mathcal{R}^2_{\mathrm{ind}\,\mathcal{S}}(X, Y)$ of

$\mathscr{R}_{\text{ind}\,\mathscr{S}}(X, Y)$ in such a way that $\sum_\alpha \varphi(\sigma\alpha)\varphi(\alpha) = 0$ for each mesh-sum $\sum_\alpha(\sigma\alpha)\alpha$ (see 8.1). We choose the $\varphi(\alpha)$ by induction on $p(Y)$.[5]

10.4. We are particularly interested in *finite* spectroids with *complete postprojective components*. They can be obtained by an inductive construction as we shall see now:

a) We first consider a finite spectroid $\mathscr{S}$ with a complete postprojective component Π. We define a partial order on the vertices of Π by setting $V \leqslant W$ if Π contains a path from V to W. We further consider a $t^\wedge$ which is maximal for this order among the projective vertices, and we denote by $\mathscr{S}'$ the full subspectroid of $\mathscr{S}$ formed by the $s \neq t$: Then the full subquiver Π_t of Π supported by $\{V \in \Pi : t^\wedge \nleqslant V\}$ is identified with a full subquiver Π_t' of $\Gamma_{\mathscr{S}'}$ which contains all the $s^\wedge$, $s \in \mathscr{S}'$, and is contained in the union Π' of the postprojective components of $\Gamma_{\mathscr{S}'}$ (Fig. 6; use the "knitting" procedure of 10.2 and induction on $p(V)$ to construct a morphism $\Pi_t \to \Gamma_{\mathscr{S}'}$ mapping V onto $V|\mathscr{S}'$). Within Π' we can describe Π_t' as follows: Set $\mathscr{R}t^\wedge = \mathscr{R}_\mathscr{S}(?, t) = \bigoplus_{i=1}^r V_i^{n_i}$, where $V_i \in \Pi$, $n_i > 0$ and $V_i \neq V_j$ if $i \neq j$. Then Π_t' is the full subquiver of Π' supported by $\{U \in \Pi' :$ $\bar{\tau}(V_i|\mathscr{S}') \nleqslant U$ for each non-injective $V_i|\mathscr{S}'\}$.

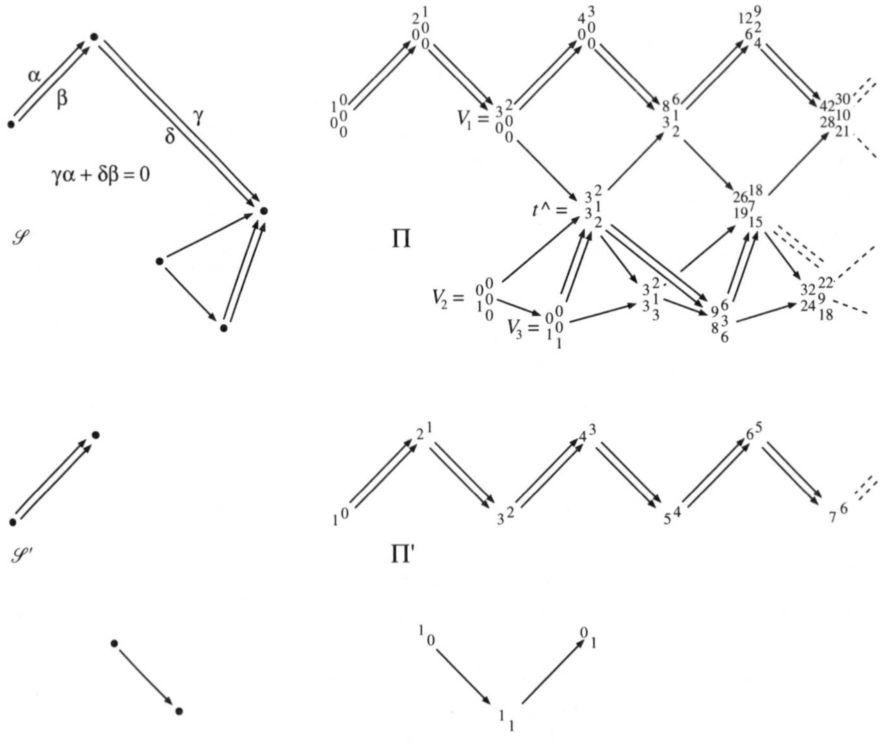

Fig. 6

b) Conversely, let $\mathcal{S}'$ be a finite spectroid such that each projective vertex of $\Gamma_{\mathcal{S}'}$ belongs to a postprojective component. Denote by Π' the union of these components, and consider vertices $W_1, \ldots, W_r$ of Π' such that $\bar{\tau}(W_i) \not\leq W_j$ for each W_j and each non-injective W_i. Then each sequence of non-zero natural numbers $n_1, \ldots, n_r$ gives rise to a finite spectroid $\mathcal{S}$ obtained from $\mathcal{S}'$ by adding a new point t such that $\mathcal{S}(s, s') = \mathcal{S}'(s, s')$, $\mathcal{S}(t, s) = 0$ and $\mathcal{S}(s, t) = W_1(s)^{n_1} \oplus \cdots \oplus W_r(s)^{n_r}$ for all $s, s' \in \mathcal{S}'$. We may assume that $\mathcal{S}$ is connected, i.e. that each connected component of Π' contains at least one W_i. Then $\mathcal{S}$ has a complete postprojective component, and we are in the situation described in a) with $W_i = V_i|\mathcal{S}'$.

10.5. If $\mathcal{S}$ admits a complete postprojective component Π, $\mathcal{S}$ is isomorphic to the full subcategory of $k_\tau\Pi$ formed by the projective vertices (10.3). Thus $\mathcal{S}$ can be described with a purely combinatorial object Π, which can be investigated for its own sake. We restrict our report to a relevant particular case.

We start with a finite *oriented tree*[6] T. A *pattern* P of the associated translation-quiver $\mathbb{N}T$ (10.2) is a set of vertices containing exactly one pair (n, s) for each $s \in T$ and among them at least one pair of the form $(0, t)$. On the vertices of $\mathbb{N}T$ we define a *dimension-function* $d_P = d$ with values in $\mathbb{N}$ so that:

a) $d(x) = 1 + \sum_{y \to x} d(y)$ if $x \in P$, where y ranges over the tails of the arrows of $\mathbb{N}T$ with head x.

b) $d(x) = -d(z) + \sum_{y \to x} d(y)$ if $x = \bar{\tau}z$ and $\sum_{y \to x} d(y) > d(z) > 0$.

c) $d(x) = 0$ in all other cases.

By Π_P we now denote the full subquiver of $\mathbb{N}T$ formed by the vertices x such that $d(x) > 0$ (Fig. 7). We equip Π_P with the translation induced by $\mathbb{N}T$: Thus the translate of (n, t) in Π_P is $(n+1, t)$ if $d(n, t) > 0$ and $d(n+1, t) > 0$; a vertex (n, t) is injective in Π_P if $d(n, t) > 0$ and $d(n+1, t) = 0$; it is projective in Π_P if it belongs to the pattern P.

In the following theorem, $\mathcal{S}_P$ denotes the full subspectroid of the mesh-category $k_\tau\Pi_P$ which is supported by P; for each vertex $x \in \Pi_P$, M_x is the $\mathcal{S}_P$-module defined by $M_x(p) = k_\tau\Pi_P(p, x)$, $p \in P$; for each $\mathcal{S}_P$-module M, we write $\underline{\dim}\, M \in \mathbb{Z}^P$ for the *dimension-function* $p \mapsto \dim M(p)$; finally, we endow $\mathbb{Z}^P$ with the quadratic form $\chi_P(\xi) = \sum_{i=0}^{i=2}(-1)^i \sum_{p,q \in P} \xi(p)\xi(q) \dim \operatorname{Ext}^i_{\mathcal{S}_P}(p^-, q^-)$, where p^- is the simple $\mathcal{S}_P$-module supported by p (3.7d).

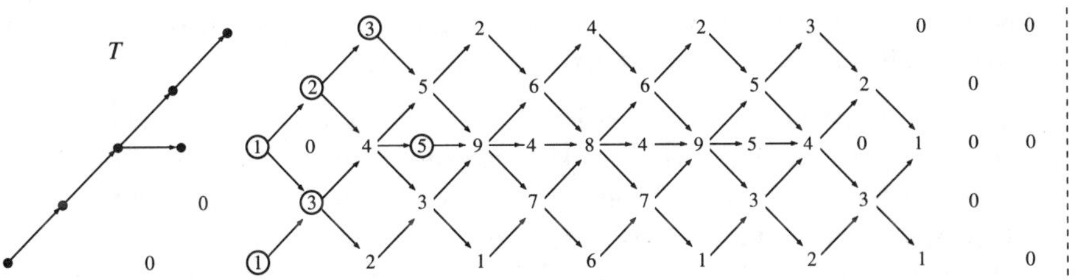

Fig. 7

Theorem. *Let T be a finite oriented tree and P a pattern of $\mathbb{N}T$. Then:*

a) *The functor* $k_\tau\Pi_P \to \text{mod}\,\mathcal{S}_P$, $x \mapsto M_x$ *is fully faithful. As a consequence, each* M_x *is indecomposable.*

b) *The map* $x \mapsto M_x$ *induces an isomorphism of the translation-quiver* Π_P *onto the union of the postprojective components of the representation-quiver* $\Gamma_{\mathcal{S}_P}$. *As a consequence,* $d_P(x) = \dim M_x := \sum_{p\in P} \dim M_x(p)$ *for each* $x \in \Pi_P$.

c) $\mathcal{S}_P$ *is finitely represented*[7] *if and only if the unit form* χ_P *is weakly positive or, equivalently, if and only if all connected components of* $\Gamma_{\mathcal{S}_P}$ *are postprojective. In this case, the map* $x \mapsto \underline{\dim}\, M_x$ *yields a bijection between the vertices of* Π_P *and the positive roots of* χ_P; *moreover, we then have* $\dim M_x(p) \leqslant 6$ *for all* $p \in P$ *and all* $x \in \Pi_P$.

The proof of the statements a) and b) is easy[8] and based on the "knitting" algorithm already used in 10.2. Statement c) generalizes Theorem 7.1 and uses central ideas of its proof[9] (the last statement follows from 6.7).

Spectroids of the form $\mathcal{S}_P$ play an eminent rôle in the investigation of finitely represented algebras. We call them *arboresque*[10].

10.6. Supportive Arboresque Spectroids. The *support* of a module $M \in \text{mod}\,\mathcal{S}$ is by definition the full subspectroid of $\mathcal{S}$ formed by the points s such that $M(s) \neq 0$. Whenever we classify representations, we may first look for the supports of the indecomposables. Accordingly, we call a spectroid *supportive* if it is the support of a finite-dimensional indecomposable representation.

Bongartz has shown that the finitely represented supportive arboresque spectroids can be arranged in 31 infinite series and a finite set of *exceptional* spectroids (whose number equals 16'325 according to new results by Dräxler).[11] We give the list of Bongartz in Fig. 8 in a slightly modified form. As it seems, the main information given by the list is that large spectroids of the considered type have long linear parts.

The spectroids of Fig. 8 are defined by their quivers and by admissible ideals (8.3). A non-oriented edge $x - y$ may be directed towards x or y. The admissible ideal is always generated by the differences of parallel paths and by the paths flanked with a dotted line.

10.7. Critical Arboresque Spectroids. An arboresque spectroid $\mathcal{S}$ is called *critical* if each full subspectroid obtained by deleting a source or a sink of the quiver $Q_{\mathcal{S}}$ is finitely represented though $\mathcal{S}$ itself is not.

Theorem.[12] *Let T be an oriented tree and P a pattern of $\mathbb{N}T$. If the associated spectroid $\mathcal{S}_P$ is critical, T is an extended Dynkin quiver.*

The theorem permits a classification of the critical arboresque spectroids: If the underlying graph of T is $\tilde{D}_n$, $n \geqslant 4$, $\mathcal{S}_P$ is isomorphic to a spectroid of Fig. 9, where non-oriented edges can be given any orientation. The defining admissible ideals (8.3) are generated by $\alpha_1\alpha_2\ldots\alpha_p - \beta_1\beta_2$ and $\gamma_1\gamma_2\ldots\gamma_q - \delta_1\delta_2$ in the first case, by $\alpha_1\alpha_2\ldots\alpha_p - \beta_1\beta_2$ in the second, by $\alpha_1\alpha_2\ldots\alpha_p + \beta_1\beta_2 + \gamma_1\gamma_2$ in the last. In the third case, they are zero.

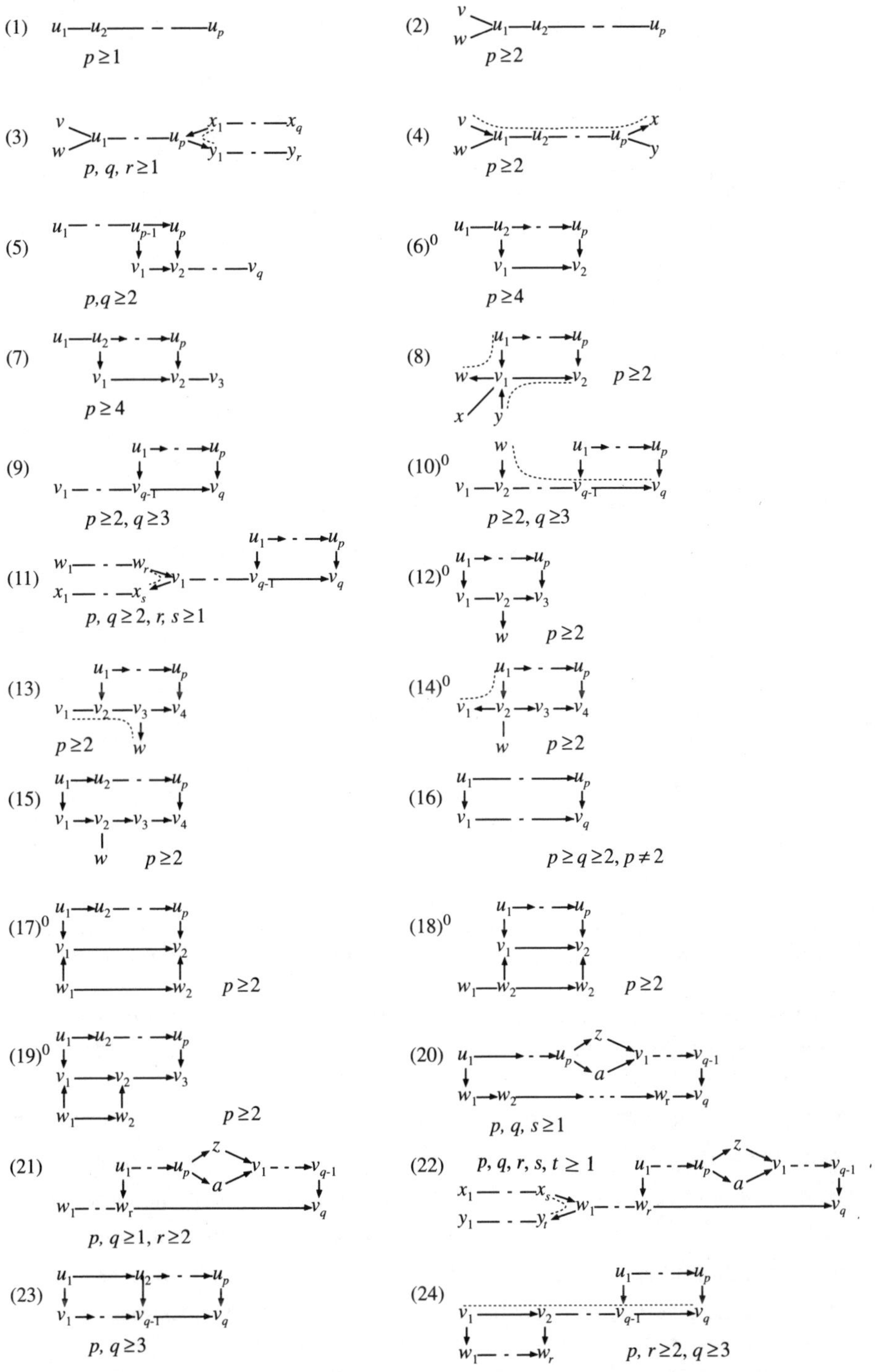

Add to this list the duals of the 7 ringed families

Fig. 8

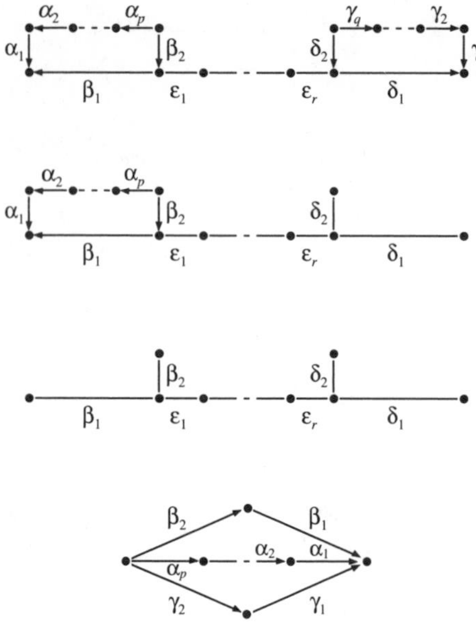

$p, q \geq 2, r \geq 0$

$p \geq 2, r \geq 0$

$r \geq 0$

$p \geq 2$

Fig. 9

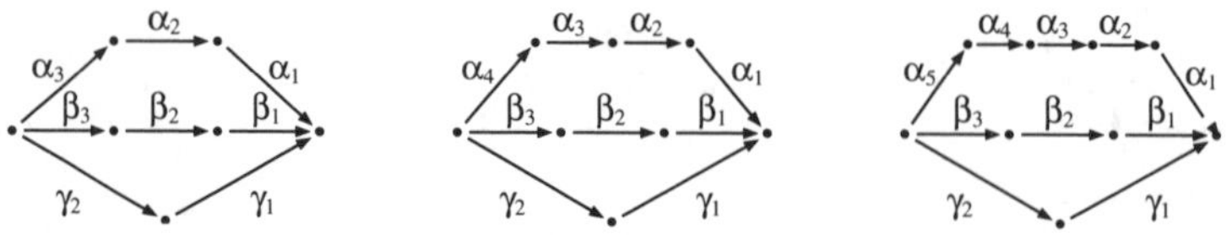

Fig. 10

If the underlying graph of T is $\widetilde{E}_n$, $6 \leqslant n \leqslant 8$, $\mathscr{S}_P$ is isomorphic to a spectroid of Fig. 10 or Fig. 11. In the case of Fig. 10, the defining admissible ideals are generated by $\alpha_1 \ldots \alpha_p + \beta_1\beta_2\beta_3 + \gamma_1\gamma_2$, $p = 3, 4, 5$.

All the spectroids of Fig. 11 are defined by quivers and "relations." Each diagram represents a family of categories obtained by orienting the edges. The admissible orientations are only subjected to the following restrictions: In a linear "subfigure" of the form  or , the dotted line points to a zero-relation; accordingly, the admissible orientations are  and  or  and , and the corresponding composed morphisms are zero. In a polygonal subfigure of the form, say 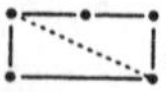, the dotted line points to a commutativity relation; accordingly, the admissible orientations are 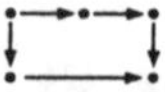 and ; in both cases, the two parallel paths give rise to the same non-zero morphism. Besides these conventions, we agree that each branch with a "free" end b of the form 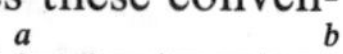 represents a connected subquiver of the quiver with relations P of Fig. 12. The

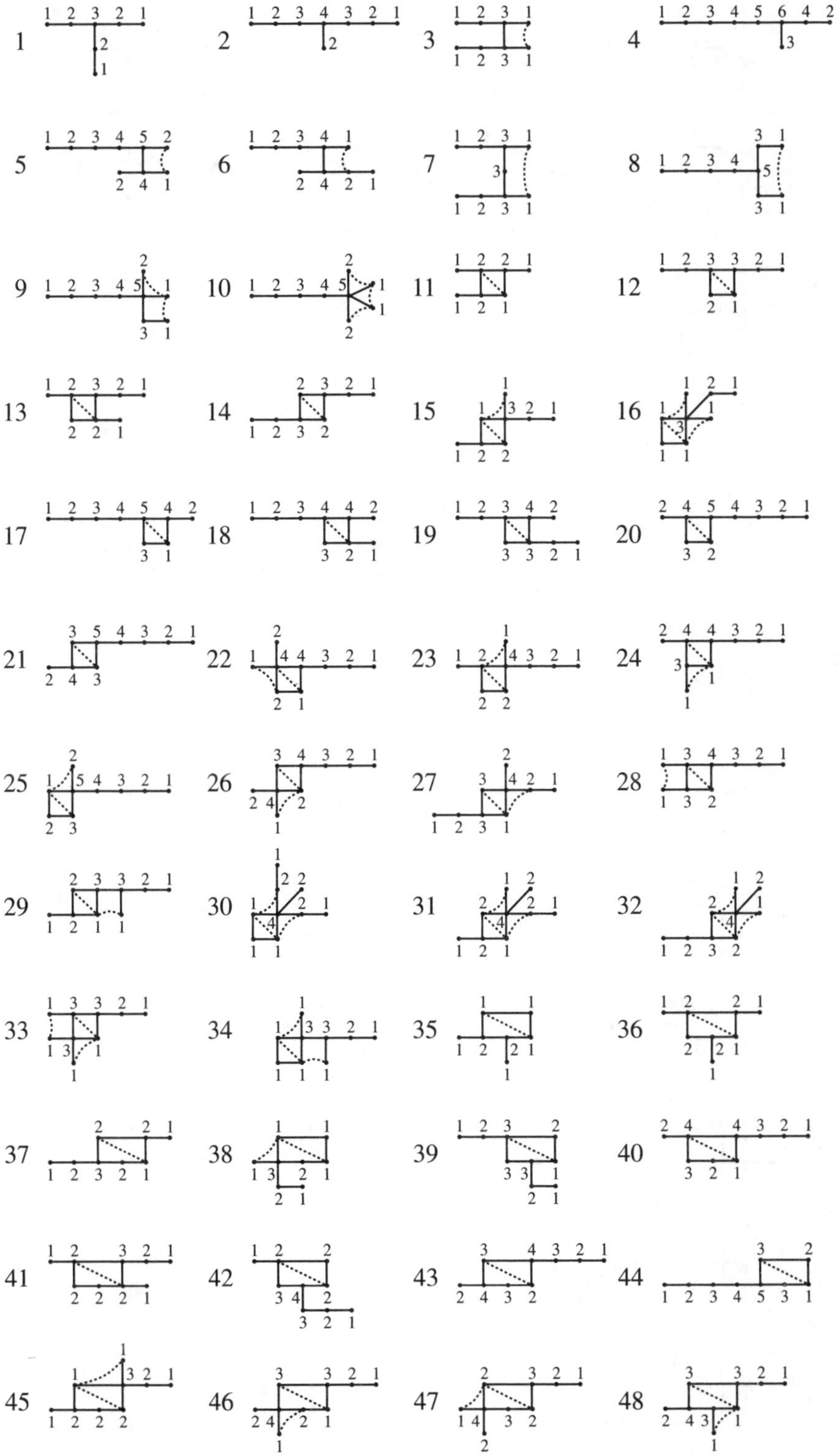

Fig. 11

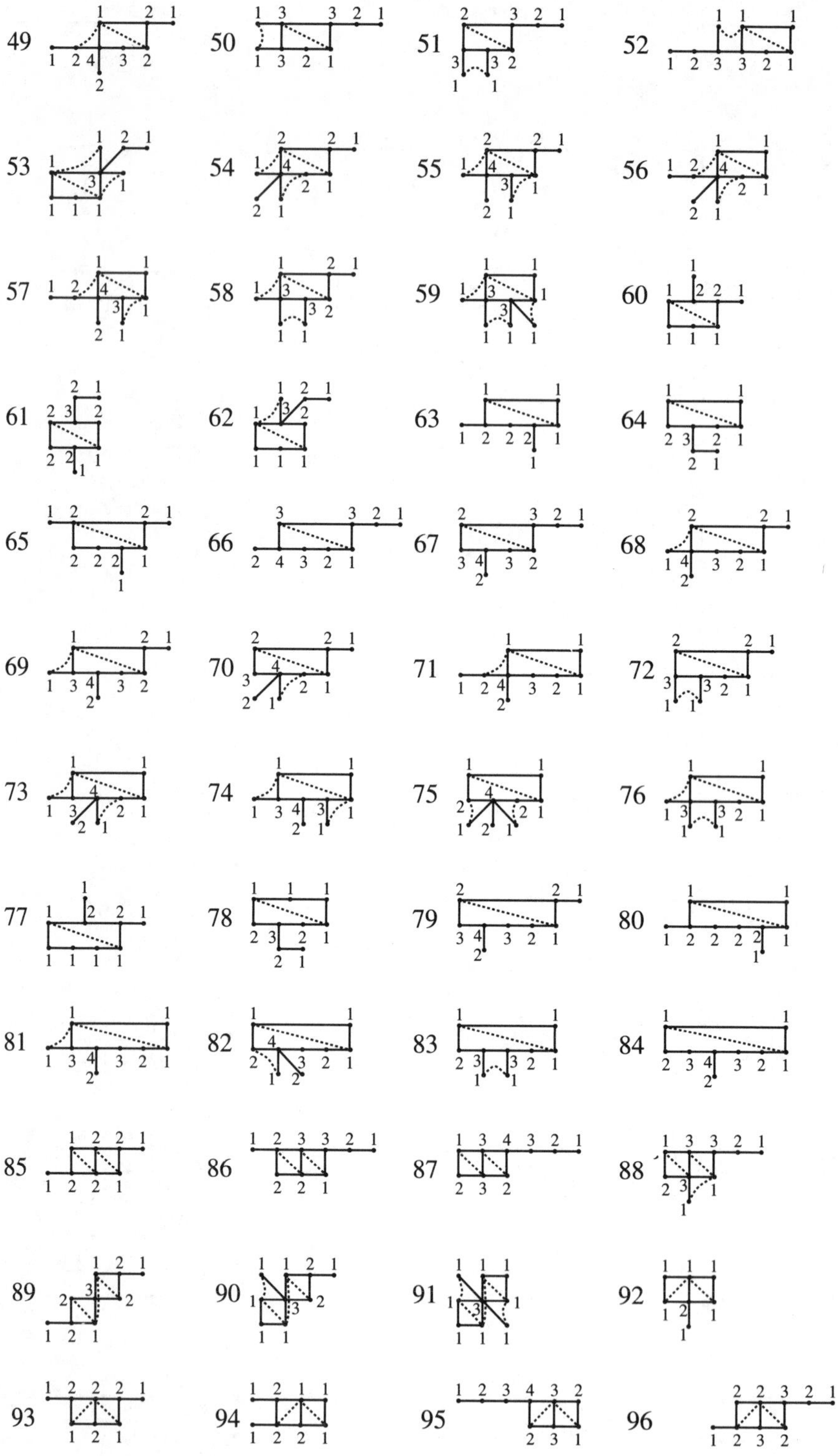
49
50
51
52
53
54
55
56
57
58
59
60
61
62
63
64
65
66
67
68
69
70
71
72
73
74
75
76
77
78
79
80
81
82
83
84
85
86
87
88
89
90
91
92
93
94
95
96

Fig. 11 (*cont.*)

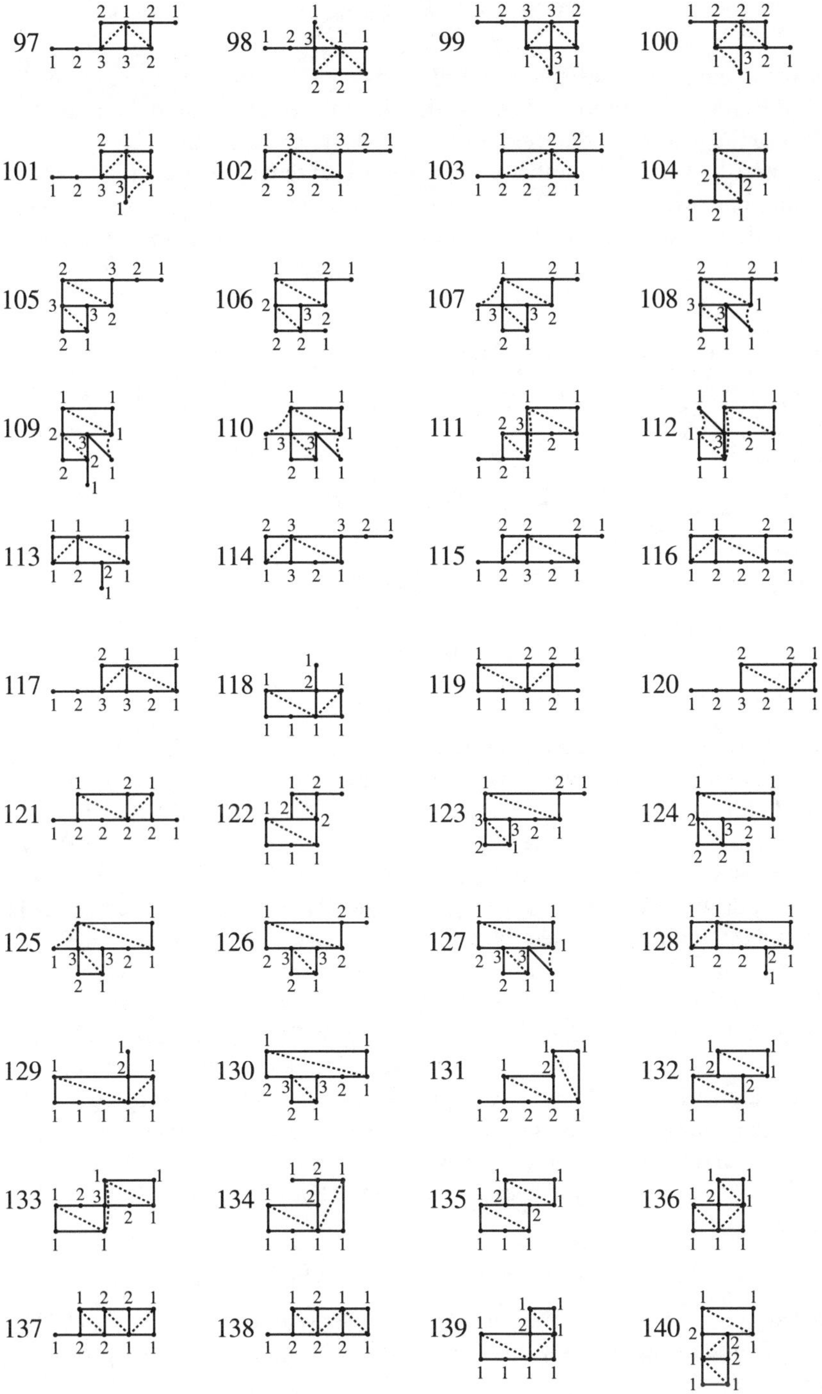

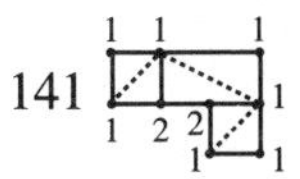

Fig. 11 (*end*)

subquiver is equipped with the induced zero-relations and has to contain the "root" of P, which is supposed to coincide with the vertex a where the branch is attached to the rest of the diagram. For instance, diagram 28 represents a family of 28 spectroids. The total number of spectroids represented by Fig. 11 is 4'299.

The numbers ascribed to the vertices of Fig. 11 are the values of the smallest dimension-functions associated with infinitely many indecomposable representations.

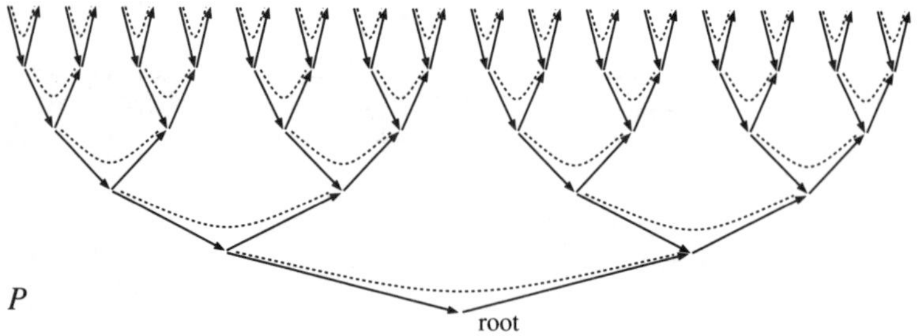

Fig. 12

***10.8. Variations on Exact Structures.** In the case of *non-abelian* exact structures, we content ourselves with two examples showing that the method used in 10.1–10.4 still works. We *first* consider the quivers L and R associated with the full arrows of Fig. 13. Subjected to the relations $\gamma\alpha - \delta\beta = \varepsilon\gamma = \zeta\delta = 0$ and $\tau\varphi - \upsilon\chi = \varphi\psi = \chi\omega = 0$, these quivers give rise to spectroids $\tilde{L}$ and $\tilde{R}$ and to the associated additive hulls $\mathcal{Q} = \oplus\tilde{L}$ and $\mathcal{J} = \oplus\tilde{R}$. By B we denote the bimodule over $\mathcal{J}$ and $\mathcal{Q}$ which is generated by the "dashed arrows"[13] $\mu \in B(c, x)$, $\nu \in B(d, v)$ subjected to the relation $\varphi\mu = \nu\delta$.

Our problem is to determine the quiver $Q_{\mathcal{A}}$ of the aggregate $\mathcal{A}$ formed by the triples (Q, b, J), where $Q \in \mathcal{Q}$, $J \in \mathcal{J}$ and $b \in B(Q, J)$. The exact structure $\mathcal{E}$ introduced in 9.1, Example 5, equips $Q_{\mathcal{A}}$ with a translation and enables us to use the "knitting" algorithm of 10.2 and 10.4, provided we know the $\mathcal{E}$-projectives (see 9.2, Example 1) and the arrows of $Q_{\mathcal{A}}$ with prescribed $\mathcal{E}$-projective head T. The tails of these arrows are the indecomposable summands of $S \in \mathcal{A}$, where $S^\wedge$ is a projective cover of $\mathcal{R}_{\mathcal{A}}(?, T)$. The determination of S is easy: In case $T = (0, 0, J)$, a projective covering

$$(0, g)^\wedge: (P, h, K)^\wedge \to \mathcal{R}_{\mathcal{A}}(?, (0, 0, J))$$

is induced by projective coverings $g^\wedge: K^\wedge \to \mathcal{R}_{\mathcal{J}}(?, J)$ and $h^\wedge: P \to \operatorname{Ker} B(?, g)$ in mod $\mathcal{Q}$ and mod $\mathcal{J}$ respectively. In case $T = (Q, p, I)$ and $Q \neq 0$ (9.2), a projective covering

$$(q, \mathbb{1}_I)^\wedge: (R, pq, I)^\wedge \to \mathcal{R}_{\mathcal{A}}(?, (Q, p, I))$$

is induced by a projective covering $q^\wedge: R^\wedge \to \mathcal{R}_{\mathcal{Q}}(?, Q)$ in mod $\mathcal{Q}$.

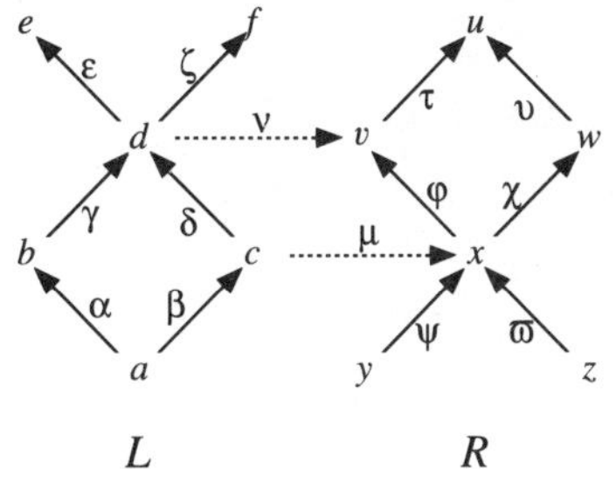

Fig. 13

The result of our knitting procedure is presented in Fig. 14, where a symbol like

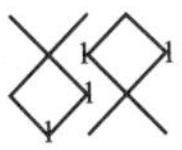

represents a triple of the form $(a \oplus c, r, v \oplus w) \in \mathscr{A}$. Each point s of L or R gives rise to an *additive* function (10.2) on $Q_{\mathscr{A}}$ whose value at $(X, r, Y) \in Q_{\mathscr{A}}$ is the multiplicity of s in X or Y. In the knitting process, the additivity of these functions was used at each "mesh". Since the starting vertices of the meshes are not $\mathscr{E}$-injective, the algorithm required the knowledge of the $\mathscr{E}$-injectives: In fact, a description of the arrows starting at an $\mathscr{E}$-injective vertex shows that additivity enforced there would have produced negative multiplicities (compare with 10.2).

In Fig. 14 each mesh is marked with a dotted edge connecting tail and head. The edges thus obtained strongly depend on the exact structure $\mathscr{E}$ imposed upon $\mathscr{A}$. As a first example, we intentionally chose an *abelian* aggregate $\mathscr{A}$ equipped with a "poor" exact structure $\mathscr{E}$. The exact structure formed by all exact pairs of $\mathscr{A}$ gives rise to a translation-quiver isomorphic to that of Fig. 15: In fact, $\mathscr{A}$ is equivalent to mod $\mathscr{S}^{op}$, where $\mathscr{S}$ is the spectroid defined by the quiver

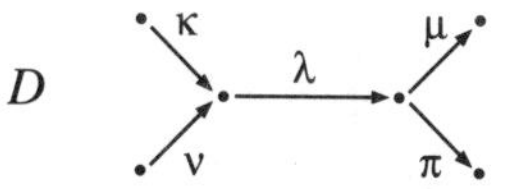

subjected to the relation $\mu\lambda\kappa = 0$.[14] The difference between Fig. 14 and Fig. 15 lies in the translations and in the corresponding meshes: We pass from $\Gamma_{\mathscr{S}^{op}}$ to the translation-quiver $Q_{\mathscr{A}}$ of $(\mathscr{A}, \mathscr{E})$ by deleting the dotted edges of $\Gamma_{\mathscr{S}}$ whose associated almost split sequences lie outside $\mathscr{E}$ *

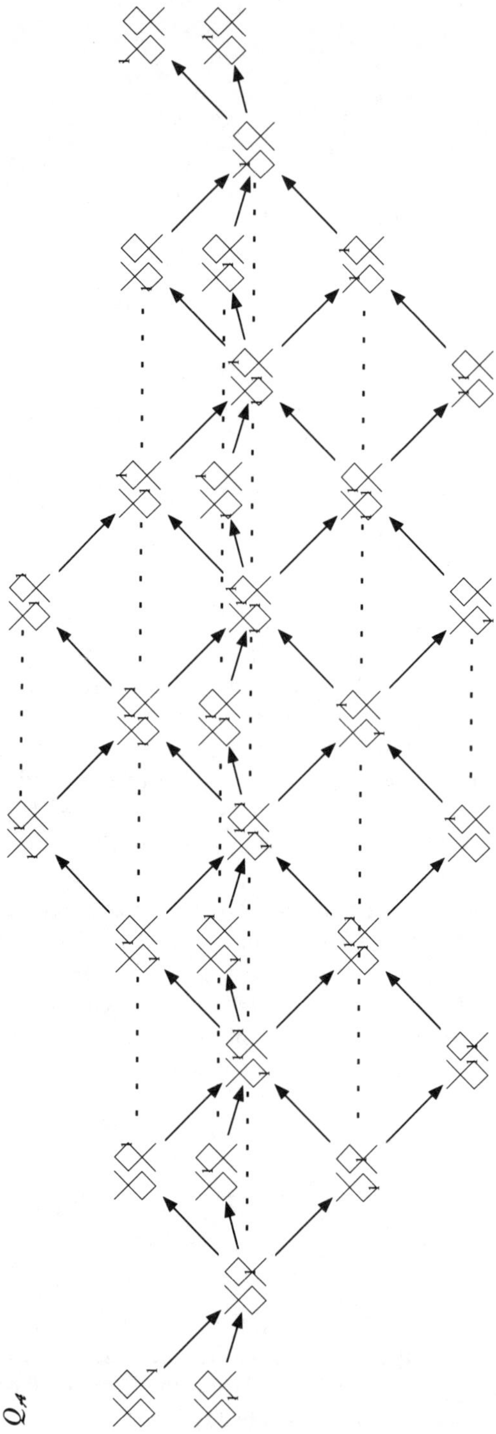

Fig. 14

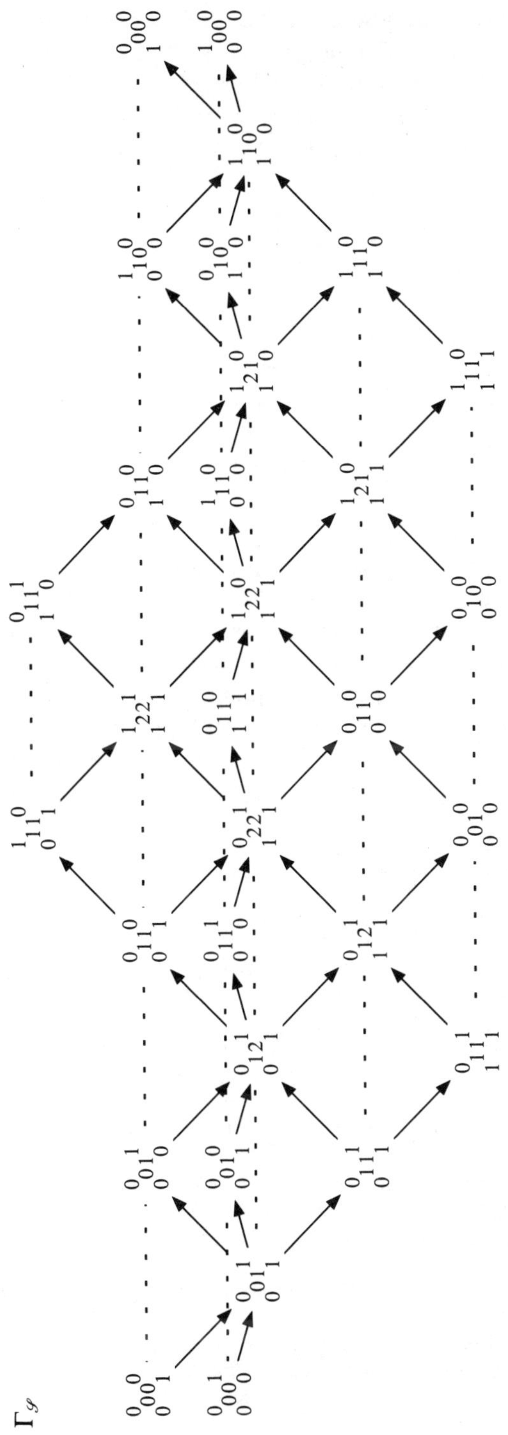

Fig. 15

***10.9. Representations of Finite Posets.** Our *second* example concerns the poset $R = \hat{4} \coprod \hat{2} \nwarrow \hat{2}$ of 5.4. The aggregate M^{Rk} of 4.1 is identified with the category of triples $\mathscr{A}$ attached to the aggregates $\mathscr{Q} = \operatorname{mod} k$, $\mathscr{J} = \bigoplus kR$ and to the bimodule B generated by 3 elements $\lambda \in B(k, a_4)$, $\mu \in B(k, b_2)$, $\nu \in B(k, c_2)$ subjected to the relation $\beta\mu = \delta\nu$ (Fig. 16; compare with 10.8).

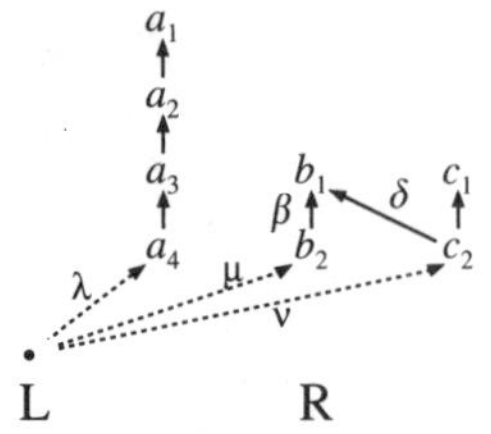

Fig. 16

The exact structure $\mathscr{E}$ equipping $\mathscr{A}$ (9.1, Example 5) defines a translation on the quiver $Q_{\mathscr{A}}$ and allows us to "knit". As a result we obtain the translation-quiver of Fig. 17. The lower part of this figure illustrates the beginning of the knitting procedure and contains all the $\mathscr{E}$-projective vertices. There each vertex is designated by the multiplicities of the involved points of L and R. After having spotted the $\mathscr{E}$-projective vertices, we replaced the 9 multiplicity-functions by their sum. This is permissible because the sum is also additive and takes values $\leqslant 2$ at all $\mathscr{E}$-injective vertices (and there only if we except the known $\mathscr{E}$-projective vertices)[15] *

10.10. Remarks and References

1. If $V_1, \ldots, V_r$ are representatives of the indecomposable modules over a finitely represented algebra A, the algebra E of endomorphisms of $\bigoplus_{i=1}^r V_i$ seems to be a natural object of study. After [3, Auslander, 1971/74] and [68, Gruson, 1974/75], an investigation of the quiver Q_E of E seemed advisable to the then adepts of diagrammatic methods. Examples were produced by Ringel in 1974. The origin of the "knitting" procedure exposed in 10.2 is controversial. Admittedly, Bautista used it in Oberwolfach in 1977. We learned it later from Ringel. The first open-cast descriptions of $\Gamma_{\mathscr{S}}$ known to us are produced in [129, Ringel, 1977/78], [58, Gabriel/Riedtmann 1977/79], [59, Gabriel, 1978/79] and in the fundamental [126, Riedtmann, 1979/80]. See also [10, Bautista, 1979/80], [103, Martinez, 1979/80], [27, Bongartz/Gabriel, 1980/82], [13, Bautista/Brenner, 1983], [31, Brenner, 1984/86].

The representation-quiver $\Gamma_{\mathscr{S}}$ is also commonly called *Auslander-Reiten quiver of* $\mathscr{S}$ (not of mod $\mathscr{S}$! See [60, Gabriel, 1979/80]). Unfortunately, new confusion is created by the increasing use of this name for the much older ordinary quiver.

2. See [60, Gabriel, 1979/80].
3. Postprojective seems more advisable than the commonly used *preprojective.*
4. See [11, Bautista, 1980], [69, Happel, 1982].
5. Compare with [126, Riedtmann, 1979/80], [103, Martinez, 1979/80].
6. oriented tree = quiver whose underlying graph is a tree (graph without cycle).
7. finitely represented = having a finite number of isoclasses of indecomposable representations.
8. [27, Bongartz/Gabriel, 1980/82].

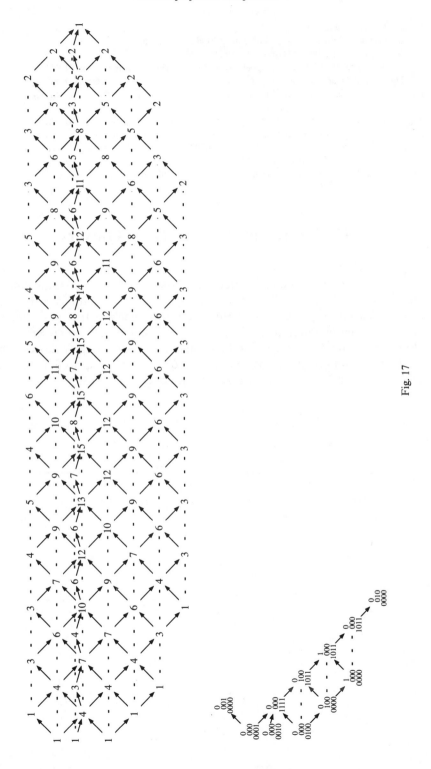

Fig. 17

9. [23, Bongartz, 1982/83].
10. arboresque = artistically tree-like (Oxford English Dictionary).
11. See [22, Bongartz, 1981/82] and [41, Dräxler, 1989].
12. [24, Bongartz, 1982/84], [72, Happel/Vossieck, 1982/83], [25, Bongartz, 1982/84].
13. The full and dashed arrows of Fig. 13 give rise to the quiver $Q_{\mathscr{T}}$ of a spectroid $\mathscr{T}$ having as points the vertices of L and R; the morphisms are such that $\mathscr{T}(l, l') = \tilde{L}(l, l')$, $\mathscr{T}(r, r') = \tilde{R}(r, r')$, $\mathscr{T}(l, r) = B(l, r)$ and $\mathscr{T}(r, l) = 0$ if $l, l' \in L$ and $r, r' \in R$.
14. In fact, each $M \in \operatorname{mod} \mathscr{S}^{op}$ admits a maximal submodule M' which vanishes on the left part > of >—<. The spectroid of the aggregate formed by these M' is identified with $\tilde{R}$, whereas $\tilde{L}$ is the spectroid of the aggregate formed by the $M'' = M/M'$, which vanish on the right part <. Moreover, $\mathrm{Ext}^1(M'', M')$ is identified with $B(M'', M')$.
15. See [132, Ringel, 1984] for the remaining supportive finitely represented posets.

11. Representations of Tame Quivers

Our objective in the present section is to classify the finite-dimensional representations of the *extended Dynkin quivers* (Fig. 1).[1] The orientations of the edges of the depicted graphs can be chosen arbitrarily. Each of the thus obtained quivers admits infinitely many isoclasses of indecomposable representations. However, for any fixed dimension-function, the corresponding isoclasses either are finite in number or can be "parametrized" by a curve up to finitely many exceptions. In this sense, the extended Dynkin quivers are *tame*.

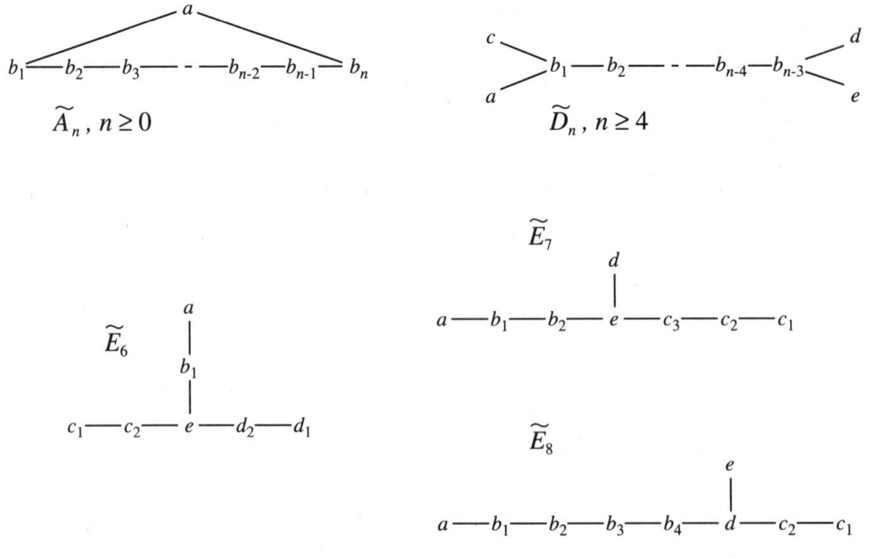

Fig. 1

Our investigation will later also provide a classification of the modules over the critical spectroids of Sect. 10. In fact, among the numerous examples of tame spectroids known by now, the critical spectroids are next to "extended Dynkin spectroids."

11.1. Representations of $\tilde{A}_n$. We suppose that each edge of $\tilde{A}_n$ has been given an orientation and consider a "*counterclockwise walk*"

$$x_0 \underset{\pi_1}{\text{——}} x_1 \underset{\pi_2}{\text{——}} x_2 ---- x_{s-1} \underset{\pi_s}{\text{——}} x_s$$

where the π_i are arrows of $\tilde{A}_n$ and $\pi_i \neq \pi_{i+1}$ if $i < s$ and $n \neq 0$. With π we associate a representation N_π of $\tilde{A}_n$ whose "stalk" at a vertex x is $N_\pi(x) = k^{\bar{x}}$ with $\bar{x} = \{i: x_i = x\}$. The map $N_\pi(\alpha)$: $k^{\bar{x}} \to k^{\bar{y}}$, $\xi \mapsto \eta$ attached to a counterclockwise oriented arrow $x \xrightarrow{\alpha} y$ is such that $\eta(0) = 0$ if $x_0 = y$ and $\eta(j) = \xi(j-1)$ if $0 \neq j \in \bar{y}$. In the case of a clockwise oriented arrow α, we set $\eta(s) = 0$ if $x_s = y$ and $\eta(j) = \xi(j+1)$ if $s \neq j \in \bar{y}$. For instance, if $\tilde{A}_n$ equals

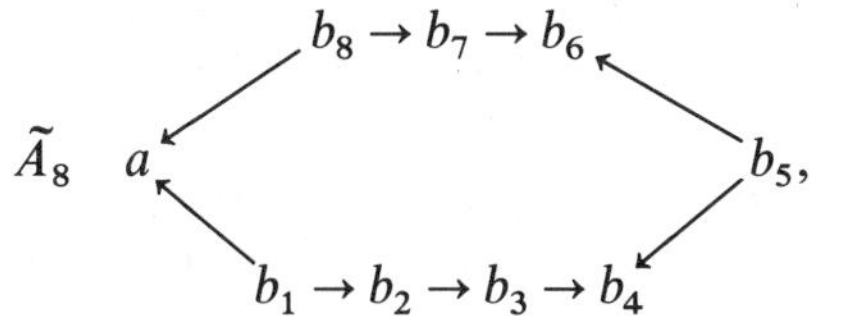

the walk

$$a \leftarrow b_1 \to b_2 \to b_3 \to b_4 \leftarrow b_5 \to b_6 \leftarrow b_7 \leftarrow b_8 \to a \leftarrow b_1$$

gives rise to the representation

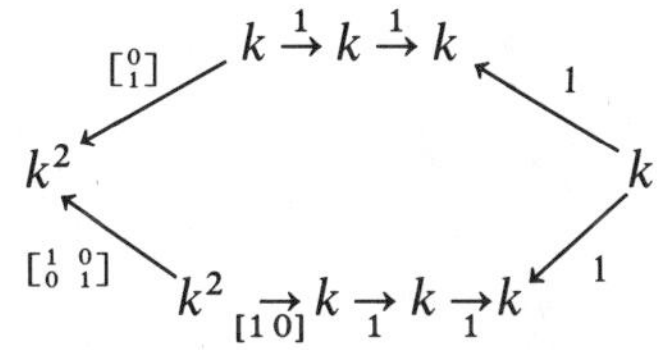

Returning to the general case, we also attach a representation

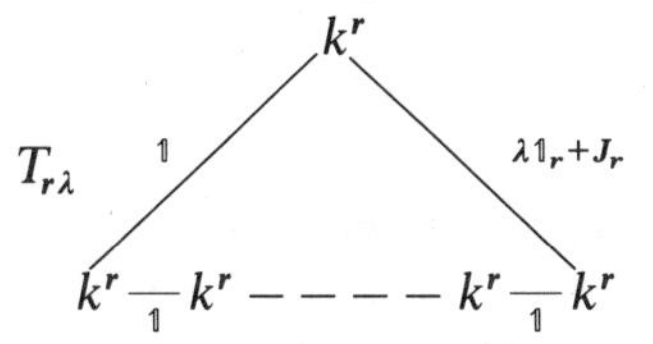

to each Jordan block $\lambda \mathbb{1}_r + J_r = \begin{bmatrix} \lambda & 1 & \\ & \ddots & \\ 0 & & \lambda \end{bmatrix} \in k^{r \times r}$, $\lambda \neq 0$ (1.7).

Theorem. *The representations N_π and $T_{r\lambda}$ constitute an exhaustive list of pairwise non-isomorphic indecomposable finite-dimensional representations of $\tilde{A}_n$.*

11.2. Quivers $\tilde{A}_n$ with Steady Orientation. If the arrows of $\tilde{A}_n$ are oriented as follows

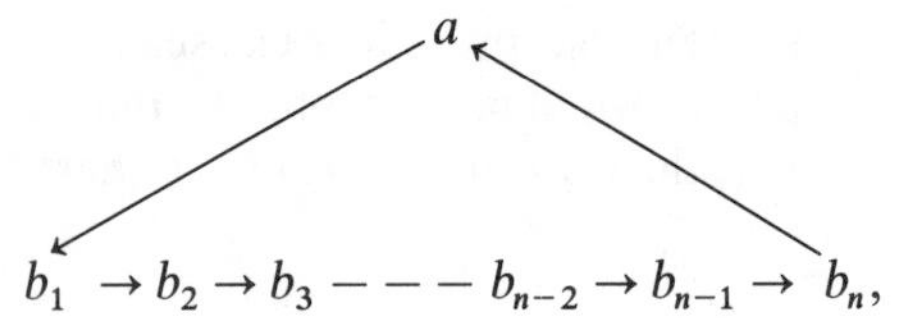

Theorem 11.1 is a slight generalization of the theorem of Jordan-Weierstrass (1.7): First one may verify that each finite-dimensional representation V is the sum of some N annihilated by large paths and some T on which the arrows act bijectively. Then one applies Jordan-Weierstrass to T and extends the observation of 2.1, Example 4 to N.

More remarkable is the existence of almost split sequences, which formally does not follow from Sect. 9 since the projectives $x^\wedge$ have infinite dimension. In fact, the investigation of the finite residue-categories of $k\tilde{A}_n$ shows that, for each $\lambda \neq 0$, the $T_{r\lambda}$ give rise to a component Γ_λ of a translation quiver $\Gamma_{\tilde{A}_n}$ whose last component Γ_0 is formed by the N_π. We refer to Fig. 2, where the left and the right borders of the vertical stripes must be identified (for $n = 3$, the identification has been carried out in the third column; in the second, yx stands for N_π, where π has origin x and terminus y).

With the same kind of arguments as for Theorem 10.3, it is easy to prove that *the spectroid formed by the representations N_π and $T_{r\lambda}$ is isomorphic to the mesh-category $k_\tau \Gamma_{\tilde{A}_n}$.*

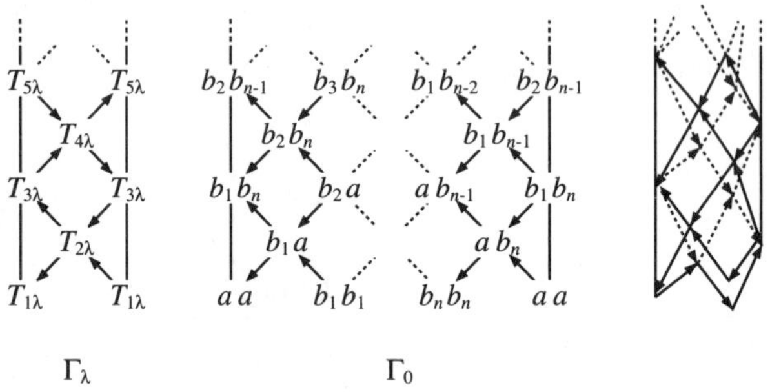

Fig. 2

11.3. Quivers $\tilde{A}_n$ with Shifting Orientation. A fairly general example of shifting orientation is given by the quiver $\tilde{A}_8$ of 11.1. For a proof of Theorem 11.1 in this case, we refer to the arguments developed below for $\tilde{E}_6$.

In contrast with the description of the indecomposables given by Theorem 11.1, the partition of the representation-quiver into connected components strongly depends on the orientations. In the case of shifting orientation, $\tilde{A}_n$ is directed (10.2) and $\Gamma_{\tilde{A}_n}$ has a postprojective component Π and a preinjective component I which are described in 10.2–3. In fact, *Π is formed by the N_π such that the two edges of $\tilde{A}_n$ which prolongate π at the origin and the terminus are counterclockwise and clockwise oriented respectively.* If these edges are clockwise

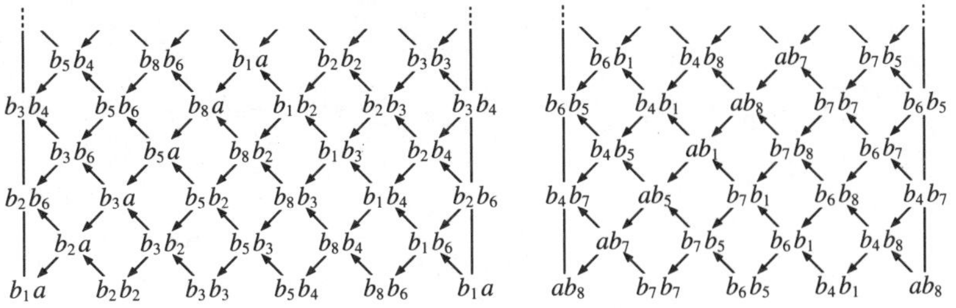

Γ_0 Γ_∞

Fig. 3

and counterclockwise oriented, N_π is preinjective. The remaining N_π form two further components Γ_0 and Γ_∞, where Γ_0 consists of the N_π such that the two edges prolongating π have the same orientation as b_n—a. This happens for instance in the case of the walk $b_1 a$ of $\tilde{A}_8$ considered in 11.1. In this case, Γ_0 and Γ_∞ are depicted in Fig. 3 (where the left and the right border of a vertical stripe must be identified).

With regard to the family $(T_{r\lambda})$, we obtain the same connected components Γ_λ as in the case of steady orientation. But the mesh-category $k_\tau\Gamma_{\tilde{A}_n}$ no longer describes the morphisms between the indecomposables. The situation now is the following:

Let us call *tubular*[2] the representations of $\tilde{A}_n$ which admit no postprojective and no preinjective summand or, equivalently, whose indecomposable summands are listed in the "*tubes*" Γ_λ, $\lambda \in k \cup \infty$. Let further Γ denote the union of *all* the tubes Γ_λ. The mesh-categories $k_\tau\Pi$, $k_\tau\Gamma$ and $k_\tau I$ are then isomorphic to the full subcategories of rep $\tilde{A}_n$ (2.3, Example 5) formed by the postprojective, tubular and preinjective vertices of $\Gamma_{\tilde{A}_n}$ respectively. If $P \in \Pi$, $T \in \Gamma$ and $J \in I$, $k_\tau\Gamma_{\tilde{A}_n}$ and rep $\tilde{A}_n$ admit only zero-morphisms from T or J to P and from J to T. However, in contrast with $k_\tau\Gamma_{\tilde{A}_n}$, rep $\tilde{A}_n$ generally admits non-zero morphisms $P \to T$, $T \to J$ and $P \to J$. If there is need for describing these morphisms by quiver and relations, we can bound the lengths of the considered representations and proceed as in 8.1, Example 6.

The preceding description of the *category* $\text{rep}_t \tilde{A}_n$ *of tubular representations* by means of $k_\tau\Gamma$ can be given the following reformulation, where for the sake of simplicity we restrict ourselves to the quiver $\tilde{A}_8$ of 11.1: Denote by Q_r the quiver $\tilde{A}_{(r-1)}$ equipped with a steady orientation, by $\text{rep}_0 Q_r$ the category formed by the finite-dimensional "nilpotent" representations (whose indecomposable summands are listed in Γ_0). Then there is an equivalence of categories[3]

$$E = [F_0 \;\; F \;\; F_\infty]\colon \text{rep}_0 Q_5 \times \text{mod}\, k_f[x, x^{-1}] \times \text{rep}_0 Q_4 \xrightarrow{\sim} \text{rep}_t \tilde{A}_8,$$

where F_0, F, F_∞ are defined as follows and $E(U, V, W) = F_0 U \oplus F V \oplus F_\infty W$:

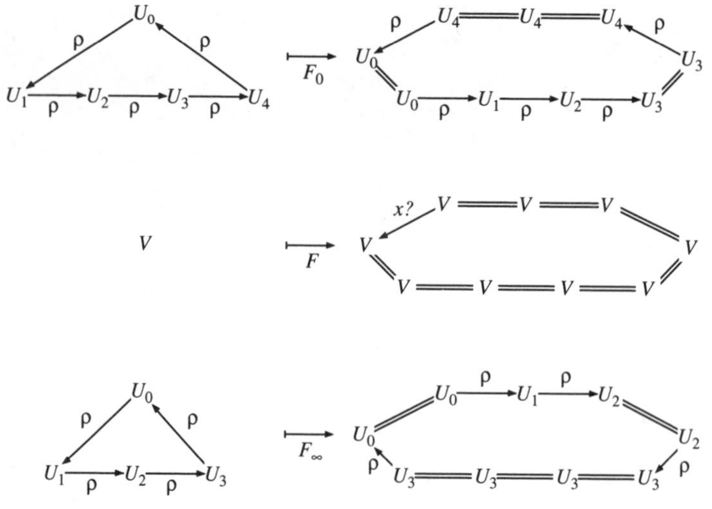

11.4. The Particular Case $\tilde{A}_1$. As we have seen in 11.3, the tubular representations form an exactly embedded full subcategory $\text{rep}_t\tilde{A}_n$ of $\text{rep}\tilde{A}_n$. In the sequel, a special rôle will be assigned to $\tilde{A}_1: a \rightrightarrows b$. For this quiver we therefore restate former results in more invariant terms (compare to 1.8).

Consider a fixed *two-dimensional vector space* Z over k. By a *Z-map* we simply mean a triple (V, f, W), where V, W are vector spaces and $f: V \to Z \otimes W$ a linear map. By a morphism from (V, f, W) to (V', f', W') we mean a pair of linear maps $v: V \to V'$ and $w: W \to W'$ such that $f'v = (Z \otimes w)f$. *The thus defined category* rep Z *of "finite-dimensional" Z-maps* is of course isomorphic to rep $\tilde{A}_1$, since each basis (z_1, z_2) of Z provides an isomorphism of categories (map $V \underset{f_2}{\overset{f_1}{\rightrightarrows}} W$ onto $V \to Z \otimes W$, $y \mapsto z_1 \otimes f_1(y) + z_2 \otimes f_2(y)$). In particular, $\text{rep}_t\tilde{A}_1$ is identified with the category $\text{rep}_t Z$ formed by the "tubular" Z-maps, whose indecomposable summands (V, f, W) satisfy $\dim V = \dim W$.

Now let L be a line (= one-dimensional subspace) of Z. We shall say that L is an *eigenline* of the Z-map (V, f, W) if there is a vector $y \in V$ such that $0 \neq f(y) \in L \otimes W$. If (V, f, W) is indecomposable and tubular, it has a *unique* eigenline and is characterized up to isomorphism by this eigenline and $\dim V$ (= *length* of (V, f, W) within $\text{rep}_t Z$).

In the sequel, T_{rL} will denote a fixed chosen tubular indecomposable with eigenline L and length r. If L is fixed, the T_{rL} then form a connected component Γ_L of the quiver Γ_Z of rep Z. For reasons of convenience, we make our choices in such a way that T_{rL} is a subrepresentation of T_{sL} if $r < s$.

11.5. Representations of $\tilde{E}_n$ ($n = 6, 7, 8$) and $\tilde{D}_n$($n \geqslant 4$). We now equip the edges of $\tilde{D}_n$ and $\tilde{E}_n$ with fixed orientations, only assuming that the edge $a - b_1$ is directed towards b_1 (using vector space duality, we may restrict ourselves to this case). By D_n and E_n we denote the quivers obtained by deleting the vertex a and the arrow $a \to b_1$.

Let Δ stand for D_n or E_n. Up to isomorphism, there is a unique indecomposable representation U of Δ such that $\dim U(b_1) = 2$ (in the case of E_6, the *dimension-function* $\underline{\dim}\, U$ of U is $12\overset{2}{3}21$ in the notation of the present section; compare with 10.2). Defining rep $U(b_1)$ as in 11.4, we thus obtain a functor

$$U^{\otimes}\colon \operatorname{rep} U(b_1) \to \operatorname{rep} \tilde{\Delta},\ (V, f, W) \mapsto R$$

such that $R(a) = V$, $R(x) = U(x) \otimes W$ if $x \in \Delta$, $R(\alpha) = f$ if α is the arrow $a \to b_1$ of $\tilde{\Delta}$ and $R(\beta) = U(\beta) \otimes W$ if β is an arrow of Δ.

Theorem. *The fully faithful functor $U^{\otimes}$ maps postprojectives onto postprojectives, preinjectives onto preinjectives and $\operatorname{rep}_t U(b_1)$ into the category $\operatorname{rep}_t \tilde{\Delta}$ of tubular representations of $\tilde{\Delta}$ (which admit no postprojective and no preinjective indecomposable summand).*

Furthermore, $U^{\otimes}$ maps each connected component of $\Gamma_{U(b_1)}$ into a connected component of $\Gamma_{\tilde{\Delta}}$ and induces a bijection between the sets of connected components of $\Gamma_{U(b_1)}$ and of $\Gamma_{\tilde{\Delta}}$.

If we except three lines of $U(b_1)$, $U^{\otimes}$ induces an isomorphism of the component Γ_L of $\Gamma_{U(b_1)}$ "with eigenline L" onto the associated connected component of $\Gamma_{\tilde{\Delta}}$.

11.6. Tubular Representations of $\tilde{E}_n (n = 6, 7, 8)$ and $\tilde{D}_n (n \geqslant 4)$. Let us first examine $\tilde{E}_6$ and direct all its edges towards e. In this case, $U(e)$ has dimension 3 and is depicted in Fig. 4 (for each arrow $x \xrightarrow{\alpha} y$ of E_6, $U(\alpha)$ is an inclusion, and x stands for $U(x)$).

Each line $L \subset U(b_1)$ provides a sequence of representations $U^{\otimes}(T_{rL})$, $r \geqslant 1$, of $\tilde{E}_6$. In particular, $U^{\otimes}(T_{1L})$ is identified with the representation of Fig. 5, which we intend to examine in detail in the three exceptional cases $L_0 = U(c_2) \cap U(b_1)$, $L_1 = (U(c_1) + U(d_1)) \cap U(b_1)$ and $L_\infty = U(d_2) \cap U(b_1)$.

$U^{\otimes}(T_{1L_0})$ has exactly 3 proper subrepresentations $0 = U_0^0 \subset U_1^0 \subset U_2^0$ which belong to $\operatorname{rep}_t \tilde{E}_6$. Their stalks at e are $\{0\} \subset L_0 \subset U(c_2)$; the dimension-functions of the associated quotients are $01\overset{\substack{1\\1}}{1}00$, $11\overset{\substack{0\\0}}{1}10$ and $00\overset{\substack{0\\1}}{1}11$. In the same way,

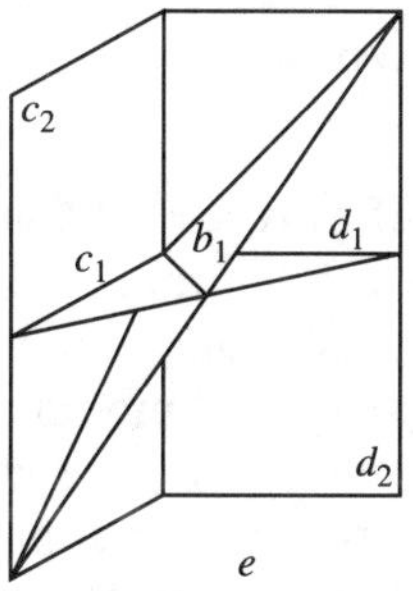

Fig. 4

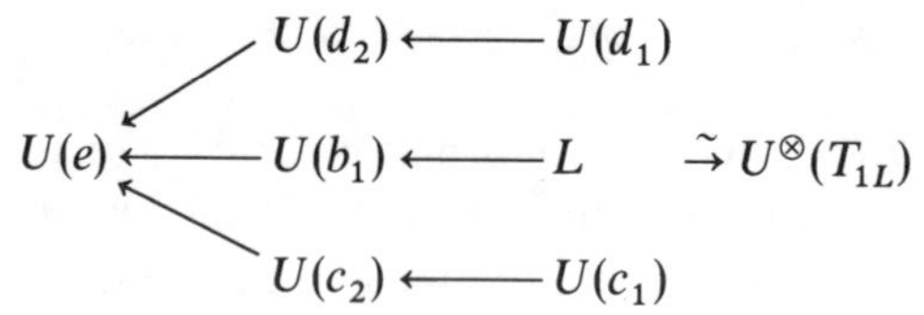

Fig. 5

$U^{\otimes}(T_{1L_\infty})$ has 3 proper subrepresentations $0 = U_0^\infty \subset U_1^\infty \subset U_2^\infty$ in $\operatorname{rep}_t \tilde{E}_6$. Their stalks at e are $\{0\} \subset L_\infty \subset U(d_2)$; the dimension-functions of the associated quotients are $00\overset{1}{\overset{1}{1}}10$, $01\overset{0}{\overset{0}{1}}11$ and $11\overset{0}{\overset{1}{1}}00$. Finally, $U^{\otimes}(T_{1L_1})$ has 2 proper subrepresentations $0 = U_0^1 \subset U_1^1$ in $\operatorname{rep}_t \tilde{E}_6$; the stalks at e are $\{0\} \subset U(c_1) + U(d_1)$ and the dimension-functions of the quotients are $11\overset{1}{\overset{1}{2}}11$ and $01\overset{0}{\overset{1}{1}}10$.

In the *general case* of a quiver $\tilde{\Delta}$ of type $\tilde{E}_n$ or $\tilde{D}_n$, we still denote by L_0, L_1, L_∞ the 3 exceptional lines (11.5). For each $i \in \{0, 1, \infty\}$, the subrepresentations of $U^{\otimes}(T_{rL_i})$ within $\operatorname{rep}_t \tilde{\Delta}$ are then linearly ordered by inclusion. We denote by $0 = U_0^i \subset U_1^i \subset \cdots \subset U_{r_i}^i = U^{\otimes}(T_{1L_i})$ the tubular subrepresentations of $U^{\otimes}(T_{1L_i})$.

Theorem. *The representations $U^{\otimes}(T_{rL})$, where L is not exceptional, and the non-zero (linearly ordered) tubular subrepresentations of*

$$U^{\otimes}(T_{1L_i})/U_j^i \subset U^{\otimes}(T_{2L_i})/U_j^i \subset \cdots \subset U^{\otimes}(T_{rL_i})/U_j^i \overset{\bullet}{\subset} \ldots,$$

where $i \in \{0, 1, \infty\}$ and $0 \leqslant j < r_i$, constitute an exhaustive list of pairwise non-isomorphic indecomposables of $\operatorname{rep}_t \tilde{\Delta}$.

If the line L of $U(b_1)$ is not exceptional, we know by Theorem 11.5 that the connected component of $\Gamma_{\tilde{\Delta}}$ formed by the indecomposables $U^{\otimes}(T_{rL})$ is a "tube" isomorphic to Γ_λ (Fig. 2). In case $L = L_i$, the connected component containing the $U^{\otimes}(T_{rL_i})$ consists of the tubular subrepresentations of $U^{\otimes}(T_{rL_i})/U_j^i$ and is a "*tube*" akin to Γ_0 (Fig. 2). The "*mouth*" of this tube consists of the r_i quotients U_j^i/U_{j-1}^i.

Concerning morphisms between indecomposable representations, we may repeat the account given in 11.3. In particular, morphisms between tubular indecomposables are described by $k_\tau \Gamma_{\tilde{\Delta}}$.

11.7. Sketch of the Proof[4] of Theorem 11.5 and Theorem 11.6. Let $\tilde{\Delta}$ denote an extended Dynkin quiver (with shifting orientation in the case of $\tilde{A}_n$) and $\mathscr{A} = \operatorname{rep} \tilde{\Delta}$ the category of its finite-dimensional representations.

a) If $\underline{\mathscr{A}}$ is the residue-category of $\mathscr{A}$ modulo injectives (9.2), the projection-functor $\mathscr{A} \to \underline{\mathscr{A}}$ has a right adjoint $\underline{\mathscr{A}} \to \mathscr{A}$ which maps N onto its largest *injective-free* (=without non-zero injective summand) quotient[5] $\underline{N}$. By duality, $\mathscr{A} \to \overline{\mathscr{A}}$ (9.2) has a left adjoint which maps $M \in \overline{\mathscr{A}}$ onto its largest *projective-free* subrepresentation $\overline{M}$.

b) By 9.6, the translation $\bar\tau$ of $\Gamma_{\tilde\Delta}$ and the inverse map τ extend to a pair of quasi-inverse functors $\bar\tau\colon \underline{\mathscr{A}} \to \overline{\mathscr{A}}$ and $\tau\colon \overline{\mathscr{A}} \to \underline{\mathscr{A}}$. We thus obtain a diagram

$$\mathscr{A} \underset{\underline{?}}{\overset{\text{can}}{\rightleftarrows}} \underline{\mathscr{A}} \underset{\tau}{\overset{\bar\tau}{\rightleftarrows}} \overline{\mathscr{A}} \underset{can}{\overset{\bar ?}{\rightleftarrows}} \mathscr{A},$$

which gives rise to a functor[6] $C\colon \mathscr{A} \to \mathscr{A}, N \mapsto \underline{\tau N}$ and to its left adjoint $\bar C\colon M \mapsto \overline{\bar\tau M}$.

c) *There is a well-defined automorphism* $c\colon \mathbb{Z}^{\tilde\Delta_v} \tilde\to \mathbb{Z}^{\tilde\Delta_v}$ *such that* $\underline{\dim}\, CM = c(\underline{\dim}\, M)$ (7.1) *for each projective-free* $M \in \operatorname{rep} \tilde\Delta$ *and* $\underline{\dim}\, \bar C N = c^{-1}(\underline{\dim}\, N)$ *for each injective-free* $N \in \operatorname{rep} \tilde\Delta$:

Indeed, the isomorphisms

$$(CM)(x) \tilde\to \mathscr{A}({}^\wedge x, CM) \tilde\to \mathscr{A}(\bar C\, {}^\wedge x, M) = \mathscr{A}(\bar\tau {}^\wedge x, M)$$

enable us to reckon the dimension of CM at $x \in \tilde\Delta$. To be precise, we examine the case of $\tilde E_6$, directing all the arrows towards e. The left end of $\Gamma_{\tilde\Delta}$ then looks as in Fig. 6 (see 10.2), where $v \in \tilde\Delta$ simply stands for ${}^\wedge v$. The "mesh" of $\Gamma_{\tilde\Delta}$ starting at e is associated with an almost split sequence which induces an exact sequence

$$0 \to \mathscr{A}(\bar\tau e, M) \to \mathscr{A}(b_1 \oplus c_2 \oplus d_2, M) \to \mathscr{A}(e, M) \to 0$$

for each projective-free M (9.3 vi). Setting $x = \underline{\dim}\, M$ and $y = \underline{\dim}\, CM$, we thus obtain $y(e) = x(b_1) + x(c_2) + x(d_2) - x(e)$. Similarly, the meshes starting at c_2 and c_1 lead to $y(c_2) = y(e) + x(c_1) - x(c_2) = x(b_1) + x(c_1) + x(d_2) - x(e)$ and $y(c_1) = y(c_2) - x(c_1) = x(b_1) + x(d_2) - x(e) \ldots.$ Thus $y(v)$ is the value at $\bar\tau v$ of the obvious additive (10.2) extension of x to the postprojective component of $\Gamma_{\tilde\Delta}$.

d) Let $d^{\tilde\Delta}$ denote the positive generator of the null space of the unit form $q_{\tilde\Delta}$ in $\mathbb{Z}^{\tilde\Delta_v}$ (7.1, 6.3). *Then we have* $c(d^{\tilde\Delta}) = d^{\tilde\Delta}$, *and there is an* $h \geqslant 1$ *and a* $\varphi\colon \mathbb{Z}^{\tilde\Delta_v} \to \mathbb{Z}$ *such that* $c^h(x) = x + \varphi(x) d^{\tilde\Delta}$, $\forall x \in \mathbb{Z}^{\tilde\Delta_v}$: This may be considered as a matter of mere computation[7]. For instance, in the case $\tilde\Delta = \tilde E_6$ considered above, we have $d^{\tilde\Delta}(e) = 3$, $d^{\tilde\Delta}(b_1) = d^{\tilde\Delta}(c_2) = d^{\tilde\Delta}(d_2) = 2$ and $d^{\tilde\Delta}(a) = d^{\tilde\Delta}(c_1) = d^{\tilde\Delta}(d_1) = 1$, hence $c(d^{\tilde\Delta})(e) = d^{\tilde\Delta}(b_1) + d^{\tilde\Delta}(c_2) + d^{\tilde\Delta}(d_2) - d^{\tilde\Delta}(e) = 2 + 2 + 2 - 3 = 3 \ \ldots.$ In the

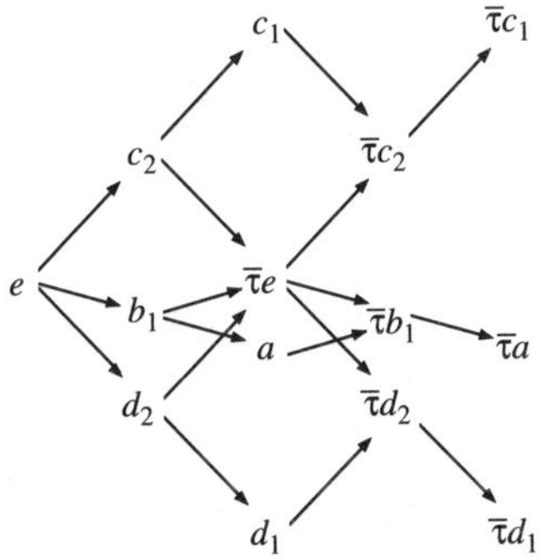

Fig. 6

same case, we further have $h = 6$, $\varphi(e^e) = -3$ and $\varphi(e^v) = 1$ for each vertex $v \neq e$ ($e^v(v) = 1$ and $e^v(w) = 0$ for $w \neq v$).

e) *An indecomposable* $V \in \operatorname{rep} \tilde{\Delta}$ *is postprojective (resp. preinjective) if and only if* $\varphi(\underline{\dim}\, V) < 0$ (resp. $\varphi(\underline{\dim}\, V) > 0$): If $V \in \Gamma_{\tilde{\Delta}}$ is postprojective, the functions $\underline{\dim}\, \bar{C}^r V$ are distinct (10.3b) and >0. It follows that $\underline{\dim}\, \bar{C}^{sh} V = c^{-sh}(\underline{\dim}\, V) = \underline{\dim}\, V - s\varphi(\underline{\dim}\, V) d^{\tilde{\Delta}}$ tends to ∞ with s, hence that $\varphi(\underline{\dim}\, V) < 0$. Conversely, $\varphi(\underline{\dim}\, V) < 0$ implies $\underline{\dim}\, V + s\varphi(\underline{\dim}\, V) d^{\tilde{\Delta}} < 0$ for large s. It follows that $\underline{\dim}\, C^{r+1} V = c(\underline{\dim}\, C^r V)$ is false for some r, hence that $C^r V$ is projective for some r.

f) *A representation* $V \in \operatorname{rep} \tilde{\Delta}$ *is tubular if and only if* $\varphi(\underline{\dim}\, V) = 0$ *and* $\varphi(\underline{\dim}\, U) \leqslant 0$ *for each subrepresentation* U, *or, equivalently, if and only if* $\varphi(\underline{\dim}\, V) = 0$ *and* $\varphi(\underline{\dim}\, W) \geqslant 0$ *for each quotient* W *of* V: The first condition is obviously sufficient since it implies that $\varphi(\underline{\dim}\, U) = 0$ if U is a summand of V. Conversely, suppose that V is tubular and that $\varphi(\underline{\dim}\, U) > 0$ for some indecomposable $U \subset V$. Then $C^{sh}(U) \subset C^{sh}(V)$ for all $s \geqslant 0$ because C is right adjoint, hence left exact. On the other hand, $\underline{\dim}\, C^{sh}(U) = \underline{\dim}\, U + s\varphi(\underline{\dim}\, U) d^{\tilde{\Delta}}$ tends to ∞, whereas $\underline{\dim}\, C^{sh}(V)$ stays constant ....

g) *The full subcategory* $\operatorname{rep}_t \tilde{\Delta}$ *formed by the tubular representations is exactly embedded*: Denote by I, K and C the image, kernel and cokernel within $\operatorname{rep} \tilde{\Delta}$ of a morphism $V \to V'$ of $\operatorname{rep}_t \tilde{\Delta}$. As a subrepresentation of V', each indecomposable summand J of I satisfies $\varphi(\underline{\dim}\, J) \leqslant 0$; as a quotient of V, it satisfies $\varphi(\underline{\dim}\, J) \geqslant 0$, hence $\varphi(\underline{\dim}\, J) = 0$. Therefore I is tubular and $\varphi(\underline{\dim}\, K) = \varphi(\underline{\dim}\, V) - \varphi(\underline{\dim}\, I) = 0$. It follows from f) that K is tubular, and so is C.

h) Thus, $\operatorname{rep}_t \tilde{\Delta}$ is an abelian pointwise finite k-category whose objects admit finite Jordan-Hölder series[8]. Let us search for the *simple objects* S of $\operatorname{rep}_t \tilde{\Delta}$ in the case $\tilde{\Delta} = \tilde{E}_6$ considered above: To this effect, we reproduce in Fig. 7 the representation-quiver of E_6 (10.2). For each vertex V of Γ_{E_6}, we indicate the numbers $\varphi(\underline{\dim}\, V^0)$ and $\dim V(b_1)$, where V^0 is the extension of V to $\tilde{E}_6$ by 0.

The restriction $S|E_6$ is isomorphic to some $\bigoplus_i V_i$, where $V_i \in \Gamma_{E_6}$. Since $V_i^0 \subset S$, we have $\varphi(\underline{\dim}\, V_i^0) \leqslant 0$ and can exclude the vertices at the right end of Γ_{E_6}. If $\varphi(\underline{\dim}\, V_i^0) = 0$, V_i^0 is tubular and $S = V_i^0$; among the 7 indecomposables of the form V_i^0, we thus obtain 5 simples of $\operatorname{rep}_t \tilde{E}_6$.

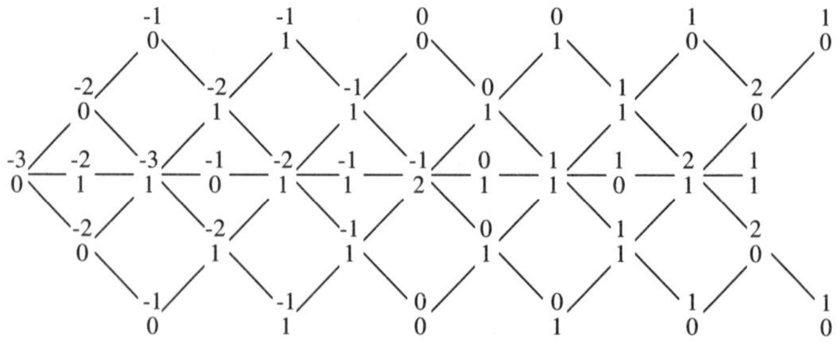

Fig. 7

Now suppose that $\varphi(\underline{\dim}\ V_i^0) < 0$ for all i, and let W_i be a maximal subrepresentation of S such that $V_i^0 \cap W_i = \{0\}$ and $W_i \supset V_j^0$ for $j \neq i$. The quotient $\overline{V}_i = S/W_i$ then satisfies $\overline{V}_i|E_6 = V_i|E_6$, $\dim \overline{V}_i(a) \leqslant \dim V_i(b_1)$ and $0 \leqslant \varphi(\underline{\dim}\ \overline{V}_i) = \varphi(\underline{\dim}\ V_i^0) + \dim \overline{V}_i(a) \leqslant \varphi(\underline{\dim}\ V_i^0) + \dim V_i(b_1)$. If $\varphi(\underline{\dim}\ V_i^0) + \dim V_i(b_1) = 0$, $\overline{V}_i$ is tubular and $S = \overline{V}_i$; with the 5 indecomposables of E_6 of the considered form V_i, we thus obtain 3 simples $\overline{V}_i$ of $\operatorname{rep}_t \tilde{E}_6$.

It remains to consider the case where $0 < -\varphi(\underline{\dim}\ V_i^0) < \dim V_i(b_1)$ for each i. In this case, all the V_i coincide with the unique vertex $U \in \Gamma_{E_6}$ such that $\dim U(b_1) = 2$. We infer that $S \xrightarrow{\sim} U^{\otimes}(k, f, k)$ for some non-zero $f: k \to U(b_1)$.

i) Finally, we apply the following lemma to the functor

$$E = [F \;\; F_0 \;\; F_1 \;\; F_\infty]: \operatorname{mod}_f A \times \operatorname{rep}_0 Q_3 \times \operatorname{rep}_0 Q_2 \times \operatorname{rep}_0 Q_3 \to \operatorname{rep}_t \tilde{E}_6$$

described by Fig. 8 ($A = k[x, x^{-1}, (1-x)^{-1}]$). By h) this functor induces a bijection between the sets of isoclasses of simples. We leave to the reader the easy homework of verifying the assumptions of the lemma below by computing the extension groups between simples of $\operatorname{rep}_t \tilde{E}_6$. $\surd$

Lemma[9]. *Let $E: \mathscr{B} \to \mathscr{C}$ be an exact functor between two abelian aggregates whose objects have finite Jordan-Hölder series. Then E is an equivalence if and only if the following two conditions are satisfied:*

a) E maps simples onto simples and induces a bijection between their sets of isoclasses.

b) For all simples $S, T \in \mathscr{B}$, the map $\operatorname{Ext}^i_{\mathscr{B}}(S, T) \to \operatorname{Ext}^i_{\mathscr{C}}(ES, ET)$ induced by E is bijective for $i = 1$ and injective for $i = 2$.

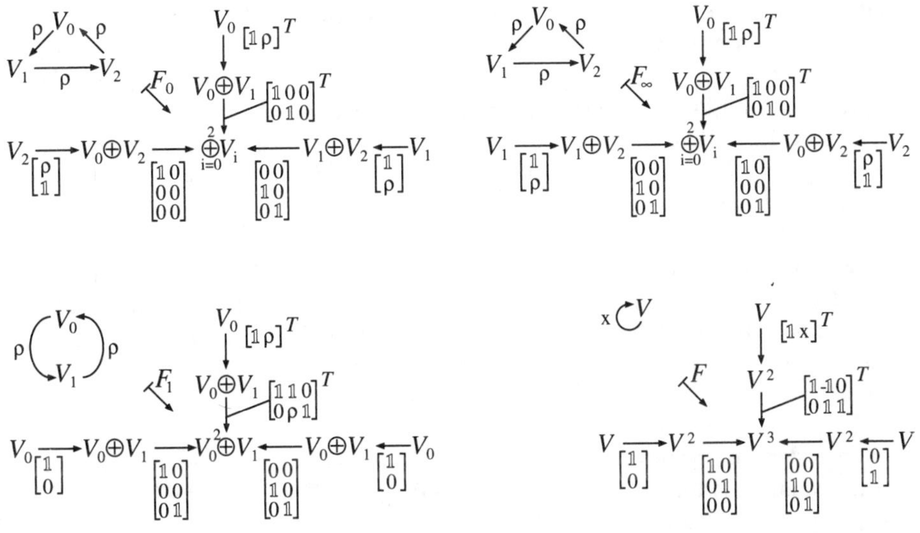

Fig. 8

11.8. Remarks and References

1. The cases $\tilde{A}_0$ and $\tilde{A}_1$ (with shifting orientation) are classical (see 1.7 and 1.8). A classification in the cases $\tilde{A}_3$ and $\tilde{D}_4$ can be found in [108, Nazarova, 1961/67] and [63, Gelfand/Ponomarev, 1970/72]. The general case is treated in [39, Donovan/Freislich, 1973] and in [109, Nazarova, 1973]. If k is not algebraically closed, see [38, Dlab/Ringel, 1973/76].
2. *tubular* = *regular* in the terminology of Gelfand-Ponomarev.
3. Let $\mathscr{A}$ be an abelian aggregate whose objects have finite lengths. With $\mathscr{A}$ we may associate a quiver $\Sigma_{\mathscr{A}}$ whose vertices are representatives of the simple objects, the number of arrows from $s \in \Sigma_{\mathscr{A}}$ to $t \in \Sigma_{\mathscr{A}}$ being equal to $\dim_k \operatorname{Ext}^1_{\mathscr{A}}(t, s)$. If the connected components of $\Sigma_{\mathscr{A}}$ are quivers $\tilde{A}_n$ with steady orientation and if $\operatorname{Ext}^2_{\mathscr{A}} = 0$, $\mathscr{A}$ is equivalent to a direct sum of categories $\operatorname{rep}_0 Q_n$ (see [50, Gabriel, 1960]).
4. The main ideas of the proof (points a) – g)) are to be found in [63, Gelfand/Ponomarev, 1970/72]. The end leans on lectures of P. Gabriel on $\tilde{D}_4$ [54, 1972].
5. The quotient is constructed within $\mathscr{A}$, which has the same objects as $\underline{\mathscr{A}}$.
6. C and $\bar{C}$ are the *Coxeter functors* of [19, Bernstein/Gelfand/Ponomarev, 1973] (see [60, Gabriel, 1979/80]).
7. In fact, c fixes the unit form $q_{\tilde{\Delta}}$ and induces a transformation $\bar{c}$ of $\mathbb{Z}^{\tilde{\Delta}_v}/\mathbb{Z}d^{\tilde{\Delta}}$ which fixes the form $\bar{q}_{\tilde{\Delta}}$ induced by $q_{\tilde{\Delta}}$ on the quotient. Since $\bar{c}$ permutes the roots of $\bar{q}_{\tilde{\Delta}}$, it satisfies $\bar{c}^h = \mathbb{1}$ for some h.
8. Compare with remark (3).
9. Such functors $E = [F \ F_0 \ F_1 \ F_\infty]$ are easily produced for all extended Dynkin quivers.

12. Derivation and Tilting[1] (by B. Keller)

Our objective is to define the (bounded) derived category of a spectroid and to show that two spectroids with the "same" derived category also share many other invariants, among which the Grothendieck group is most important. This naturally leads to the question of when two spectroids are "derivatively equivalent". The answer is provided in 12.6 by a theorem relating equivalences between derived categories and "tilting spectroids". As an application, we determine the representation quiver of a critical arboresque spectroid.

In Sect. 12.1–12.3, $\mathscr{A}$ will denote an *exact category with splitting idempotents, enough projectives and enough injectives*[2]. From 12.5 onwards, *k is an algebraically closed field.*

12.1. *The homotopy category* $\mathscr{H}\mathscr{A}$ is the residue category of the category $\mathscr{C}\mathscr{A}$ of *differential complexes*[3] modulo the ideal of *null-homotopic*[4] morphisms. The *suspension functor* S: $\mathscr{H}\mathscr{A} \to \mathscr{H}\mathscr{A}$ is defined by $(SK)_n = K_{n-1}$, $d_n^{SK} = -d_{n-1}^K$. A *triangle* of $\mathscr{H}\mathscr{A}$ is an *S-sequence*[5] isomorphic to some

$$K \xrightarrow{\bar{f}} L \xrightarrow{\bar{g}} M \xrightarrow{\bar{h}} SK,$$

where $\bar{m}$ denotes the residue class of a morphism m, all (f_n, g_n) are split conflations (9.4, Example 1) and h satisfies $h_n = r_{n-1} d_n^L s_n$ for all n and some[6] s_n, r_n with $g_n s_n = \mathbb{1}$, $r_n f_n = \mathbb{1}$ and $r_n s_n = 0$. Thus, the triangles of $\mathscr{H}\mathscr{A}$ keep track of the pointwise split conflations of $\mathscr{C}\mathscr{A}$. The *mapping cone construction*[7] shows that each morphism $\bar{e}: K \to N$ can be completed to a triangle $K \xrightarrow{\bar{e}} N \to Ce \to SK$. We shall need the following property of the "triangulated[8] category" $\mathscr{H}\mathscr{A}$:

Lemma. *For each triangle $K \to L \to M \to SK$ and each $C \in \mathscr{H}\mathscr{A}$, the induced sequences*

$$\cdots \to (\mathscr{H}\mathscr{A})(C, S^nK) \to (\mathscr{H}\mathscr{A})(C, S^nL) \to (\mathscr{H}\mathscr{A})(C, S^nM) \to (\mathscr{H}\mathscr{A})(C, S^{n+1}K) \to \cdots$$

$$\cdots \leftarrow (\mathscr{H}\mathscr{A})(S^nK, C) \leftarrow (\mathscr{H}\mathscr{A})(S^nL, C) \leftarrow (\mathscr{H}\mathscr{A})(S^nM, C) \leftarrow (\mathscr{H}\mathscr{A})(S^{n+1}K, C) \leftarrow \cdots$$

are exact.

It follows immediately that a morphism $\bar{e}$ is invertible if and only if Ce is null-homotopic[4].

12.2. Resolution. Let $\mathscr{P}$ (resp. $\mathscr{I}$) denote the full subcategory of the projectives (resp. the injectives) of $\mathscr{A}$ and $\mathscr{H}_+\mathscr{A}$ (resp. $\mathscr{H}_-\mathscr{A}$) the full subcategory of the *right bounded*[9] (resp. *left bounded*) complexes of $\mathscr{H}\mathscr{A}$. Let A be an *acyclic* complex, i.e. a complex admitting conflations

$$Z_n \xrightarrow{i_n} A_n \xrightarrow{p_n} Z_{n-1}$$

such that $d_n^A = i_{n-1}p_n$, $\forall n$.[10]

Lemma[11]. a) *We have $(\mathscr{H}\mathscr{A})(P, A) = 0 = (\mathscr{H}\mathscr{A})(A, I)$ for all $P \in \mathscr{H}_+\mathscr{P}$, $I \in \mathscr{H}_-\mathscr{I}$.*

b) *There is*[12] *a functor which maps each complex $K \in \mathscr{H}_+\mathscr{A}$ to a triangle $\mathbf{p}K \to K \to \mathbf{a}_+K \to S\mathbf{p}K$ in $\mathscr{H}_+\mathscr{A}$ such that $\mathbf{a}_+K$ is acyclic, and $\mathbf{pp}K = \mathbf{p}K \in \mathscr{H}_+\mathscr{P}$, $\mathbf{p}SK = S\mathbf{p}K$.*

c) *There is*[12] *a functor which maps each complex $K \in \mathscr{H}_-\mathscr{A}$ to a triangle $\mathbf{a}_-K \to K \to \mathbf{i}K \to S\mathbf{a}_-K$ in $\mathscr{H}_-\mathscr{A}$ such that $\mathbf{a}_-K$ is acyclic, and $\mathbf{ii}K = \mathbf{i}K \in \mathscr{H}_-\mathscr{I}$, $\mathbf{i}SK = S\mathbf{i}K$.*

12.3. The *derivative* $\mathscr{D}\mathscr{A}$ is the category with the same objects as $\mathscr{H}_b\mathscr{A} := \mathscr{H}_+\mathscr{A} \cap \mathscr{H}_-\mathscr{A}$ whose morphism spaces $(\mathscr{D}\mathscr{A})(K, L)$ are identified with $(\mathscr{H}\mathscr{A})(\mathbf{p}K, \mathbf{i}L)$. The composition of $\bar{a}: \mathbf{p}K \to \mathbf{i}L$ and $\bar{b}: \mathbf{p}L \to \mathbf{i}M$ is $\overline{ba'} = \overline{b'}\bar{a}$, where $\overline{a'}$ and $\overline{b'}$ belong to the commutative diagram

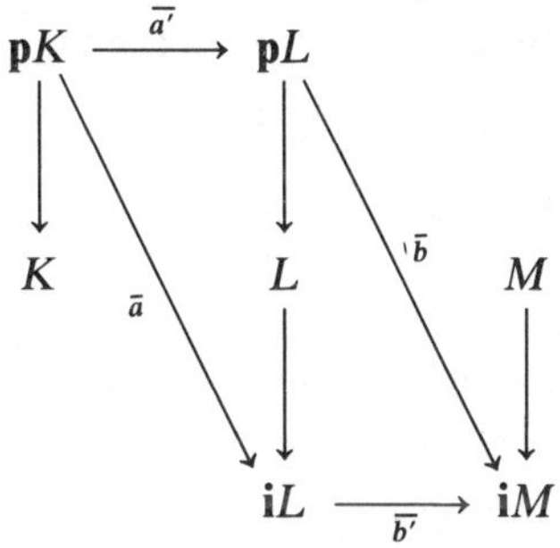

provided by 12.2 and 12.1. We call *triangles* of $\mathscr{D}\mathscr{A}$ the S-sequences isomorphic to images of triangles under *can*: $\mathscr{H}_b\mathscr{A} \to \mathscr{D}\mathscr{A}$. Via the fully faithful (!) embedding

$A \mapsto (\dots 0 \to A \to 0 \dots)$, we identify $\mathcal{A}$ with a subcategory of $\mathcal{D}\mathcal{A}$ and set

$$\mathrm{Ext}^n_{\mathcal{A}}(A, B) = (\mathcal{D}\mathcal{A})(A, S^n B), \qquad \forall A, B \in \mathcal{A}, \qquad n \in \mathbb{N}.$$

Remarks a) By construction, $\mathbf{p}|\mathcal{H}_b\mathcal{A}$ admits a unique and fully faithful factorization $\bar{\mathbf{p}}$: $\mathcal{D}\mathcal{A} \to \mathcal{H}_+\mathcal{A}$ through *can*: $\mathcal{H}_b\mathcal{A} \to \mathcal{D}\mathcal{A}$. In particular, an object $X \in \mathcal{H}_b\mathcal{A}$ vanishes in $\mathcal{D}\mathcal{A}$ if and only if $\mathbf{p}X = 0$, which by 12.1 and 12.2 means that X is acyclic.

b) A morphism $\bar{s}$ of $\mathcal{H}_b\mathcal{A}$ becomes invertible in $\mathcal{D}\mathcal{A}$ if and only if $(\mathcal{H}\mathcal{A})(P, \bar{s})$ is invertible for all $P \in \mathcal{H}_+\mathcal{P}$, which by 12.2 and 12.1 means that Cs is acyclic.

c) Let $X \xrightarrow{j} Y \xrightarrow{q} Z$ be a pair of morphisms of bounded complexes such that (j_n, q_n) is a conflation of $\mathcal{A}$ for each n. It is then easy to see that the mapping cone over the morphism

$$s\colon Cj \to Z, \qquad s_n = [q_n \;\; 0]\colon Y_n \oplus X_{n-1} \to Z_n$$

is acyclic. Hence $\bar{s}$ becomes invertible in $\mathcal{D}\mathcal{A}$ and (j, q) gives rise to a triangle of $\mathcal{D}\mathcal{A}$, the *derived triangle*.

d) We say that $A \in \mathcal{A}$ has *projective dimension at most n* (pdim $A \leqslant n$) if there is an acyclic complex

$$0 \to P_n \to P_{n-1} \to \cdots \to P_1 \to P_0 \to A \to 0$$

with projective P_i. If each $A \in \mathcal{A}$ has finite projective dimension, the embedding $\mathcal{H}_b\mathcal{P} \to \mathcal{D}\mathcal{A}$ is an equivalence.

12.4. The Hereditary Case. Suppose that $\mathcal{A}$ is a hereditary (8.2) abelian category (9.1, Example 2). Let P be a bounded complex of projectives with $P_n = 0$ for all $n < 0$. Since $\mathrm{Im}\, d_1^P \subset P_0$ is projective, the morphism of complexes

$$\begin{array}{ccccccccc}
\cdots \longrightarrow & P_2 & \longrightarrow & P_1 & \longrightarrow & P_0 & \longrightarrow & 0 & \longrightarrow \cdots \\
 & \downarrow & & \downarrow & & \downarrow & & \downarrow & \\
\cdots \longrightarrow & 0 & \longrightarrow & \mathrm{Im}\, d_1^P & \subset & P_0 & \longrightarrow & 0 & \longrightarrow \cdots
\end{array}$$

is a retraction. By induction on the number of non-vanishing components of P, we see that P is a direct sum of complexes of the form

$$R = (\cdots \to 0 \to R_{n+1} \subset R_n \to 0 \cdots) \in \mathcal{H}_b\mathcal{P}.$$

In $\mathcal{D}\mathcal{A}$, we have $R \xrightarrow{\sim} S^n H_n R$[13] by 12.3 b). Since $\mathcal{H}_b\mathcal{P} \xrightarrow{\sim} \mathcal{D}\mathcal{A}$ (12.3 d), we infer that each indecomposable of $\mathcal{D}\mathcal{A}$ is of the form $S^n U$ for some indecomposable U of $\mathcal{A}$ and some $n \in \mathbb{Z}$. In the following examples, we consider categories of the form $\mathcal{A} = \mathrm{mod}\, kQ/\mathcal{L}$, where Q is a quiver, $\mathcal{L}$ an admissible ideal and k a field. If $\mathcal{L} = 0$, $\mathcal{A}$ is hereditary (8.2).

Example 1. For $\mathcal{L} = 0$ and

$$Q = \vec{A}_n\colon 1 \to 2 \to \cdots \to n-1 \to n,$$

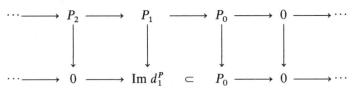

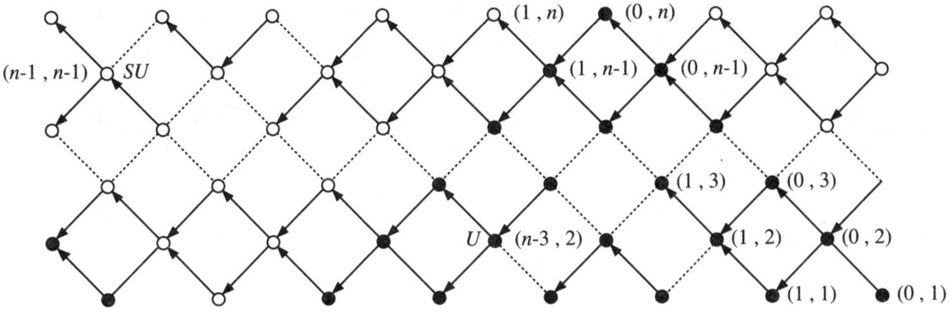

Fig. 1. $\mathbb{Z}\vec{A}_n$

we represent the quiver of $\mathscr{D}\mathscr{A}$ by the infinite stripe $\mathbb{Z}\vec{A}_n$ of Fig. 1. The point $(0, h)$ represents the indecomposable projective $h^\wedge$. The points (g, h) with $g \geqslant 0$, $h \geqslant 1$ and $g + h \leqslant n$ represent the indecomposables $(g + h)^\wedge/g^\wedge$ (we set $0^\wedge = 0$). The suspension functor of $\mathscr{D}\mathscr{A}$ induces the "shifting reflection" $(g, h) \mapsto (g + h, n + 1 - h)$. The indecomposables $S^n U$, $U \in \mathscr{I}\mathscr{A}$, are marked with • for even n, by ∘ for odd n.

Example 2. For $\mathscr{L} = 0$ and

$$Q: 1 \rightarrow 2 \rightarrow 3 \leftarrow 4 \leftarrow 5 \rightarrow 6,$$

we obtain the "same" derivative as for $\vec{A}_6$. In general, changing the direction of the arrows in a Dynkin quiver yields an equivalent derivative[14], as we shall see in 12.8, Example 1. Of course, $\mathscr{I}\mathscr{A}$ here corresponds to a different "piece" of $\mathbb{Z}\vec{A}_6$. See Fig. 2, where the indecomposable projectives are numbered.

Example 3. We consider the quiver

$$Q: 3 \leftarrow 2 \leftarrow 1 \xrightarrow{\alpha} 4 \xrightarrow{\beta} 5$$

and the ideal $\mathscr{L}$ generated by $\beta\alpha$. We obtain the "same" derivative as for $\vec{A}_5$ (cf. 12.6, Example). Here, however, the objects $S^n U$, $U \in \mathscr{I}\mathscr{A}$, $n \in \mathbb{Z}$, do not exhaust

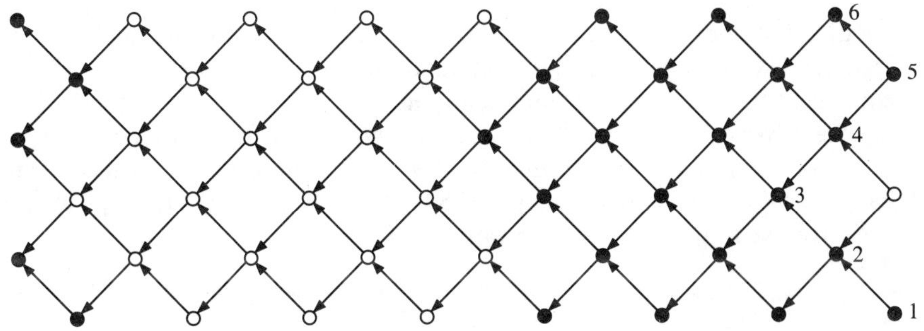

Fig. 2. $\mathbb{Z}\vec{A}_6$

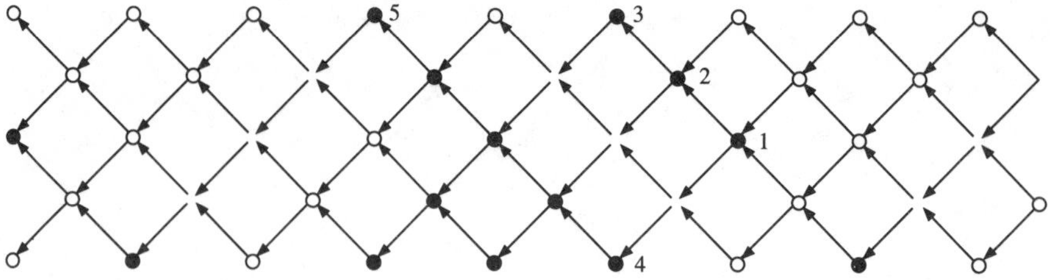

Fig. 3. $\mathbb{Z}\vec{A}_5$

the indecomposables of $\mathcal{D}\mathcal{A}$. The unmarked points in Fig. 3 represent indecomposables with non-vanishing homology in two degrees.

12.5. Subordinate Invariants. We call two spectroids $\mathcal{S}$, $\mathcal{T}$ (over the fixed algebraically closed field k) *derivatively equivalent* if the derivatives $\mathcal{D}\mathcal{S} = \mathcal{D} \bmod \mathcal{S}$ and $\mathcal{D}\mathcal{T}$ are S-equivalent[15]. In the sequel, we suppose that $\mathcal{S}$ and $\mathcal{T}$ are finite and denote by $G\colon \mathcal{D}\mathcal{T} \leftrightarrows \mathcal{D}\mathcal{S}\colon F$ a pair of quasi-inverse S-equivalences.

a) *G gives rise to an S-equivalence* $\mathcal{H}_b \operatorname{pro} \mathcal{T} \to \mathcal{H}_b \operatorname{pro} \mathcal{S}$ [16]: We use the following characterization[17] of $\mathcal{H}_b \operatorname{pro} \mathcal{S}$ within $\mathcal{D}\mathcal{S}$: An object $P \in \mathcal{D}\mathcal{S}$ is isomorphic to a bounded complex of projectives if and only if for each $L \in \mathcal{D}\mathcal{S}$, there is an n_0 with $(\mathcal{D}\mathcal{S})(P, S^n L) = 0$, $\forall n \geqslant n_0$: Indeed, in order to see that this condition suffices, we take L to be the sum of the simple modules X^-, $X \in \mathcal{S}$. The assumption then implies that, for all large n, the map $\operatorname{Hom}(Q_n/B_n, L) \leftarrow \operatorname{Hom}(Q_{n-1}, L)$ is surjective, where $Q = \mathbf{p}P$ and $B_n = \operatorname{Im} d^Q_{n+1}$. Hence $Q_n/B_n \to Q_{n-1}$ splits, which clearly implies that Q is isomorphic in $\mathcal{H} \operatorname{pro} \mathcal{S}$ to a bounded complex of projectives. $\surd$

b) By a), we may assume that $GX^\wedge$ is a bounded complex of projectives for each $X \in \mathcal{T}$ and moreover, since $\mathcal{T}$ is finite, that $(GX^\wedge)_n = 0$ for all $n < 0$. Let w be maximal among the n such that $(GX^\wedge)_n \neq 0$ for some $X \in \mathcal{T}$. *Then* gldim $\mathcal{S}$[18] *is finite if and only if so is* gldim $\mathcal{T}$. *In this case,* $|\operatorname{gldim} \mathcal{S} - \operatorname{gldim} \mathcal{T}| \leqslant w$: The isomorphisms

$$(H_n FK)(X) \tilde{\to} (\mathcal{D}\mathcal{T})(S^n X^\wedge, FK) \tilde{\to} (\mathcal{D}\mathcal{S})(S^n GX^\wedge, K), \quad K \in \mathcal{D}\mathcal{S}, \quad X \in \mathcal{T},$$

show that $F\mathcal{D}_{\geqslant w} \subset \mathcal{D}_{\geqslant 0}$ and $F\mathcal{D}_{<0} \subset \mathcal{D}_{<0}$ [19]. By forming the orthogonal subcategories[20] we obtain[21] $G\mathcal{D}_{<0} \subset \mathcal{D}_{<w}$ and $G\mathcal{D}_{\geqslant 0} \subset \mathcal{D}_{\geqslant 0}$. Since gldim $\mathcal{S} \leqslant d$ is equivalent to $\mathcal{D}_{\geqslant d} \subset \mathcal{D}^{\perp}_{<0}$ [22], our claim follows. $\surd$

c) The *Grothendieck group*[23] $K_0\mathcal{S} = K_0 \bmod \mathcal{S}$ is free with basis $[X^-]$, $X \in \mathcal{S}$. The canonical embedding $\bmod \mathcal{S} \to \mathcal{D}\mathcal{S}$ induces an isomorphism $K_0\mathcal{S} \to K_0\mathcal{D}\mathcal{S}$, whose inverse is given by $[K] \mapsto \sum_{n\in\mathbb{Z}} (-1)^n [H_n K]$. Thus, F and G give rise to mutually inverse isomorphisms

$$K_0\mathcal{S} \to K_0\mathcal{T}, [M] \mapsto \sum_{n\in\mathbb{Z}} (-1)^n [H_n FM]$$

$$K_0\mathcal{T} \to K_0\mathcal{S}, [N] \mapsto \sum_{n\in\mathbb{Z}} (-1)^n [H_n GN].$$

Now suppose that gldim $\mathcal{S} < \infty$. Then the bilinear form

$$\langle ?, ? \rangle: K_0\mathcal{S} \times K_0\mathcal{S} \to \mathbb{Z}, ([A], [B]) \mapsto \sum_{n \in \mathbb{Z}} (-1)^n \dim_k \operatorname{Ext}^n_{\mathcal{S}}(A, B)$$

is well defined. It is associated with the form

$$K_0\mathcal{D}\mathcal{S} \times K_0\mathcal{D}\mathcal{S} \to \mathbb{Z}, ([K], [L]) \mapsto \sum_{n \in \mathbb{Z}} (-1)^n \dim_k(\mathcal{D}\mathcal{S})(K, S^n L).$$

Therefore, the isomorphisms induced by F and G respect the bilinear forms on $K_0\mathcal{S}$ and $K_0\mathcal{T}$. The fundamental importance of these invariants is already apparent in the case where $\mathcal{S}$ is the k-category of paths of a quiver Q: Then $\langle [V], [V] \rangle$ is identified with the form $q_Q(\underline{\dim}\, V)$ of 7.1.

d) By c), the cardinality $|\mathcal{S}|$ of the set of points of $\mathcal{S}$ equals the rank of $K_0\mathcal{S}$. We conclude that $|\mathcal{S}| = |\mathcal{T}|$.

12.6. Derivatively Equivalent Spectroids. Let $\mathcal{S}$, $\mathcal{T}$ be finite spectroids over k. By 12.5 a), an S-equivalence $G: \mathcal{D}\mathcal{T} \to \mathcal{D}\mathcal{S}$ gives rise to an S-equivalence $\bar{G}$: $\mathcal{H}_b \operatorname{pro} \mathcal{T} \to \mathcal{H}_b \operatorname{pro} \mathcal{S}$, hence to a fully faithful $E: \mathcal{T} \to \mathcal{H}_b \operatorname{pro} \mathcal{S}$, $X \mapsto \bar{G}X^\wedge$, which obviously enjoys the following properties:

a) $(\mathcal{D}\mathcal{S})(EX, S^n EY) = 0$, $\forall n \neq 0$, $\forall X, Y \in \mathcal{T}$.

b) $\mathcal{H}_b \operatorname{pro} \mathcal{S}$ is contained in the smallest full subcategory of $\mathcal{D}\mathcal{S}$ containing the EX, $X \in \mathcal{T}$, stable under S, S^{-1} and closed under extensions[24].

Theorem.[25] *If $E: \mathcal{T} \to \mathcal{H}_b$ pro $\mathcal{S}$ is a fully faithful embedding satisfying* a) *and* b) *above, there is an S-equivalence $G: \mathcal{D}\mathcal{T} \to \mathcal{D}\mathcal{S}$ and a natural isomorphism $GX^\wedge \xrightarrow{\sim} EX$, $X \in \mathcal{T}$.*

Since an S-equivalence induces an isomorphism of the Grothendieck groups, conditions a) and b) imply $|\mathcal{T}| = |\mathcal{S}|$. Therefore each spectroid $\mathcal{T}$ "sharing" its derivative with $\mathcal{S}$ is given by a *tilting spectroid*, i.e. a full subcategory $\mathcal{X}$ of $\mathcal{D}\mathcal{S}$ formed by $|\mathcal{S}|$ pairwise non-isomorphic indecomposables of $\mathcal{D}\mathcal{S}$ such that $(\mathcal{D}\mathcal{S})(X, S^n Y) = 0$, $\forall n \neq 0$, $\forall X, Y \in \mathcal{X}$. Conversely, under the hypothesis that $\mathcal{S}$ is hereditary, one can show[26] that each tilting spectroid of $\mathcal{D}\mathcal{S}$ is derivatively equivalent to $\mathcal{S}$.

Example.[27] We describe the tilting spectroids in $\mathcal{D}k\vec{A}_n$ as sets of vertices of the quiver $\mathbb{Z}\vec{A}_n$ (12.4). We call *bifurcate quiver* an oriented finite tree K whose set of arrows is decomposed into a class of *α-arrows* and a class of *β-arrows* such that each vertex of K is the head of at most one α-arrow and one β-arrow and the tail of at most one α-arrow and one β-arrow. For each vertex x of such a K, let x_α (resp. ${}^\alpha x$) be the number of vertices y of K such that the shortest "walk" from x to y begins with an α-arrow with head (resp. tail) x. Correspondingly, we define x^β and ${}_\beta x$. Then there is exactly one map of the underlying sets of vertices $K_0 \to (\mathbb{Z}\vec{A}_n)_0$, $x \mapsto (gx, hx)$ such that

a) $\min_{x \in K} gx = 0$,

b) $(gy, hy) = (gx, hx + x^\beta + {}_\beta y + 1)$ for each α-arrow $x \xrightarrow{\alpha} y$ and

c) $(gy, hy) = (gx + x_\alpha + {}^\alpha y + 1, hx - x_\alpha - {}^\alpha y - 1)$ for each β-arrow $x \xrightarrow{\beta} y$. This map is well defined because $hx = 1 + {}_\beta x + x_\alpha$. Let $\mathcal{X}_K$ denote its image. *The assignment* $K \mapsto \mathscr{X}_K$ *induces a bijection from the isoclasses of bifurcate quivers with n points to the tilting spectroids* $\mathscr{X}$ *in* $\mathscr{I}\mathscr{D}k\vec{A}_n = \mathbb{Z}\vec{A}_n$ *with* $\min_{(g,h)\in\mathscr{X}} g = 0$. *Moreover, the spectroid associated with* $\mathscr{X}_K$ *is defined by the quiver* K *and all possible relations* $\alpha\beta = 0$ *and* $\beta\alpha = 0$. An example of a bifurcate quiver is

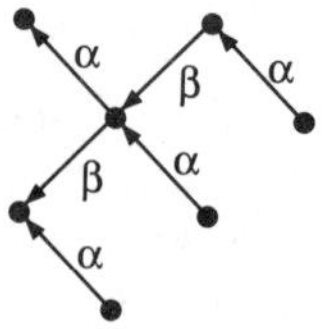

The points of $\mathscr{X}_K$ are marked with • in Fig. 4.

12.7. Equivalences Induced on Pieces. We continue the analysis begun in 12.5 using the same notations. Set $F^n = H_{-n}F|\text{mod}\,\mathscr{S}$ and $G_n = H_nG|\text{mod}\,\mathscr{T}$. Clearly, $F^n = 0 = G_n$ for $n < 0$ or $n > w$. By 12.3 a), an $M \in \text{mod}\,\mathscr{S}$ (resp. $N \in \text{mod}\,\mathscr{T}$) vanishes if $F^nM = 0$ (resp. $G_nN = 0$), $\forall n$. An exact sequence $0 \to M' \to M \to M'' \to 0$ of mod $\mathscr{S}$ gives rise to a triangle $M' \to M \to M'' \to SM'$ of $\mathscr{D}\mathscr{S}$ (12.3c), hence to a long exact sequence $\cdots \to F^nM' \to F^nM \to F^nM'' \to F^{n+1}M' \to \cdots$ and similarly for exact sequences of mod $\mathscr{T}$ and the G_n. This makes it clear that *the subcategories*

$$\mathscr{A}_n = \{M \in \text{mod}\,\mathscr{S} \colon F^iM = 0, \forall i \neq n\}$$

$$\mathscr{B}_n = \{N \in \text{mod}\,\mathscr{T} \colon G_iN = 0, \forall i \neq n\}$$

are closed under extensions, that *they vanish for* $n < 0$ or $n > w$ and that $\mathscr{A}_w$, $\mathscr{B}_0$ *are closed under submodules and* $\mathscr{A}_0$, $\mathscr{B}_w$ *under quotients*. Moreover, since $F^n|\mathscr{A}_n \xrightarrow{\sim} S^nF|\mathscr{A}_n$ and $G_n|\mathscr{B}_n \xrightarrow{\sim} S^{-n}G|\mathscr{B}_n$, *the functors* F^n *and* G_n *induce quasi-inverse equivalences* $\mathscr{A}_n \xrightarrow{\sim} \mathscr{B}_n$, $\forall n$.

Let us suppose that $\mathscr{S}$ *is hereditary*[18]. By 12.4, an indecomposable of $\mathscr{D}\mathscr{S}$ has non-zero homology in just one degree. Therefore, *each indecomposable of*

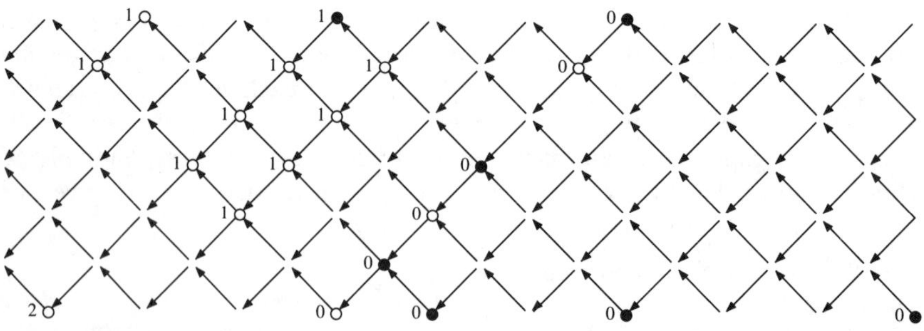

Fig. 4. A tilting spectroid in $\mathbb{Z}\vec{A}_7$

mod $\mathcal{T}$ *is contained in one of the* $\mathcal{B}_n \tilde{\rightarrow} \mathcal{A}_n$. Thus, as an additive category, mod $\mathcal{T}$ is made up of "pieces" of the category mod $\mathcal{S}$. Whence the terminology that mod $\mathcal{T}$ is *piecewise hereditary*. Observe that $\mathcal{T}$ is finitely represented if and only if there occur only finitely many isoclasses of indecomposables in each $\mathcal{A}_n$; for example, if $\mathcal{S}$ is finitely represented.

Example. In the example of 12.6, $\mathcal{T}$ is given by K. The points corresponding to indecomposables of $\mathcal{B}_n$ are marked with n in Fig. 4.

12.8. Tilting Modules. In addition to the assumptions of 12.6, we *suppose for simplicity that* gldim $\mathcal{S} < \infty$. We consider a tilting spectroid $\mathcal{T}$ in $\mathcal{D}\mathcal{S}$ such that each $X \in \mathcal{T}$ is an $\mathcal{S}$-module with pdim $X \leqslant 1$. One can show[28] that for each $Y \in \mathcal{S}$ there is an exact sequence

$$0 \rightarrow Y^\wedge \rightarrow \bigoplus_{X \in \mathcal{T}} X^{\mu(X)} \rightarrow \bigoplus_{X \in \mathcal{T}} X^{\nu(X)} \rightarrow 0, \quad \mu(X), \nu(X) \in \mathbb{N}.$$

By 12.6 it follows that $\mathcal{T}$ is derivatively equivalent to $\mathcal{S}$. In this case, a pair of S-equivalences $\mathcal{D}\mathcal{T} \leftrightarrows \mathcal{D}\mathcal{S}$ can be obtained as follows: For $M \in \text{mod}\, \mathcal{S}$ we define the $\mathcal{T}$-module $H_{\mathcal{T}}M$ by $X \mapsto \text{Hom}(X, M)$, whereas for $N \in \text{mod}\, \mathcal{T}$, the $\mathcal{S}$-module $T_{\mathcal{T}}N$ is the cokernel of the morphism

$$v_N\colon \coprod_{X, Z \in \mathcal{T}} N(Z) \otimes_k \mathcal{T}(X, Z) \otimes_k X \rightarrow \coprod_{X \in \mathcal{T}} N(X) \otimes_k X$$

given by $v_n(n \otimes \varphi \otimes \sigma) = n\varphi \otimes \sigma - n \otimes \varphi(\sigma)$ (cf. 9.1). Since both, gldim $\mathcal{S}$ and gldim $\mathcal{T}$, are finite (12.5 b), there are well defined total derived functors[8] $G = LT_{\mathcal{T}}\colon \mathcal{D}\mathcal{T} \rightarrow \mathcal{D}\mathcal{S}$ and $F = RH_{\mathcal{T}}\colon \mathcal{D}\mathcal{S} \rightarrow \mathcal{D}\mathcal{T}$ sending K to $T_{\mathcal{T}}(\mathbf{p}K)$ and L to $H_{\mathcal{T}}(\mathbf{i}L)$ (apply the functors to each component) respectively. By a well-known induction argument[29], G is an S-equivalence. It is easy to see that F is right adjoint, hence quasi-inverse, to G. By definition, $\mathcal{A}_1$ consists of the $\mathcal{S}$-modules M with $H_{\mathcal{T}}M = 0$. It is not hard to show[30] that $\mathcal{A}_0$ is the subcategory of all quotients of sums of modules in $\mathcal{T}$. Moreover, $(\mathcal{A}_1, \mathcal{A}_0)$ and $(\mathcal{B}_0, \mathcal{B}_1)$ are *torsion pairs* in the respective module categories, i.e. $\mathcal{A}_0^\perp = \mathcal{A}_1$ and $\mathcal{A}_0 = {}^\perp\mathcal{A}_1$.[20]

Example 1. Suppose there is an $X \in \mathcal{S}$ such that $\mathcal{S}(Y, X) = 0$, $\forall Y \neq X$, but $\mathcal{S}(X, Y) \neq 0$ for some $Y \neq X$. Then $\bar{\tau}X^\wedge$ and the $Y^\wedge$, $Y \neq X$, form a tilting spectroid by 9.7.[31]

Example 2. We continue Example 3 of 12.4. Let $\mathcal{S} = kQ/\mathcal{L}$. The indecomposables $U \in \mathcal{I}\mathcal{D}\mathcal{S}$ with $\text{Hom}(4^\wedge, U) \neq 0$ and $\text{Hom}(U, 5^\wedge) \neq 0$ (cf. Fig. 3) form a tilting spectroid isomorphic to $k\vec{A}_5$. The subcategory $\mathcal{A}_1$ of "torsion free" modules consists of the direct sums of copies of $1^\wedge$, $2^\wedge$ and $3^\wedge$. The subcategory $\mathcal{A}_0$ of "torsion" modules consists of the direct sums of copies of indecomposables lying "to the left" of the tilting spectroid.

12.9. Tilted Spectroids. Let Q be a finite directed quiver and $\mathcal{S} = kQ$ its k-category of paths. A spectroid $\mathcal{T}$ is *tilted of type* Q if it is isomorphic to a tilting spectroid lying in mod $\mathcal{S} \subset \mathcal{D}\mathcal{S}$. We contend that this is the situation described in statement i) of the following theorem. Indeed, if $\mathscr{T}$ is a tilting spectroid in mod $\mathscr{S}$, we may take $\mathscr{S}'$ to be the image of the embedding

$$\mathscr{S} \to \operatorname{mod} \mathscr{T},\ X \mapsto F^0 X^\vee = (Y \mapsto \operatorname{Hom}_{\mathscr{S}}(Y, X^\vee)),$$

and use the formula $Y(X) \xrightarrow{\sim} D \operatorname{Hom}_{\mathscr{S}}(Y, X^\vee)$ to identify the points of $\mathscr{T}$ with the modules $M \mapsto DM(Y)$. Statement (ii) characterizes within mod $\mathscr{T}$ the spectroids $\mathscr{S}'$ arising in this way.

Theorem.[32] *Let $\mathscr{T}$ be a finite spectroid and $\mathscr{S}'$ a full finite subspectroid of* ind $\mathscr{T} = \mathscr{I}$ mod $\mathscr{T}$. *The following statements are equivalent*:

(i) *The spectroid $\mathscr{S}'$ is hereditary and the functor mapping $Y \in \mathscr{T}$ to the $\mathscr{S}'$-*module *$V \mapsto DV(Y)$ is an isomorphism of $\mathscr{T}$ onto a tilting spectroid of $\mathscr{D}\mathscr{S}'$.*

(ii) *$\mathscr{S}'$ is a tilting spectriod of $\mathscr{D}\mathscr{T}$ whose points are $\mathscr{T}$-modules of projective dimension $\leqslant 1$. Moreover, if $V \in \mathscr{S}'$ and $W \to V$ is an arrow of the representation quiver $\Gamma_{\mathscr{T}}$, then $W \xrightarrow{\sim} U$ or $W \xrightarrow{\sim} \tau U$ for some $U \in \mathscr{S}'$.*

Corollary.[33] *Let $\tilde{\Delta}$ be an extended Dynkin quiver and $\mathscr{T} = \mathscr{S}_P$ a critical arboresque spectroid associated with a pattern P of $\mathbb{N}\tilde{\Delta}$. Then $\mathscr{T}$ is tilted of type $\tilde{\Delta}$. In particular,* mod $\mathscr{T}$ *is piecewise hereditary.*

We prove the corollary and describe the representation quiver $\Gamma_{\mathscr{T}}$ when $\mathscr{T}$ is defined by the following quiver and relations (10.7, Fig. 12, entry 11)

$$\begin{array}{ccccccc} b & \longleftarrow & q & \xrightarrow{\beta} & a & \longleftarrow & d \\ & & \Big\downarrow{\scriptstyle \delta} & & \Big\downarrow{\scriptstyle \alpha} & & \\ f & \longrightarrow & c & \xrightarrow{\gamma} & e & & \end{array} \quad , \qquad \alpha\beta = \gamma\delta.$$

The same technique applies to all the spectroids listed in Figures 9, 10 and 12 of 10.7.

We start by "knitting" the postprojective component of $\Gamma_{\mathscr{T}}$ from the left to the right until we discover a full subquiver Q such that Q^{op} is isomorphic to $\tilde{E}_6$ with the orientation of Fig. 1 of 10.2. In Fig. 5, projectives are marked with ○; points of Q with □. Let $\mathscr{S}'$ be the full subspectroid of ind $\mathscr{T}$ supported by the points of Q. Using 9.4 and 10.3 we check the condition of (ii). The second condition is clear by our choice of Q. The assertion of the corollary now follows from the theorem.

For $Y \in \mathscr{T}$, denote by T_Y the $\mathscr{S}'$ module $M \mapsto DM(Y)$, $M \in \mathscr{S}'$. To determine $\Gamma_{\mathscr{T}}$, we recall from 12.8 that each indecomposable of Mod $\mathscr{T}$ is contained in one of the subcategories $\mathscr{B}_0 \xleftarrow{\sim} \mathscr{A}_0$ or $\mathscr{B}_1 \xleftarrow{\sim} \mathscr{A}_1$, where $\mathscr{A}_0 = {}^{\perp}\mathscr{A}_1$ and $\mathscr{A}_1$ consists of the $M \in \operatorname{mod} \mathscr{S}'$ with $\operatorname{Hom}(T_Y, M) = 0$, $\forall Y \in \mathscr{T}$. We read off the dimension functions of the T_Y in Fig. 5 and find that they are all preinjective. They are marked with ○ in Fig. 6. Under the equivalence $\mathscr{B}_1 \xleftarrow{\sim} \mathscr{A}_1$, the points of Q correspond to the $M^\vee$, $M \in \mathscr{S}'$, marked with □. By 11.6 the indecomposables of $\mathscr{A}_1$ are represented by $\tau e^\vee$, by the preinjective vertices of $\Gamma_{\mathscr{S}'}$ lying " to the left" of all T_Y and by all tubular and all postprojective vertices. The unmarked vertices

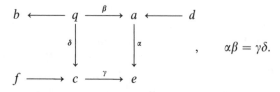

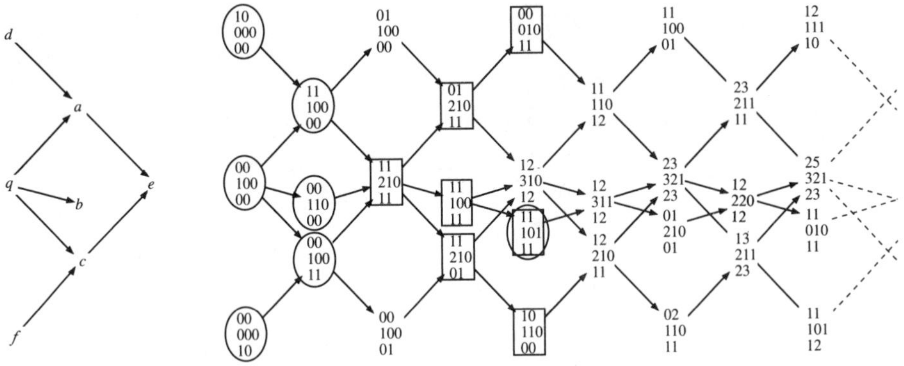

Fig. 5. The postprojective component of $\Gamma_{\mathscr{T}}$

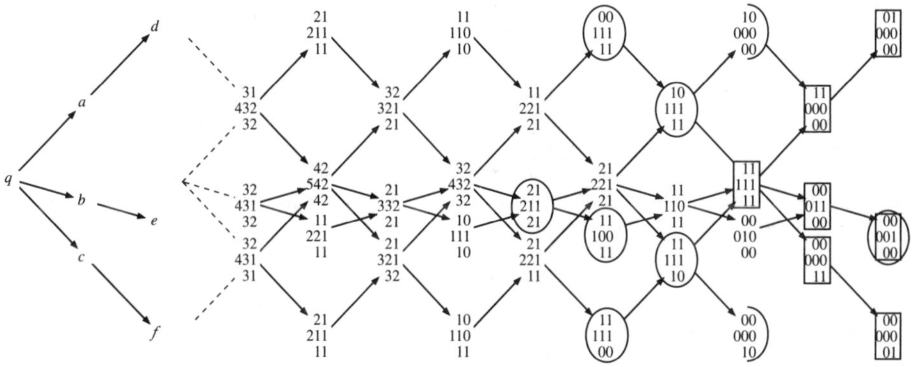

Fig. 6. The preinjective component of $\Gamma_{\mathscr{S}'}$

in Fig. 6 except $\tau q^{\vee}$ and $\tau b^{\vee}$ lie in $\mathscr{A}_1$. The indecomposables of $\mathscr{A}_0$ are marked with ○, ⊃ or □ in Fig. 6. The relations between $\Gamma_{\mathscr{S}'}$ and $\Gamma_{\mathscr{T}}$ are schematically depicted in Fig. 7. Apart from the postprojective and the preinjective component, $\Gamma_{\mathscr{T}}$ contains an isomorphic copy of the quiver of $\mathrm{rep}_t \tilde{E}_6$. The isomorphism is induced by F^1. Composing F^1 with the functor $U^{\otimes}$ of 11.5, we obtain the following functor from the category rep $\tilde{A}_1$ of pairs of maps $f_1, f_2 \colon V \rightrightarrows W$ to mod $\mathscr{T}$

$$
(f_1, f_2) \mapsto \quad
\begin{array}{ccccccc}
W & \xrightarrow{[\mathbb{1}\,0]^T} & W \oplus W & \xleftarrow{\mathbb{1}} & W \oplus W & \xrightarrow{[\mathbb{1}\,\mathbb{1}]} & W \\
 & & \Big\uparrow{\scriptstyle \mathbb{1}} & & \Big\uparrow{\scriptstyle \begin{bmatrix} f_1 \\ f_2 \end{bmatrix}} & & \\
W & \xleftarrow{[\mathbb{1}\,0]} & W \oplus W & \xleftarrow{[f_1\,f_2]^T} & V & &
\end{array}
$$

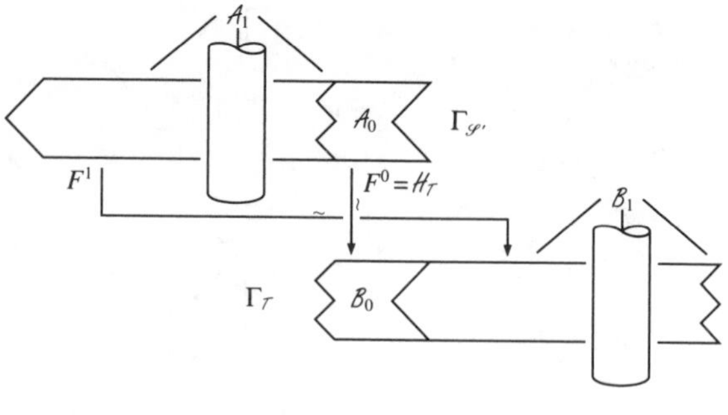

Fig. 7. Equivalent pieces of $\Gamma_{\mathcal{S}'}$ and $\Gamma_{\mathcal{T}}$

12.10. Remarks and References

1. Tilting theory is the reincarnation of Grothendieck-Roos-duality in the world of finite-dimensional algebras [90, Keller/Vossieck, 1988]. It originated with [63, Gelfand/Ponomarev, 1970/72] and subsequently became the object of thorough investigation ([19, Bernstein/Gelfand/Ponomarev, 1973], [6, Auslander/Platzeck/Reiten, 1979], [32, Brenner/Butler, 1979/80], [102, Marmaridis, 1979/80], [71, Happel/Ringel, 1980/82], [21, Bongartz, 1981]). The first work devoted to the relation between tilting and derived categories is [69, Happel, 1987]. In a wider context, derived categories serve to establish better communications to the mainland of mathematics. For instance, the coherent sheaves on Grassmann varieties and quadrics give rise to the "same" derived categories as suitable finite-dimensional algebras ([18, Bernstein/Gelfand/Gelfand, 1978], [16, Beilinson, 1978], [83, Kapranov, 1983], [84, Kapranov, 1986].

2. We suppose that $\mathcal{A}$ is equipped with a fixed exact structure $\mathcal{E}$. By "projectives" we mean $\mathcal{E}$-projectives. The category $\mathcal{A}$ has *enough projectives* if for each object A there is a deflation $P \to A$ with projective P.

3. We denote a differential complex K by

$$\cdots \longrightarrow K_{n+1} \xrightarrow{d_{n+1}^K} K_n \xrightarrow{d_n^K} K_{n-1} \longrightarrow \cdots, \quad K_n \in \mathcal{A}, \quad d_n^K d_{n+1}^K = 0, \quad n \in \mathbb{Z}$$

4. A morphism $f\colon K \to L$ is *null-homotopic*, if there is a sequence of $r_n \in \mathcal{A}(K_n, L_{n+1})$ with $f_n = d_{n+1}^L r_n + r_{n-1} d_n^K$, $\forall n$. A complex K is *null-homotopic* if so is $\mathbb{1}_K$.

5. Whenever K is a category with an endo-functor $S\colon K \to K$, an *S-sequence* is a sequence of the form $X \xrightarrow{u} Y \xrightarrow{v} Z \xrightarrow{w} SX$ and a *morphism of S-sequences* is a commutative diagram

$$\begin{array}{ccccccc} X & \xrightarrow{u} & Y & \xrightarrow{v} & Z & \xrightarrow{w} & SX \\ \downarrow{\scriptstyle a} & & \downarrow{\scriptstyle b} & & \downarrow{\scriptstyle c} & & \downarrow{\scriptstyle Sa} \\ X' & \xrightarrow{u'} & Y' & \xrightarrow{v'} & Z' & \xrightarrow{w'} & SX'. \end{array}$$

6. In fact, the homotopy class $\bar{h}$ is uniquely determined by the conflation (f, g).

7. The *mapping cone Ce* over a morphism of complexes $e\colon K \to N$ is the complex

$$(Ce)_n = N_n \oplus K_{n-1}, \qquad d_n^{Ce} = \begin{bmatrix} d_n^N & e_{n-1} \\ 0 & -d_{n-1}^K \end{bmatrix}.$$

In $\mathcal{H}\mathcal{A}$, the sequence

$$K \xrightarrow{\bar{e}} N \xrightarrow{\bar{b}} Ce \xrightarrow{\bar{c}} SK, \quad b_n = [\mathbb{1}\ \ 0]^T, \quad c_n = [0\ \ -\mathbb{1}], \quad n \in \mathbb{Z},$$

becomes isomorphic to the triangle produced by $K \xrightarrow{f} L \xrightarrow{g} Ce$, where

$$L_n = N_n \oplus K_n \oplus K_{n-1}, \quad d_n^L = d_n^N \oplus \begin{bmatrix} 0 & \mathbb{1} \\ 0 & 0 \end{bmatrix}, \quad g_n = \begin{bmatrix} \mathbb{1} & -e_n & 0 \\ 0 & d_n^K & -\mathbb{1} \end{bmatrix}, \quad f_n = \begin{bmatrix} e_n \\ \mathbb{1} \\ d_n^K \end{bmatrix}.$$

In this terminology, a morphism $f: K \to M$ of $\mathscr{C}\mathscr{A}$ is null-homotopic if it factors through the canonical morphism $K \to C\mathbb{1}_K$. If K is isomorphic to L in $\mathscr{H}\mathscr{A}$, it follows that K is isomorphic to a summand of $L \oplus C\mathbb{1}_K$ in $\mathscr{C}\mathscr{A}$.

8. [145, Verdier, 1961/77].
9. i.e. $K_n = 0$ for all $n \leqslant b$ and some $b \in \mathbb{Z}$.
10. The mapping cone $C\mathbb{1}_K$ considered in (7) is acyclic for each K. If K is isomorphic to an acyclic L in $\mathscr{H}\mathscr{A}$, it follows that K is acyclic.
11. a) is classical [36, Cartan/Eilenberg, V 1.1], b) is classical for complexes K with $K_n = 0$, $\forall n \neq 0$. In this case, we can choose as $\mathbf{a}_+ K$ an acyclic complex

$$\cdots \to P_2 \to P_1 \to K_0 \to 0 \to 0 \to \cdots$$

with projective P_n, which amounts to choosing a projective resolution for K_0. (For a proof in the general case cf. 4.1 in [85, Keller, 1990].)
12. It is the left (resp. right) adjoint of the functor $(X \to K \to Z \to SX) \mapsto K$ defined on the category of triangles of $\mathscr{H}_+\mathscr{A}$ (resp. $\mathscr{H}_-\mathscr{A}$) with acyclic Z (resp. X).
13. By definition, $H_n K = \operatorname{Ker} d_n / \operatorname{Im} d_{n+1}$ for $K \in \mathscr{C}\mathscr{A}$.
14. [120, Parthasarathy, 1987].
15. By an *S-category*, we mean an additive category $\mathscr{C}$ with an additive endo-functor $S: \mathscr{C} \to \mathscr{C}$ and a class of distinguished S-sequences called triangles. If $\mathscr{C}$ and $\mathscr{D}$ are S-categories, an *S-functor* $(F, \varphi): \mathscr{C} \to \mathscr{D}$ consists of an additive functor $F: \mathscr{C} \to \mathscr{D}$ and an isomorphism $\varphi: FS \xrightarrow{\sim} SF$ such that

$$FX \xrightarrow{Fu} FY \xrightarrow{Fv} FZ \xrightarrow{(\varphi X)(Fw)} SFX$$

is a triangle whenever (u, v, w) is (cf. [88, Keller/Vossieck, 1987]). A *morphism of S-functors* $(F, \varphi) \to (F', \varphi')$ is given by a morphism of functors $\mu: F \to F'$ such that $\varphi'(\mu S) = (S\mu)\varphi$. An S-functor (F, φ): $\mathscr{C} \to \mathscr{D}$ is an *S-equivalence* if there is an S-functor $(G, \gamma): \mathscr{D} \to \mathscr{C}$ such that the *composed S-functors* $(GF, (\gamma F)(G\varphi))$ and $(FG, (\varphi G)(F\gamma))$ are isomorphic to the *identical* S-functors $(1_{\mathscr{C}}, 1_S)$ and $(1_{\mathscr{D}}, 1_S)$ respectively. Of course, F is an equivalence if (F, φ) is an S-equivalence; the converse also holds (cf. 6.7 in [86, Keller, 1990/91]).

For instance, if $\mathscr{S}$ is a finite spectroid and $\mathscr{S}^{\mathbb{Z}}$ its repetition (8.3, Example 3), the residue-category $\underline{\operatorname{mod} \mathscr{S}^{\mathbb{Z}}}$ of 9.2 is equipped with a natural S-category structure. Moreover, there is a fully-faithful S-functor $E: \mathscr{D}\mathscr{S} \to \underline{\operatorname{mod} \mathscr{S}^{\mathbb{Z}}}$, which is an S-equivalence if $gl \dim \mathscr{S} < \infty$ [70, Happel, 1987].
16. We regard $\mathscr{H}_b$ pro $\mathscr{S}$ as an S-category with the suspension functor induced by S and the triangles of $\mathscr{H}$ pro $\mathscr{S}$ having their terms in $\mathscr{H}_b$ pro $\mathscr{S}$.
17. Cf. 8.3 in [125, Rickard, 1989].
18. By definition, gldim $\mathscr{S} \leqslant d$ if and only if $\operatorname{Ext}^n_{\mathscr{S}}(?, ?) = 0$ for all $n > d$. In case gldim $\mathscr{S} = 1$, we say that $\mathscr{S}$ and mod $\mathscr{S}$ are *hereditary.*
19. Whenever $\mathscr{A}$ is abelian and $n \in \mathbb{Z}$, we denote by $\mathscr{D}_{\geqslant n}$ (resp. $\mathscr{D}_{<n}$) the full subcategory of $\mathscr{D}\mathscr{A}$ consisting of the complexes K with $H_i K = 0$, $\forall i < n$ (resp. $\forall i \geqslant n$).
20. Whenever $\mathscr{C}$ is an additive category and $\mathscr{U}$ a class of objects of $\mathscr{C}$, we denote by $\mathscr{U}^{\perp}$ (resp. ${}^{\perp}\mathscr{U}$) the right (resp. left) orthogonal subcategory of $\mathscr{U}$, i.e. the full subcategory of $\mathscr{C}$ consisting of the objects X such that $\mathscr{C}(Y, X) = 0$ (resp. $\mathscr{C}(X, Y) = 0$) for all $Y \in \mathscr{U}$.
21. $\mathscr{D}^{\perp}_{\geqslant 0} = \mathscr{D}_{<0}$ and $\mathscr{D}_{\geqslant 0} = {}^{\perp}\mathscr{D}_{<0}$ by 1.3.4 in [17, Beilinson/Bernstein/Deligne, 1982].
22. This condition is clearly sufficient. To see that it is necessary, we use that each $X \in \mathscr{D}\mathscr{S}$ may be built up from the $S^n H_n X$, $n \in \mathbb{Z}$ by successive extensions (cf. 1.3.13 in loc. cit. (21)).
23. The *Grothendieck group* $K_0\mathscr{C}$ of an exact (resp. an S-) category $\mathscr{C}$ is the quotient of the free abelian group on the isomorphism classes $cl(X)$, $X \in \mathscr{C}$, by all the relations $cl(X) - cl(Y) + cl(Z)$ provided by the conflations $X \to Y \to Z$ (resp. the triangles $X \to Y \to Z \to SX$) of $\mathscr{C}$. The image of $cl(X)$ in $K_0\mathscr{C}$ is denoted by $[X]$ ([67, Grothendieck, 1977]).
24. An extension of X by Z is an object Y occurring in a triangle $X \to Y \to Z \to SX$.

25. 8.3 of loc. cit (17); see also 2.1 in [85, Keller, 1990].
26. Cf. Sect. 4 in [89, Keller/Vossieck, 1988].
27. Cf. 2.2 of loc. cit. (26) and [1, Assem/Happel, 1981/82]. The tilting spectroids of $\mathscr{D}kD_n$ can be found in [2, Assem/Skowronski, 1988] and [87, Keller, 1991].
28. 2.1 in [21, Bongartz, 1981].
29. [16, Beilinson, 1978].
30. [21, Bongartz, 1981].
31. [6, Auslander/Platzeck/Reiten, 1979].
32. 2.4 in loc. cit. (28); cf. 7.2 in [71, Happel/Ringel, 1980/82].
33. [72, Happel/Vossieck, 1983].

13. Multiplicative Bases

We say that a k-category $\mathscr{R}$ is *finitely represented*[1] if it is svelte and pointwise finite, if its spectroid is locally bounded and if, for each $X \in \mathscr{R}$, there are only finitely many isoclasses of indecomposables $M \in \text{mod}\,\mathscr{R}$ such that $M(X) \neq 0$. Algebras of Dynkin quivers are finitely represented and so are in case char $k = p > 0$ the algebras of the finite groups whose p-subgroups are cyclic.[2] Out of these two examples, a general theory has been developed which we present in Sect. 13 and Sect. 14. We devote Sect. 13 to some combinatorial invariants attached to finitely represented algebras.[3] In particular, we shall see that each dimension only allows of finitely many isoclasses of finitely represented algebras.

13.1 Lemma. *If M is a pointwise finite module over a spectroid $\mathscr{S}$, the following two statements are equivalent:*

(i) *The lattice of submodules of M is distributive.*

(ii) *For each $x \in \mathscr{S}$, the $\mathscr{S}(x, x)$-submodules of $M(x)$ are linearly ordered.*

If the dimension of M is finite, these statements amount to saying that the number of submodules of M is finite.

13.2. Lemma. *If $\mathscr{S}$ is a spectroid, the following two statements are equivalent:*

(i) *The lattice of ideals of $\mathscr{S}$ is distributive.*

(ii) *For each $x \in \mathscr{S}$, $\mathscr{S}(x, x)$ is isomorphic to a truncated polynomial algebra $k[t]/t^{n_x}$, and for all $x, y \in \mathscr{S}$, $\mathscr{S}(x, y)$ is a cyclic*[2] *module over $\mathscr{S}(x, x)$ or over $\mathscr{S}(y, y)$.*

If $\mathscr{S}$ is finite, these statements amount to saying that the number of ideals of $\mathscr{S}$ is finite.

Proof. Consider the spectroid $\mathscr{C} = \mathscr{S} \otimes \mathscr{S}^{op}$ whose set of objects is $\mathscr{C}_0 = \mathscr{S}_0 \times \mathscr{S}_0$, whose morphism-spaces are the $\mathscr{C}((x, y), (x', y')) = \mathscr{S}(x, x') \otimes_k \mathscr{S}(y', y)$. Then $\mathscr{S}$ gives rise to a pointwise finite $\mathscr{C}$-module whose value at (x, y) is $\mathscr{S}(x, y)$, and our lemma follows from 13.1. $\surd$

In the sequel, we call a spectroid *distributive* if it is locally bounded and satisfies the conditions (i) and (ii) of the preceding lemma. In this terminology, if A is a finite-dimensional k-algebra, the following three conditions are equivalent: a) The number of (two-sided) ideals of A is finite; b) the lattice of ideals of A is distributive; c) the spectroid of A is distributive. The investigation of such "*ideal-finite*" algebras has some interest in itself.

13.3. Lemma. *Let $\mathscr{R}$ be a finitely represented k-category. Then all residue-categories $\mathscr{R}/\mathscr{I}$ and all full subcategories $\mathscr{Q}$ of $\mathscr{R}$ are finitely represented; moreover, the spectroid $\mathscr{S}$ of $\mathscr{R}$ is distributive.*

Proof. The modules over $\mathscr{R}/\mathscr{I}$ are identified with the $\mathscr{R}$-modules annihilated by $\mathscr{I}$. On the other hand, the functor which maps $N \in \operatorname{mod} \mathscr{Q}$ onto $x \mapsto \operatorname{Hom}(x^\wedge | \mathscr{Q}, N)$ provides a full embedding $\operatorname{mod} \mathscr{Q} \to \operatorname{mod} \mathscr{R}$. This implies the first two statements. Now, if $\mathscr{Q}$ is finite, the ideals $\mathscr{K}$ of $\mathscr{Q}$ give rise to pairwise non-isomorphic $\mathscr{Q}$-modules $\bigoplus_{x \in \mathscr{Q}} \mathscr{Q}(?, x)/\mathscr{K}(?, x)$ of bounded dimension. Since $\mathscr{Q}$ is finitely represented, there are only finitely many such $\mathscr{Q}$-modules, hence only finitely many $\mathscr{K}$. It then follows from 13.2 that each finite full subcategory of $\mathscr{S}$ is distributive, hence that so is $\mathscr{S}$. $\surd$

13.4. *Let the spectroid $\mathscr{S}$ be distributive.* We say that two morphisms $\mu, \nu \in \mathscr{S}(x, y)$ are *equivalent* if $\nu = \sigma\mu\tau$ for some automorphisms τ, σ. For instance, suppose that $\mathscr{S}(x, y)$ is cyclic over $\mathscr{S}(x, x)$; denote by ρ a generator of the maximal ideal of $\mathscr{S}(x, x)$, by μ a generator of $\mathscr{S}(x, y)$ over $\mathscr{S}(x, x)$, by m the smallest natural number such that $\mu\rho^m = 0$. Then $\mu, \mu\rho, \dots, \mu\rho^{m-1}$ and $\mu\rho^m = 0$ are representatives of the equivalence classes of $\mathscr{S}(x, y)$

In general, we denote by $\vec{\mu}$ the equivalence class of $\mu \in \mathscr{S}(x, y)$, and we call $\vec{\mu}$ a *ray* from x to y. These rays behave as follows with respect to composition: Given a second morphism $\nu \in \mathscr{S}(y, z)$, we have[3] either $\nu\mathscr{R}_{\mathscr{S}}(y, y)\mu \subset \mathscr{R}_{\mathscr{S}}(z, z)\nu\mu + \nu\mu\mathscr{R}_{\mathscr{S}}(x, x)$ or $\nu\mu \in \nu\mathscr{R}_{\mathscr{S}}(y, y)\mu \neq 0$. In the first case, $D := \{\nu'\mu' : \nu' \in \vec{\nu}, \mu' \in \vec{\mu}\}$ is equal to $\overrightarrow{\nu\mu}$, and we set $\vec{\nu}\vec{\mu} := \overrightarrow{\nu\mu}$. In the second case, $D = \nu\mathscr{R}_{\mathscr{S}}(y, y)\mu$ is a non-zero subbimodule of $\mathscr{S}(x, z)$; we then set $\vec{\nu}\vec{\mu} := \vec{0}$. For the surprising composition thus defined, the rays are the morphism of a *ray-category*, i.e. of a category P with the following properties:

a) the objects of P form a set and are pairwise not isomorphic.

b) There is a (necessarily unique) family $({}_x0_y)_{x, y \in P}$ of *zero-morphisms* $0 = {}_y0_x : x \to y$ such that $\mu 0 = 0 = 0\nu$ for all μ, ν.

c) For each $x \in P$, $P(x, y) = \{0\} = P(y, x)$ for almost all $y \in P$.

d) For each $x \in P$, $P(x, x)$ is isomorphic to the semi-group $H_n = \{1, \sigma, \sigma^2, \dots, \sigma^{n-1} \neq \sigma^n = \sigma^{n+1} = \cdots = 0\}$ for some $n \geqslant 1$ depending on x.

e) For all $x, y \in P$, the set $P(x, y)$ is cyclic[2] under the action of $P(x, x)$ or of $P(y, y)$.

f) If λ, μ, ν, π are morphisms of P, $\mu\nu = \mu\pi \neq 0$ implies $\nu = \pi$, and $\mu\nu = \lambda\nu \neq 0$ implies $\mu = \lambda$ (*cancellation property*).

In the sequel, we denote by $\vec{\mathscr{S}}$ *the ray-category attached to* $\mathscr{S}$. Within the class of all ray-categories, a special rôle is played by the ray-categories P such that $P(x, x)$ is isomorphic to H_1 for all $x \in P$. This condition implies that

$|P(x, y)| \leqslant 1$ for all $x, y \in P$, and that P is determined up to isomorphism by its set E of objects and the set $F = \{(x, y, z) \in E \times E \times E: P(y, z)P(x, y) \neq \{0\}\}$. For instance, each finite poset M gives rise to such a P: Set $E = M$ and $F = \{(x, y, z) \in M \times M \times M: x \leqslant y \leqslant z\}$.

13.5. Reversely, each ray-category P gives rise to a distributive spectroid, the *linearization* $k(P)$ of P which has the same objects as P and whose morphism-spaces are the quotients $k(P)(x, y) = kP(x, y)/k_y 0_x$, where $kP(x, y)$ denotes the vector space freely generated by $P(x, y)$; the composition of $k(P)$ is induced by that of P. Thus, $P \setminus \{0\}$ provides a *multiplicative basis* [14] of $k(P)$. The ray-category $\overrightarrow{k(P)}$ of $k(P)$ is identified with P. Therefore, each ray-category is isomorphic to the ray-category attached to some distributive spectroid.

We say that a distributive spectroid is *standard* if it is isomorphic to the linearization of a ray-category. If $\mathscr{S}$ is standard, then so is each full subcategory and each residue-spectroid of $\mathscr{S}$. In fact, the ideals of any distributive $\mathscr{S}$ can be described in terms of $\vec{\mathscr{S}}$: Define an *ideal* of a ray-category P to be a family $\mathscr{J}$ of subsets $\mathscr{J}(x, y) \subset P(x, y)$ which contain the zero-morphisms and are closed under external composition ($\mu \in \mathscr{J}(x, y)$ implies $\nu\mu\lambda \in \mathscr{J}(w, z)$ for all $\lambda \in P(w, x)$ and $\nu \in P(y, z)$). In case $P = \vec{\mathscr{S}}$, we then have a natural bijection $\mathscr{I} \mapsto \vec{\mathscr{I}}$ between the ideals of $\mathscr{S}$ and those of P; the ideal $\vec{\mathscr{I}}$ consists of the rays $\vec{\mu}$ of the morphisms μ belonging to $\mathscr{I}$; it gives rise to an isomorphism $\vec{\mathscr{S}}/\vec{\mathscr{I}} \xrightarrow{\sim} \overrightarrow{\mathscr{S}/\mathscr{I}}$ if $\mathscr{I} \subset \mathscr{R}_{\mathscr{S}}$ (the symbol $P/\mathscr{J}$ denotes the residue-category of P obtained by annihilating the morphisms lying in $\mathscr{J}$).

One of the objectives of the present section is to give an insight into the proof of the following

Theorem. *Each finitely represented spectroid is standard if* char $k \neq 2$.

13.6. The ray-categories provide an appropriate language for the description of the finitely represented algebras. They have a combinatorial nature with a slight algebraic touch. We must try and understand them.

We first remark that each morphism $\neq 0$ of a ray-category P is a finite (in general not uniquely determined) composition of *irreducible* morphisms (irreducible here means non-identical, non-zero and without proper factorization). With P we can therefore associate a quiver Q_P which has the objects of P as vertices and the irreducible morphisms as arrows. In case $P = \vec{\mathscr{S}}$, Q_P is identified with the quiver $Q_{\mathscr{S}}$ of $\mathscr{S}$: The reason is that the bijection $\mathscr{I} \mapsto \vec{\mathscr{I}}$ of 13.5 is compatible with the multiplication of ideals, hence that $\mathscr{R}_{\mathscr{S}}(x, y) \neq \mathscr{R}^2_{\mathscr{S}}(x, y)$ is equivalent to $\vec{\mathscr{R}}_{\mathscr{S}}(x, y) \neq \overrightarrow{\mathscr{R}^2_{\mathscr{S}}}(x, y) = \vec{\mathscr{R}}^2_{\mathscr{S}}(x, y)$.

Reversely, each quiver Q determines a category $\mathscr{P}Q$ whose objects are the vertices of Q, whose morphisms are the *paths*, to which we adjoin formal zero-morphisms. In case $Q = Q_P$ we obtain a canonical functor $\mathscr{P}Q \to P$ which maps zero-morphisms onto zero-morphisms and formal compositions $p = \alpha_n|\ldots|\alpha_2|\alpha_1$ of arrows α_i onto the corresponding compositions $\vec{p} = \alpha_n \ldots \alpha_2\alpha_1$ in P. Thanks to this functor we can interpret each ray-category as a residue-category of some $\mathscr{P}Q$.

The equivalence relation on $\mathscr{P}Q_P$ which defines P can be described as follows: First we look at *zero-paths*, i.e. at paths p of Q_P such that $\vec{p} = 0$. We say that the zero-path p is *minimal* if it cannot be written as formal composition $p = v|q|u$ where u or v is not identical and $\vec{q} = 0$.

Next, we call two paths q, r *interlaced* if (q, r) belongs to the equivalence relation generated by the following relation R: $(q, r) \in R$ if and only if there are paths v, q_1, r_1, u such that u, v are not both identical and $q = vq_1u$, $r = vr_1u$, $\vec{q}_1 = \vec{r}_1 \neq 0$. We call *contour* of P a set $\{q, r\}$ of two non-interlaced paths such that $\vec{q} = \vec{r} \neq 0$. Finally, we say that two contours are *equivalent* if each path of one contour is interlaced with a path of the other.

The ray-category P is then identified with $\mathscr{P}Q_P/E_P$, where E_P denotes the smallest equivalence relation on the morphisms of $\mathscr{P}Q_P$ which is *stable* under composition – $(x, y) \in E_P$ implies $(txz, tyz) \in E_P$ – and contains the pairs $(p, 0)$, (q, r) for each minimal zero-path p and each element $\{q, r\}$ of a set of representatives of the equivalence classes of contours. This description of P easily implies that a distributive spectroid $\mathscr{S}$ with ray-category $\vec{\mathscr{S}} = P$ is standard if and only if there are representatives $\bar{\alpha} \in \alpha \subset \mathscr{S}(x, y)$ of the arrows $x \xrightarrow{\alpha} y$ of Q_P such that $\bar{\alpha}_n \ldots \bar{\alpha}_2\bar{\alpha}_1 = 0$ for each minimal zero-path $p = \alpha_n|\ldots|\alpha_2|\alpha_1$ and $\bar{\beta}_m \ldots \bar{\beta}_1 = \bar{\gamma}_l \ldots \bar{\gamma}_1$ for each representative contour $\{q = \beta_m|\ldots|\beta_1, r = \gamma_l|\ldots|\gamma_1\}$.

13.7. Example. The first non-trivial example is provided by the quiver

$$B \qquad {}^{\sigma}\circlearrowright y \xrightarrow{\nu} x \circlearrowleft^{\tau}$$

In order to describe the ray-categories with quiver B, we consider the set F of all $(m, n, s, t) \in \mathbb{N}^4$ such that $1 \leqslant m \leqslant n \leqslant t$, $n \leqslant sm$, $2 \leqslant s$ and $2 \leqslant t$. Given $(m, n, s, t) \in F$, we denote by E the stable[4] equivalence relation generated by the following pairs of morphisms of $\mathscr{P}B$: $(\sigma^s, 0)$, $(\tau^t, 0)$, $(\tau^n\nu, 0)$ and $(\nu\sigma, \tau^m\nu)$. The residue-category $P = \mathscr{P}B/E$ then is a ray-category, and each ray-category with quiver B is isomorphic or antiisomorphic to P for some $(m, n, s, t) \in F$.

We now want to investigate the distributive spectroids $\mathscr{S}$ such that $\vec{\mathscr{S}} = P$: Denote by $\bar{\sigma}$, $\bar{\nu}$ and $\bar{\tau}$ morphisms of $\mathscr{S}$ whose rays are σ, ν and τ respectively. Then we have $\mathscr{S}(x, y) = 0$,

$$\mathscr{S}(y, y) = k\mathbb{1}_y \oplus k\bar{\sigma} \oplus \cdots \oplus k\bar{\sigma}^{s-1} \quad \text{with } \bar{\sigma}^s = 0,$$

$$\mathscr{S}(y, x) = k\bar{\nu} \oplus k\bar{\tau}\bar{\nu} \oplus \cdots \oplus k\bar{\tau}^{n-1}\bar{\nu} \quad \text{with } \bar{\tau}^n\bar{\nu} = 0,$$

and

$$\mathscr{S}(x, x) = k\mathbb{1}_x \oplus k\bar{\tau} \oplus \cdots \oplus k\bar{\tau}^{t-1} \quad \text{with } \bar{\tau}^t = 0.$$

We therefore "know" all the morphism-spaces, and the composition of $\mathscr{S}$ is determined by the module structure of $\mathscr{S}(y, x)$ over $\mathscr{S}(y, y)$. This is defined by an algebra-homomorphism $h: \mathscr{S}(y, y) \to \mathscr{S}(x, x)/\bar{\tau}^n$ such that $h(\bar{\sigma}) = a_m\bar{\tau}^m + \cdots + a_{n-1}\bar{\tau}^{n-1} \bmod \bar{\tau}^n$, where $a_m \neq 0$ and $\bar{\nu}\bar{\sigma} = h(\bar{\sigma})\bar{\nu}$. Moreover, two homomorphisms h and h' give rise to isomorphic spectroids if and only if $h' = \xi h \eta^{-1}$ for some automorphisms ξ of $\mathscr{S}(x, x)/\bar{\tau}^n$ and η of $\mathscr{S}(y, y)$. Equivalently, let us call a subalgebra of $\mathscr{S}(x, x)/\bar{\tau}^n$ *cyclic of order m* if it is generated by an element of the form $\bar{\tau}^m + a\bar{\tau}^{m+1} + \cdots \bmod \bar{\tau}^n$; associating $\operatorname{Im} h$ with h, we obtain a *bijection between the isoclasses of spectroids with ray-category P and the orbits of cyclic subalgebras of order m under the action of the automorphisms of* $\mathscr{S}(x, x)/\bar{\tau}^n$. If char $k = 0$, there is only one orbit because $\bar{\tau}^m + a\bar{\tau}^{m+1} + \cdots = \left(\bar{\tau} + \frac{1}{m}a\bar{\tau}^2 + \cdots\right)^m$. If char $k = p > 0$, some further computations lead to the following results:

a) Each spectroid with ray-category P is standard if p does not divide m or if $n \leqslant m + 1$.

b) The number of isoclasses of non-standard spectroids with ray-category P is finite and $\geqslant 1$ in the following cases:

– $m = p < n - 1$. As representatives of the orbits of cyclic subalgebras of order m we can choose the $k[\bar{\tau}^p + \bar{\tau}^u]$ where p does not divide u and $p < u < n$.

– $p = 2$, $m = 2(2r + 1) < n - 1 \leqslant m + 4$, $r \geqslant 1$. As representatives we can choose $k[\bar{\tau}^m]$, $k[\bar{\tau}^m + \bar{\tau}^{m+1}]$, and $k[\bar{\tau}^m + \bar{\tau}^{m+3}]$ if $m + 3 < n$.

– $p > 2$, $m = pq < n - 1 \leqslant m + p + 1$, where q is >1 and prime to p. As representatives we can choose $k[\bar{\tau}^m]$ and the $k[\bar{\tau}^m + \bar{\tau}^u]$ where $m < u < n$ and $u \neq m + p$.

– p^2 divides $m < n - 1 \leqslant m + p$. As representatives we can choose $k[\bar{\tau}^m]$ and the $k[\bar{\tau}^m + \bar{\tau}^u]$ where $m < u < n$.

c) In all the other cases, there are infinitely many isoclasses of spectroids with ray-category P. The following families consist of pairwise non-equivalent cyclic subalgebras of order m:

– $p = 2$, $m = 2(2r + 1)$, $r \geqslant 1$, $m + 5 < n$: $k[\bar{\tau}^m + \bar{\tau}^{m+3} + c\bar{\tau}^{m+5}]$, $c \in k$.

– $p > 2$, $m = pq$, $m + p + 2 < n$, where q is >1 and prime to p: $k[\bar{\tau}^m + \bar{\tau}^{m+p+1} + c\bar{\tau}^{m+p+2}]$, $c \in k$.

– p^2 divides m, $m + p + 1 < n$: $k[\bar{\tau}^m + \bar{\tau}^{m+p} + c\bar{\tau}^{m+p+1}]$, $c \in k$.

13.8. The preceding example shows that a small ray-category can give rise to a multitude of non-standard distributive spectroids. By way of contrast, ray-categories of finitely represented spectroids give rise to few spectroids, as we intend to show in the sequel.

For this purpose, we identify the modules over the linearization of a ray-category P with the functors $P^{op} \to \operatorname{Mod} k$ which map zero-morphisms onto zero-maps. And we simply call them *modules over P*. Moreover, we say that P is *finitely represented* if so is $k(P)$. Our strategy is to investigate finitely represented ray-categories in a first stage and then to use the acquired knowledge for proving that a distributive spectroid $\mathscr{S}$ is finitely represented if and only if so is $\vec{\mathscr{S}}$.

In order to exclude pathological examples, we need a handy criterion for infinite representment. Therefore, given two ray-categories P and Σ, we say that a functor $C\colon \Sigma \to P$ is *cleaving* if the following a) and b) are true: a) $C\mu = 0$ if and only if $\mu = 0$; b) if $\alpha \in \Sigma(x, z)$ (resp. $\beta \in \Sigma(z, y)$) is irreducible and $C\mu\colon Cx \to Cy$ factors through $C\alpha$ (resp. $C\beta$), then μ factors through α (resp. β).

Lemma. *Let $C: \Sigma \to P$ be a cleaving functor between two ray-categories. If Σ is not finitely represented, then neither is P.*

Proof. The conditions a) and b) mean that the morphism $C(x, y): \Sigma(x, y) \to P(Cx, Cy)$ of functors in 2 variables x, y admits a (functorial) retraction which maps the complement of $\operatorname{Im} C(x, y)$ onto 0. If we set $\Sigma' = k(\Sigma)$, $P' = k(P)$, $C' = k(C)$, it follows that the linear morphism $C'(x, y): \Sigma'(x, y) \to P'(C'x, C'y)$ of functors in x, y admits a linear retraction.

Now C' induces a restriction-functor $C^{\bullet}: \operatorname{Mod} P' \to \operatorname{Mod} \Sigma'$ and its right adjoint $C_{\bullet}: \operatorname{Mod} \Sigma' \to \operatorname{Mod} P'$ which is defined by $(C_{\bullet}M)(z) = \operatorname{Hom}(P'(C'?, z), M)$. The functor $C_{\bullet}$ maps $x^{\vee}$ onto $(Cx)^{\vee}$, hence finite-dimensional modules onto finite-dimensional ones. Moreover, the adjunction morphisms

$$(C^{\bullet}C_{\bullet}M)(y) = \operatorname{Hom}(P'(C'?, C'y), M) \to M(y) \tilde{\to} \operatorname{Hom}(\Sigma'(?, y), M)$$

are induced by the linear maps $C'(x, y)$ considered above. It follows that the adjunction morphism $C^{\bullet}C_{\bullet} \to \mathbb{1}_{\operatorname{Mod} \Sigma'}$ admits a section and that each indecomposable $M \in \operatorname{mod} \Sigma'$ is a summand of $C^{\bullet}C_{\bullet}M$, i.e. of $C^{\bullet}N$ for some indecomposable summand N of $C_{\bullet}M$.

Now suppose P finitely represented. Denote by $N_1, \ldots, N_r$ representatives of the indecomposable $L \in \operatorname{mod} P'$ such that $L(Cx) \neq 0$. If $M_{i1}, \ldots, M_{is}$ are pairwise non-isomorphic finite-dimensional indecomposable summands of $C^{\bullet}N_i$, the canonical morphism $\bigoplus_j M_{ij} \to C^{\bullet}N_i$ is mono (this follows from 3.3c) by restriction to a finite subcategory of Σ'). Therefore we have $\sum_j \dim M_{ij}(x) \leqq \dim N_i(Cx)$ and $s \leqslant \dim N_i(Cx)$ if all $M_{ij}(x)$ are $\neq 0$. We infer that there are at most $\sum_i \dim N_i(Cx)$ pairwise non-isomorphic indecomposable $M \in \operatorname{mod} \Sigma'$ such that $M(x) \neq 0$. This provides the wanted contradiction. $\surd$

13.9. We first apply Lemma 13.8 to the case $\Sigma = \mathscr{P}\mathbb{A}_{\infty}$, where $\mathbb{A}_{\infty}$ is the infinite quiver

$$0 \overset{\sigma_1}{\leftarrow} 1 \overset{\rho_1}{\rightarrow} 2 \overset{\sigma_2}{\leftarrow} 3 \overset{\rho_2}{\rightarrow} 4 \overset{\sigma_3}{\leftarrow} 5 \overset{\rho_3}{\rightarrow} \cdots$$

A cleaving functor $C: \mathscr{P}\mathbb{A}_{\infty} \to P$ will be called *zigzag* of P. In this terminology, Lemma 13.8 implies that finitely represented ray-categories are *zigzag-free*.

Zigzags occur in relation with a special class of distributive spectroids: Let us start with an arbitrary ray-category P and simply call *cocycle* a function f with values in $k^{\bullet} = k \setminus \{0\}$ which is defined on the pair (τ, σ) of composable morphisms such that $\tau\sigma \neq 0$ and which satisfies $f(\sigma, \rho)f(\tau, \sigma\rho) = f(\tau\sigma, \rho)f(\tau, \sigma)$ whenever $\tau\sigma\rho$ makes sense and is $\neq 0$. Among the cocycles, we have the *coboundaries* δg with the values $(\delta g)(\tau, \sigma) = g(\sigma)g(\tau\sigma)^{-1}g(\tau)$, where g is defined on the non-zero morphisms and has its values in $k^{\bullet}$. The cocycles and the coboundaries form commutative groups which we denote by $Z(P)$ and $B(P)$ respectively. The residue-group $H(P) = Z(P)/B(P)$ is the *cohomology group* of P.

If $\mathscr{S}$ is a distributive spectroid, $H(\vec{\mathscr{S}})$ contains a *canonical class*, which is defined as follows: Choose a representative $g(\rho) \in \mathscr{S}(x, y)$ in each non-zero ray $\rho \in \vec{\mathscr{S}}(x, y)$. Then, each pair (τ, σ) of rays with non-zero composition $\tau\sigma: z \to t$

provides us with a coefficient $c(\tau, \sigma) \in k^{\cdot}$ such that $g(\tau)g(\sigma) - c(\tau, \sigma)g(\tau\sigma)$ belongs to the radical of the bimodule $\mathscr{S}(t, t)g(\tau\sigma)\mathscr{S}(z, z)$. The class $\bar{c} \in H(\vec{\mathscr{S}})$ of the so defined cocycle c does not depend on the choice of the representatives $g(\rho)$ and is the announced canonical class.

Reversely, we associate a distributive spectroid $k^f(P)$ with each cocycle f: The objects and the morphisms of $k^f(P)$ are the same as in $k(P)$. In particular, the set of non-zero morphisms $\rho \in P(x, y)$ is identified with a basis of $k^f(P)(x, y)$. The composition $(\nu, \mu) \mapsto \nu \overset{f}{.} \mu$ of $k^f(P)$ is defined on these bases by $\tau \overset{f}{.} \sigma = f(\tau, \sigma)\tau\sigma$ if $\tau\sigma$ is $\neq 0$ in P, by $\tau \overset{f}{.} \sigma = 0$ if $\tau\sigma = 0$. The associativity of the composition of $k^f(P)$ results from the cocycle condition. The k-category $k^f(P)$ is a distributive spectroid with ray-category P and canonical class $\bar{f} = f \bmod B(P)$. It is isomorphic to $k^{f(\delta g)}(P)$ for each coboundary δg.

The following statement is an important link in the proof of Theorem 13.5.

Theorem. $H(P) = \{1\}$ *whenever the ray-category* P *is zigzag-free.*

This is a purely combinatorial statement which will therefore not be proved here.[3] The statement makes sense even in the case where P is the ray-category attached to a finite poset (13.4). In this case, $H(P)$ is a classical invariant of the poset, even though the statement surprises.

Example.[5] Consider the quiver Q of Fig. 1 and the ray-category $P = \mathscr{P}Q/E$, where E is the stable[4] equivalence relation generated by $(\eta|\gamma|\alpha, 0)$, $(\gamma|\alpha, \varepsilon|\beta)$, $(\delta|\alpha, \zeta|\beta)$, $(\eta|\gamma, \vartheta|\delta)$ and $(\eta|\varepsilon, \vartheta|\zeta)$. The category P admits a zigzag C such that $\sigma_{2p-1} = \gamma$, $\rho_{2p-1} = \delta$, $\sigma_{2p} = \zeta$ and $\rho_{2p} = \varepsilon$. The cohomology group $H(P)$ is isomorphic to $k^{\cdot}$.

13.10. Next we are to apply Lemma 13.8 to cleaving functors whose source is crucial. For the convenience of the proof, we here call *crucial* the ray-categories listed in Fig. 2, their duals and the $\mathscr{P}\tilde{\Delta}$, where $\tilde{\Delta}$ is extended Dynkin. The squares of the last seven categories of Fig. 2 are commutative and the dotted lines point to zero-paths.

In order to apply Lemma 13.8, we have to make sure that the crucial categories are not finitely represented. In fact, their linearizations are critical arboresque spectroids, which are infinitely represented by Sect. 12.

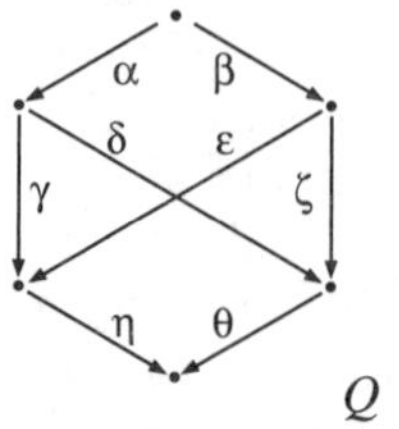

Fig. 1

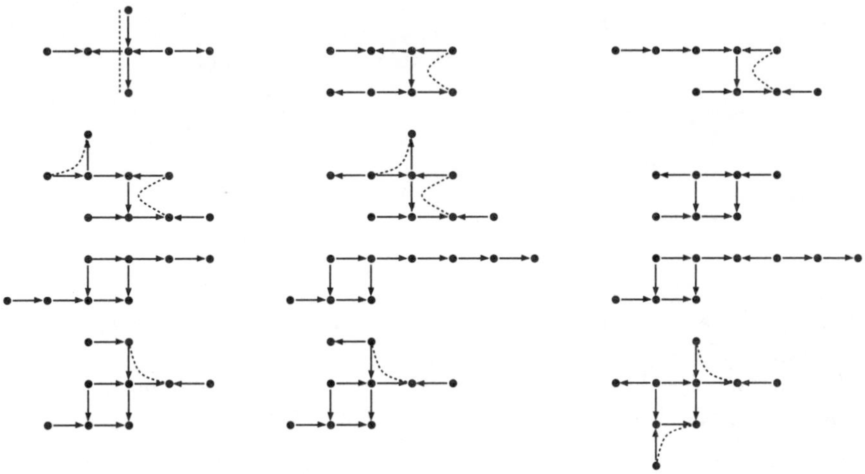

Fig. 2

Theorem. *Let P be a ray-category. Suppose that* char $k \neq 2$ *and that there is no cleaving functor $C: \Sigma \to P'/\mathscr{I}$ where Σ is crucial and $\mathscr{I}$ a non-zero ideal of a full subcategory P' of P having at most 5 objects. Then each distributive spectroid $\mathscr{S}$ with ray-category P is isomorphic to $k^f(P)$ for some cocycle f.*

Although this theorem is purely combinatorial, we will comment on its proof. Together with Theorem 13.9 and Lemma 13.8, it implies Theorem 13.5. It also applies to the case where P is *minimal representation-infinite*, i.e. where $P/\mathscr{I}$ is finitely represented for each ideal $\mathscr{I} \neq 0$ though P itself is not (each ideal $\mathscr{I}'$ of P' extends to an ideal $\mathscr{I}$ of P). In fact, the proof will show that the assumption char $k \neq 2$ is superfluous if P is minimal representation-infinite.

Remark. As we shall see later, the conditions imposed on P in the theorem above are satisfied whenever P is the ray-category of a finitely represented or minimal representation-infinite distributive spectroid.

13.11. Proof of Theorem 13.10. First Part. We intend to show here that representatives $\bar{\alpha} \in \alpha$ of the arrows α of $Q_{\mathscr{S}} = Q_P$ can be chosen in such a way that $\bar{\alpha}_n \dots \bar{\alpha}_2\bar{\alpha}_1 = 0$ for each minimal zero-path

$$p\colon x_0 \xrightarrow{\alpha_1} x_1 \xrightarrow{\alpha_2} x_2 \to \cdots \to x_{n-1} \xrightarrow{\alpha_n} x_n$$

of Q_P (13.6). The method used for this purpose is typical of our cleaving technique.

First we notice that there are two classes of minimal zero-paths. Either $\varphi_n \dots \varphi_1 = 0$ for each sequence of morphisms $\varphi_i \in \alpha_i$, a case which will not tease us further! Or p is a *critical path*, i.e. the composition $\vec{p} = \alpha_n \dots \alpha_1$ is zero and there are endomorphisms ρ_i such that $\alpha_n \dots \alpha_{i+1}\rho_i\alpha_i \dots \alpha_1 \neq 0$ if $0 < i < n$: indeed, in the case considered now, there are morphisms $\varphi_i \in \alpha_i$ such that $\varphi_n \dots \varphi_1 \neq 0$; setting $\mu_i = \varphi_i \dots \varphi_1$ and $\nu_i = \varphi_n \dots \varphi_{i+1}$, we then have $\nu_i\mu_i \neq 0$ and $\vec{\nu}_i\vec{\mu}_i = 0$,

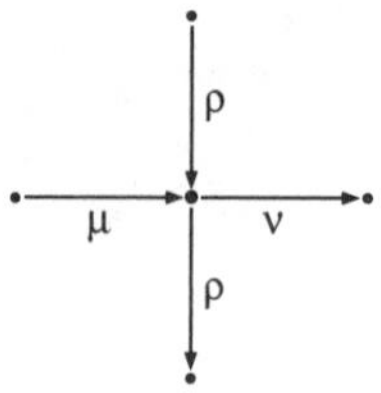

Fig. 3

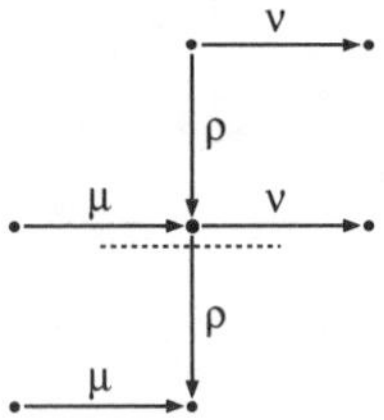

Fig. 4

hence $v_i\mu_i \in v_i\mathscr{R}_{\mathscr{S}}(x_i, x_i)\mu_i \neq 0$ by 13.4 and $\vec{v}_i\vec{\rho}_i\vec{\mu}_i \neq 0$ where $\rho_i = \mathscr{R}_{\mathscr{S}}^{m_i}(x_i, x_i)\setminus \mathscr{R}_{\mathscr{S}}^{m_i+1}(x_i, x_i)$ and $m_i = \sup\{e\colon v_i\mathscr{R}_{\mathscr{S}}^{e}(x_i, x_i)\mu_i \neq 0\}$.

In order to prove the existence of representatives $\bar{\alpha} \in \alpha$ which annihilate all the minimal zero-paths, it therefore suffices to attain to Lemma 3 below.

Lemma 1. *Suppose that $P(a, c)$ and $P(c, b)$ are not cyclic under the actions of $P(a, a)$ and $P(b, b)$ respectively. Then we have $|P(a, c)| \leqslant 3$ or $|P(c, b)| \leqslant 3$.*

Proof-sketch. Let ρ, μ and v be generators of $P(c, c)$, $P(a, c)$ and $P(c, b)$ respectively. Suppose that $\rho^2\mu \neq 0 \neq v\rho^2$. Denote by P' the full subcategory of P having a, b, c as objects, by $\mathscr{I}$ the ideal of P' generated by $\rho^2\mu$ and $v\rho^2$. In case $v\mu \neq 0$, we consider the functor $C\colon \Sigma \to P'/\mathscr{I}$ illustrated by Fig. 3 (the picture describes Σ, which is extended Dynkin; the letters denote the images of the arrows of Σ under C). It is easy to show that C is cleaving. In case $v\mu = 0$, Fig. 4 illustrates a cleaving functor with range $P'/\mathscr{I}$ and crucial domain.

Lemma 2. *All critical paths have length* 2.

Proof-sketch. Suppose that the path p above is critical and that $n \geqslant 3$. Denote by P' the full subcategory of P consisting of $x = x_0, y = x_1, z = x_{n-1}$ and $t = x_n$, by σ and τ the generators of $P(y, y)$ and $P(z, z)$. Set $\alpha = \alpha_1, \beta = \alpha_{n-1} \ldots \alpha_2, \gamma = \alpha_n$. By Lemma 1, we have $\gamma\tau^2 = 0$ or $\tau^2 P(x, z) = 0$, hence $\gamma\tau^t P(x, z) = 0$ for $t \geqslant 1$. Since $\gamma\tau^t\beta\alpha \neq 0$ for some $t \geqslant 1$, we infer that $\gamma\tau\beta\alpha \neq 0$ and that $\beta\alpha$ generates $P(x, z)$. Similarly, we have $P(y, t)\sigma^t\alpha = 0$ for $t \geqslant 1$, $\gamma\beta\sigma\alpha \neq 0$, and $\gamma\beta$ generates $P(y, t)$. It follows that β is irreducible in P' and satisfies $\beta\sigma = \tau\beta$.

Now we claim that σ and τ are irreducible in P': Otherwise, we would have $\sigma = \lambda\beta$ and $\tau = \beta\lambda$ for some irreducible $\lambda \in P'(z, y)$ and Fig. 5 would provide a cleaving functor with range $P'/\mathscr{I}$, where $\mathscr{I}$ is generated by $\sigma\alpha$ and $\gamma\tau$.

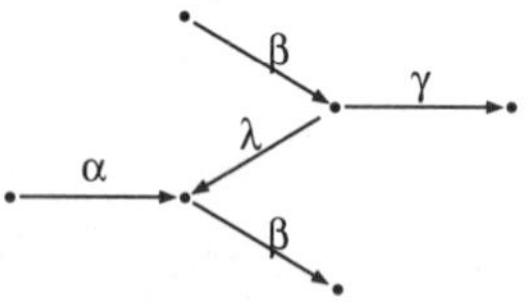

Fig. 5

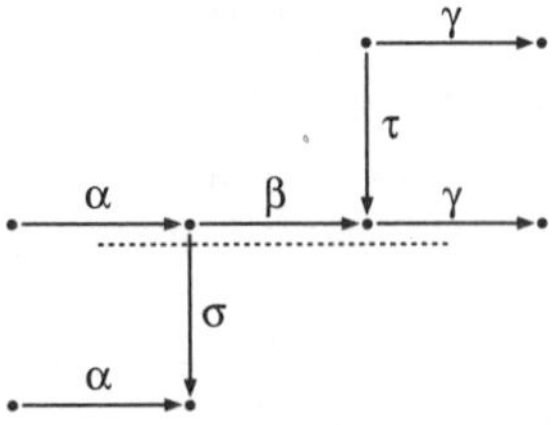

Fig. 6

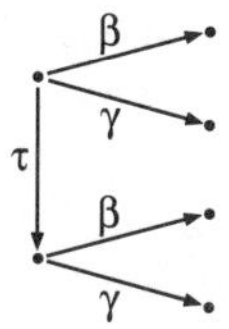

Fig. 7

Finally, we obtain the required contradiction with Fig. 6, which illustrates a cleaving functor with range $P'/\tau\beta$ $\surd$

Lemma 3. *Two distinct critical paths have no common arrow.*

Proof-sketch. Suppose that the arrow $a \xrightarrow{\alpha} m$ belongs to two distinct critical paths of length 2. Up to duality, we may suppose that $P(a, m)$ is cyclic under $P(m, m)$. The given critical paths then have the form $a \xrightarrow{\alpha} m \xrightarrow{\beta} b$ and $a \xrightarrow{\alpha} m \xrightarrow{\gamma} c$. If τ generates $P(m, m)$, it easily follows that Fig. 7 provides a cleaving functor with range $P'/\{\beta\tau\alpha, \gamma\tau\alpha\}$, where P' denotes the full subcategory of P whose objects are a, m, b, c.

13.12. Proof of Theorem 13.10. Second Part. Our next objective is to choose the representatives $\bar{\alpha}$ of the arrows α of $Q_{\mathscr{S}}$ in such a way that $\bar{\beta}_m \ldots \bar{\beta}_1$ and $\bar{\gamma}_l \ldots \bar{\gamma}_1$ are *proportional* for each representative contour $\{q = \beta_m| \ldots |\beta_1, r = \gamma_l| \ldots |\gamma_1\}$ (13.6). Here again, we meet with two cases. In the first, we have $\vec{q}\sigma = \vec{r}\sigma = 0 = \tau\vec{q} = \tau\vec{r}$ for all non-identical $\sigma \in P(a, a)$ and $\tau \in P(z, z)$, where a is the source of β_1, γ_1 and z the head of β_m, γ_l; in this case, which will not tease us further, $\bar{\beta}_m \ldots \bar{\beta}_1$ and $\bar{\gamma}_l \ldots \bar{\gamma}_1$ lie in the one-dimensional socle of the bimodule $\mathscr{S}(a, z)$ for each choice of the $\bar{\alpha}$. In the opposite case, we say that the contour $\{q, r\}$ is *non-deep*. We give 3 examples of possible non-deep contours supported by *subquivers* of Q_P (Fig. 8).

In the first example, we suppose that the full subcategory of P with objects x, y is identified with $\mathscr{P}B/E_B$, where E_B is the stable[4] equivalence generated by $(v|\sigma, \tau|v)$, $(\sigma^s, 0)$ and $(\tau^t, 0)$ where $\inf(s, t) = 3$ and $\sup(s, t) \leqslant 5$. Then $\{v|\sigma, \tau|v\}$ is a non-deep contour of P; we call it a *dumb-bell*.

In the second example, we suppose that the full subcategory of P with objects $x_0, \ldots, x_{n-1}$ is identified with $\mathscr{P}F/E_F$; here, E_F is generated by $(\alpha_n| \ldots |\alpha_1, \rho^2)$,

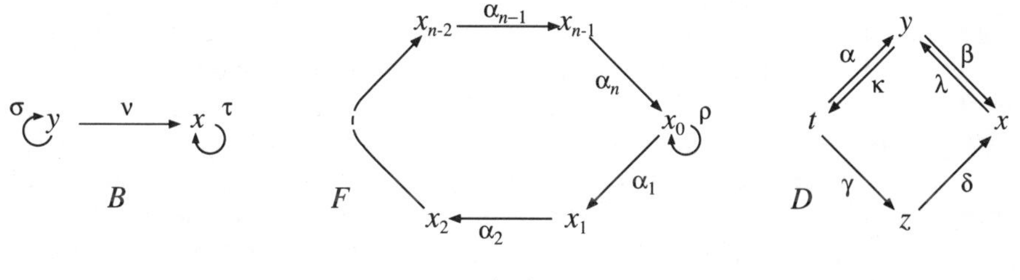

Fig. 8

$(\alpha_1|\alpha_n, 0)$ and $(\alpha_{ei}|\ldots|\alpha_1|\rho|\alpha_n|\ldots|\alpha_{i+1}, 0)$ for $1 \leqslant i \leqslant n-1$, where e is a given non-decreasing function $\{1, \ldots, n-1\} \to \{1, \ldots, n\}$. We call the non-deep contour $\{\alpha_n|\ldots|\alpha_1, \rho^2\}$ a *penny-farthing*.

In the third example, we suppose that the full subcategory of P with objects x, y, z, t is identified with $\mathscr{P}D/E_D$, where E_D is generated by $(\beta|\alpha, \delta|\gamma)$, $(\alpha|\kappa, \lambda|\beta)$ and $(\kappa|\lambda, 0)$. The non-deep contour $\{\beta|\alpha, \delta|\gamma\}$ is called a *diamond*.

Theorem. *Let P be a ray-category such that there is no cleaving functor $C\colon \Sigma \to P'/\mathscr{I}$ where Σ is crucial and $\mathscr{I}$ a non-zero ideal of a full subcategory P' of P having at most 5 objects. Then:*

a) *Each non-deep contour of P is a dumb-bell, a penny-farthing or a diamond.*
b) *Two non-deep contours with share an arrow are equal.*

The proof[3] of this theorem involves the techniques used in the lemmas of 13.11 above. It is lengthy.

13.13. Proof of Theorem 13.10. Conclusion. We choose representatives $\bar{\xi} \in \xi \subset \mathscr{S}(u, v)$ of the arrows $u \xrightarrow{\xi} v$ of $Q_{\mathscr{S}}$ which annihilate all minimal zero-paths (13.11). For each dumb-bell $\{v|\sigma, \tau|v\}$, we then set $\tilde{\tau} = \upsilon\bar{\tau}$ where υ is defined by $\bar{v}\bar{\sigma} = \upsilon\bar{\tau}\bar{v}$. For each diamond $\{\beta|\alpha, \delta|\gamma\}$, we set $\tilde{\beta} = \eta\bar{\beta}$ where $\bar{\delta}\bar{\gamma} = \eta\bar{\beta}\bar{\alpha}$. Now, if $\{\alpha_n|\ldots|\alpha_1, \rho^2\}$ is a penny-farthing and if $a, b \in k$ are such that $\bar{\alpha}_n \ldots \bar{\alpha}_1 = \bar{\rho}^2(a^2 + b\bar{\rho})$, we set $\tilde{\rho} = \bar{\rho}(a + \frac{1}{2}a^{-1}b\bar{\rho})$. These definitions are compatible because of 13.12b. We further set $\tilde{\xi} = \bar{\xi}$ if ξ is not of the form τ, β or ρ. In this way, we get a new choice of representatives $\tilde{\xi}$ which still annihilate the minimal zero-paths because no critical path contains an arrow τ, ρ, or β ($P(x, x)\beta = \beta P(y, y)$!). In addition, $\tilde{\beta}_m \ldots \tilde{\beta}_1$ and $\tilde{\gamma}_l \ldots \tilde{\gamma}_1$ are proportional for each contour $\{\beta_m|\ldots|\beta_1, \gamma_l|\ldots|\gamma_1\}$ (they are even equal if the contour is non-deep). This clearly means that the display-functor $kQ_{\mathscr{S}} \to \mathscr{S}$, $\xi \mapsto \tilde{\xi}$ induces an isomorphism $k^f(P) \xrightarrow{\sim} \mathscr{S}$, where f lies in the canonical class of $\mathscr{S}$ (13.9) and is constructed as follows: For each non-zero ray φ we choose a decomposition $\varphi = \varphi_r \ldots \varphi_1$ of φ into irreducible morphisms φ_i; we then set $g(\varphi) = \tilde{\varphi}_r \ldots \tilde{\varphi}_1$ and define f by $g(\psi)g(\varphi) = f(\psi, \varphi)g(\psi\varphi)$. $\surd$

13.14. The Case char $k = 2$. In this case, the term $a + \frac{1}{2}a^{-1}b\bar{\rho}$ of 13.13 is not defined. In fact, if $e(n-1) \neq 1$, the category $P_e = \mathscr{P}F/E_F$ of 13.12 is the ray-category of 2 non isomorphic distributive spectroids: the standard spectroid $k(P_e)$ and the non-standard $k_{\bullet}(P_e)$ which is defined by the quiver F of 13.12 and the relations $0 = \alpha_{ei} \ldots \alpha_1 \rho \alpha_n \ldots \alpha_{i+1} = \alpha_1 \alpha_n - \alpha_1 \rho \alpha_n = \alpha_n \ldots \alpha_1 - \rho^2$, $1 \leqslant i < n$.

In the more general case of a ray-category P satisfying the assumption of Theorem 13.12, we denote by $\mathscr{P}\!f_P$ the set of penny-farthings such that $\alpha_1 \rho \alpha_n \neq 0$ (with the notations of 13.12). To each cocycle f and each subset $\mathscr{N} \subset \mathscr{P}\!f_P$ we can then attach a new distributive spectroid $k^f_{\mathscr{N}}(P)$ with ray-category P: The objects and the morphisms are the same as in $k^f(P)$ and $k(P)$. As in 13.9, we define the composition $(\tau, \sigma) \mapsto \tau^f_{\mathscr{N}}\sigma$ of $k^f_{\mathscr{N}}(P)$ on the non-zero morphisms of P which we identify with "basis-morphisms" of $k^f_{\mathscr{N}}(P)$. For this definition, we first consider the case $\tau = \alpha_j \ldots \alpha_i$, $\sigma = \alpha_n \ldots \alpha_{i+1}$, where the α_p are arrows of a penny-farthing of $\mathcal{N}$ and $1 \leqslant i \leqslant n-1$, $j < e(i)$ (notations as in 13.12); we then set $\tau^f_{\mathcal{N}}\sigma = \tau^f_{\bullet}\rho^f_{\bullet}\sigma = f(\rho, \sigma) f(\tau, \rho\sigma)\tau\rho\sigma$, where $0 \neq \tau\rho\sigma \in P(x_i, x_j)$. In all other cases, $\tau^f_{\mathcal{N}}\sigma = \tau^f_{\bullet}\sigma$ (13.9). The proof of the associativity of the new composition uses the following statement and its dual[6]: If x lies on the penny-farthing and y outside, and if $P(y, x_0) \neq 0$, then $\rho P(y, x_0) = 0 = P(x, y)$.

Theorem. *Suppose that* char $k = 2$, *that P is a ray-category and that there is no cleaving functor $C: \Sigma \to P'/\mathcal{I}$ where Σ is crucial and $\mathcal{I}$ a non-zero ideal of a full subcategory P' of P having at most 5 objects. Then each distributive spectroid $\mathcal{S}$ with ray-category P is isomorphic to $k^f_{\mathcal{N}}(P)$ for some cocycle $f \in Z(P)$ and some $\mathcal{N} \subset \mathcal{P}\!f_P$. Moreover, $k^f_{\mathcal{N}}(P)$ is isomorphic to $k^{f\delta g}_{\mathcal{N}}(P)$ for each coboundary δg.*

Of course, each penny-farthing of $\mathcal{N}$ gives rise to a full subcategory of $k^f_{\mathcal{N}}(P)$ which is isomorphic to some non-standard $k_{\bullet}(P_e)$.

Proof. We start as in 13.13 by choosing representatives $\bar{\xi} \in \xi \subset \mathcal{S}(u, v)$ of the arrows ξ of $Q_{\mathcal{S}}$. For each dumb-bell and each diamond, we replace $\bar{\tau}$ and $\bar{\beta}$ by $\tilde{\tau}$ and $\tilde{\beta}$ as in 13.13. However, if $\{\alpha_n|\ldots|\alpha_1, \rho^2\}$ is a penny-farthing, we now set $\tilde{\alpha}_n = \zeta\bar{\alpha}_n$, where $\bar{\rho}^2 = \zeta\bar{\alpha}_n \ldots \bar{\alpha}_1$. We further set $\tilde{\xi} = \bar{\xi}$ if ξ is not of the form τ, β or α_n. In this way, we obtain a new choice of representatives such that $\tilde{\beta}_m \ldots \tilde{\beta}_1$ and $\tilde{\gamma}_l \ldots \tilde{\gamma}_1$ are proportional for each contour $\{\beta_m|\ldots|\beta_1, \gamma_l|\ldots|\gamma_1\}$. Unfortunately, if our first choice $\xi \mapsto \bar{\xi}$ annihilates all minimal zero-paths, this may be no longer true for the new choice $\xi \mapsto \tilde{\xi}$ and the paths $\alpha_1|\alpha_n$.

Denote by $\mathcal{N}$ the set of penny-farthings of P such that $\tilde{\alpha}_1\tilde{\alpha}_n \neq 0$, i.e. $\tilde{\alpha}_1\tilde{\alpha}_n = u\tilde{\alpha}_1\tilde{\rho}\tilde{\alpha}_n$, where $u \in k^{\bullet}$. With the help of $\mathcal{N}$, we construct a category $P_{\mathcal{N}}$ which has the same objects and morphisms as P, but has a new composition $(\tau, \sigma) \mapsto \tau_{\mathcal{N}}\sigma$: For each penny-farthing of $\mathcal{N}$, we set $\tau_{\mathcal{N}}\sigma = \tau\rho\sigma$ for $\tau = \alpha_j \ldots \alpha_1$ and $\sigma = \alpha_n \ldots \alpha_{i+1}$ (notations as above); in all other cases $\tau_{\mathcal{N}}\sigma = \tau\sigma$. The category $P_{\mathcal{N}}$ satisfies the conditions a) – e) of 13.4, but not f) if $\mathcal{N} \neq \emptyset$. This permits us to define contours, cocycles and coboundaries of $P_{\mathcal{N}}$ as we did with P (13.6 and 13.9). In fact, the contours of $P_{\mathcal{N}}$ are those of P plus some contours $\{\alpha_1|\alpha_n, \alpha_1|\rho|\alpha_n\}$, one for each penny-farthing of $\mathcal{N}$. The first part of our proof therefore implies that the map $\xi \mapsto \tilde{\xi}$ extends to an isomorphism $k^c(P_{\mathcal{N}}) \xrightarrow{\sim} \mathcal{S}$, where c is a suitable cocycle on $P_{\mathcal{N}}$.

So it remains to examine $Z(P_{\mathcal{N}})$: A cocycle $c \in Z(P_{\mathcal{N}})$ is defined for all composable pairs (τ, σ) of P such that $\tau\sigma \neq 0$ and for all pairs $(\alpha_j \ldots \alpha_1, \alpha_n \ldots \alpha_{i+1})$ as above. We therefore have a restriction $Z(P_{\mathcal{N}}) \to Z(P)$, whose kernel is identified with $k^{\bullet\mathcal{N}}$ and lies in $B(P_{\mathcal{N}})$: A function $h: \mathcal{N} \to k^{\bullet}$ is identified with a cocycle having at $(\alpha_j \ldots \alpha_1, \alpha_n \ldots \alpha_{i+1})$ the same value as h at $\{\alpha_n|\ldots|\alpha_1, \rho^2\} \in \mathcal{N}$. Reversely, each cocycle f on P has a natural extension to $P_{\mathcal{N}}$: Set $f(\tau, \sigma) = f(\rho, \sigma) f(\tau, \rho\sigma)$ if $\tau = \alpha_j \ldots \alpha_1$ and $\sigma = \alpha_n \ldots \alpha_i$! We infer that $Z(P_{\mathcal{N}})$ is identified with $k^{\bullet\mathcal{N}} \times Z(P)$, $H(P_{\mathcal{N}})$ with $H(P)$ and $k^f(P_{\mathcal{N}})$ with $k^f_{\mathcal{N}}(P)$ if $f \in Z(P)$ $\surd$

13.15. For the determination of the representation type of $k^f(P)$ and $k^f_{\mathcal{N}}(P)$ we need the following lemma.

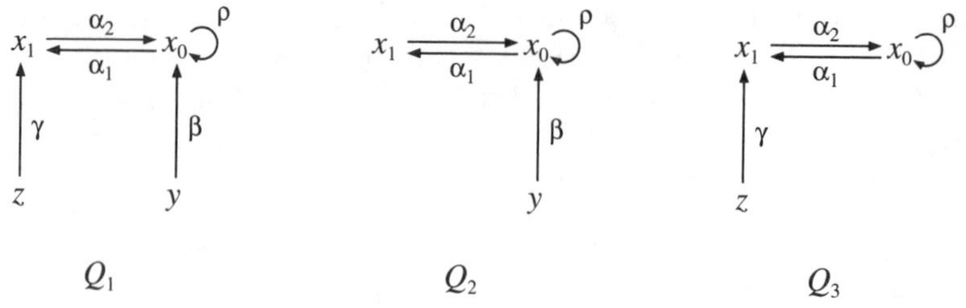

Fig. 9

Lemma. *Let P be a ray-category such that each full subcategory with at most 5 objects is finitely represented or minimal representation-infinite. Let F be the full subquiver of Q_P supporting a penny-farthing whose loop-vertex x_0 satisfies $P(y, x_0) \neq 0$ for some $y \notin F$. Then F has 2 objects, and the full subquiver of Q_P consisting of the arrows having at least one extremity in F has the form Q_1, Q_2 or Q_3 of Fig. 9. In all 3 cases, no further arrow of Q_P starts at y or z.*

a) *In case Q_1, $0 = \alpha_1\beta = \alpha_2\gamma = \rho\beta = \beta\delta = \gamma\varepsilon$ for all δ, ε stopping at y or z.*

b) *In case Q_2, $0 = \rho\beta = \beta\delta$ for each δ stopping at y.*

c) *In case Q_3, $0 = \rho\alpha_2\gamma = \gamma\varepsilon$ for each ε stopping at z.*

The proof[7] of the lemma is an easy application of the cleaving technique. Besides the crucial ray-categories of 13.10, it needs 11 further ray-categories of critical arboresque spectroids. In fact, the assumption really needed is that there is no cleaving functor $C: \Sigma \to P'/\rho^3$, where Σ belongs to the enlarged crucial list and P' is a full subcategory of P containing at most 4 objects besides x_0.

13.16. Let P satisfy the assumption of Lemma 13.15. For each $\mathcal{N} \subset \mathcal{P}\!f_P$, we denote by $\mathcal{F}_{\mathcal{N}}$ an ideal of P whose non-zero morphisms correspond bijectively to the penny-farthings of $\mathcal{N}$: Each penny-farthing of $\mathcal{N}$ which gives rise to case Q_3 of 13.15 or to its dual contributes to $\mathcal{F}_{\mathcal{N}}$ the endomorphism $\alpha_1\rho\alpha_2$ of its *crux* x_1. For any other penny-farthing of $\mathcal{N}$, the *crux* will denote the loop-vertex x_0, and the contribution to $\mathcal{F}_{\mathcal{N}}$ will be $\rho^3 \in P(x_0, x_0)$.

If f is a cocycle on P, we denote by $k\mathcal{F}_{\mathcal{N}}$ the ideal of $k^f_{\mathcal{N}}(P)$ associated with $\mathcal{F}_{\mathcal{N}}$. The residue-spectroid $k^f_{\mathcal{N}}(P)/k\mathcal{F}_{\mathcal{N}}$ then is isomorphic to $k^{\bar{f}}(P/\mathcal{F}_{\mathcal{N}})$, where $\bar{f}$ is the cocycle on $P/\mathcal{F}_{\mathcal{N}}$ induced by f.

Theorem.[7] *Let $M \in \operatorname{mod} k^f_{\mathcal{N}}(P)$ be an indecomposable which is not annihilated by the ideal $k\mathcal{F}_{\mathcal{N}}$ above. Then M is isomorphic to $x^\wedge$ or to $x^\vee$, where x is the crux of a suitable penny-farthing of $\mathcal{N}$.*

Proof. Consider some $F \in \mathcal{N}$. Denote by $\mathcal{F}_x$ the ideal of P which is generated by ρ^3 if the crux x of F coincides with the loop-vertex, or else by $\alpha_1\rho\alpha_n$. It then suffices to prove that for each x, $x^\wedge$ and $x^\vee$ are the only indecomposables not annihilated by $\mathcal{F}_x$. We consider 3 cases:

a) *$P(x_0, y) = 0 = P(y, x_0)$ for each $y \notin F$.* Then $x^\vee$ is isomorphic to $x^\wedge$, and the proof is antique: Each $m \in M(x_0)$ such that $m\rho^3 \neq 0$ gives rise to a morphism

$m^\wedge: x^\wedge \to M$ with non-zero restriction to the socle, which is generated by $\rho^3 \in x^\wedge(x_0)$. Therefore, $m^\wedge$ provides a monomorphism $x^\vee \to M$, which must be invertible because $x^\vee$ is injective and M indecomposable.

b) *Q_P contains an arrow starting outside F and stopping at x_0.* We then obtain case Q_1 or Q_2 of 13.15 and the new demonstration is a perestroika of the antique: First suppose that $\operatorname{Ker} M(\beta) \subset \operatorname{Ker} M(\rho)^3$ and choose an $m \in \operatorname{Ker} M(\beta) \setminus \operatorname{Ker} M(\rho)^3$. The associated morphism $m^\wedge: x^\wedge \to M$ then factors through $x^\wedge/\beta \xrightarrow{\sim} x^\vee$ and induces a morphism $x^\vee \to M$ with non-zero restriction to the socle, which is generated by $\rho^3 \in x^\vee(x_0) \xleftarrow{\sim} x^\wedge(x_0)$. We conclude as in case a).

Now suppose that $\operatorname{Ker} M(\beta) \subset \operatorname{Ker} M(\rho)^3 \neq M(x_0)$. Then we choose an $m \in M(x_0) \setminus \operatorname{Ker} M(\rho)^3$ and a linear form $f: M(x_0) \to k$ such that $f(m) = 1$ and $f | \operatorname{Ker} M(\rho)^3 = 0$. The last equality implies the existence of linear forms g and h on $M(x_0)$ and $M(y)$ such that $gM(\rho)^3 = f = hM(\beta)$. The forms g and h define morphisms $M \to x^\vee$ and $M \to y^\vee$. Since $x^\wedge$ is naturally embedded into $x^\vee \oplus y^\vee$, the equality $gM(\rho)^3 = hM(\beta)$ is equivalent to the existence of a morphism $v: M \to x^\wedge$ such that $v(\xi_0) = g(a\rho^3)\mathbb{1}_{x_0} + g(a\rho^2)\rho + g(a\rho)\rho^2 + g(a)\rho^3 \in x^\wedge(x_0)$ for each $a \in M(x_0)$, $v(b) = g(b\alpha_1\rho)\alpha_2 + g(b\alpha_1)\rho\alpha_2$ for each $b \in M(x_1)$ and $v(c) = h(c)\beta$ for each $c \in M(y)$. The equality $g(m\rho^3) = f(m) = 1$ means that $v(m)$ generates $x^\wedge$. Hence v is an epimorphism which must be invertible because $x^\wedge$ is projective and M indecomposable.

c) *$P(z, x_0) \neq 0$ for some $z \notin F$, but there is no arrow of Q_P outside F which stops at x_0.* This is case Q_3 of 13.15. We then have $x = x_1$. The argument is similar to that of b).

The remaining cases are dual to b) or c).

13.17. A spectroid $\mathscr{S}$ is called *minimal representation-infinite* if it is locally bounded and not finitely represented though $\mathscr{S}/\mathscr{I}$ is finitely represented for each ideal $\mathscr{I} \neq 0$.

Theorem. *Let $\mathscr{S}$ be a distributive spectroid. Then:*

a) *$\mathscr{S}$ is minimal representation-infinite if and only if so is $\vec{\mathscr{S}}$. Under these conditions, $\mathscr{S}$ is isomorphic to $k^f(\vec{\mathscr{S}})$ for some cocyle f; it is standard if $\vec{\mathscr{S}}$ is zigzag-free.*

b) *$\mathscr{S}$ is finitely represented if and only if so is $\vec{\mathscr{S}}$. Under these conditions, $\mathscr{S}$ is standard if* char $k \neq 2$; *it is isomorphic to some $k^1_{\mathscr{N}}(\vec{\mathscr{S}})$ if* char $k = 2$.

Proof.[8] 1) First we suppose *$\vec{\mathscr{S}}$ finitely represented*, hence zigzag-free (13.9): If char $k \neq 2$, it follows that $\mathscr{S} \xrightarrow{\sim} k(\vec{\mathscr{S}})$ (13.9 and 13.10), hence that $\mathscr{S}$ is finitely represented. If char $k = 2$, $\mathscr{S}$ is isomorphic to some $k^1_{\mathscr{N}}(\vec{\mathscr{S}})$ by 13.9 and 13.14; by 13.16, almost all indecomposables of $k^1_{\mathscr{N}}(\vec{\mathscr{S}})$ are annihilated by some ideal $k\mathscr{F}_{\mathscr{N}}$; since $k^1_{\mathscr{N}}(\vec{\mathscr{S}})/k\mathscr{F}_{\mathscr{N}} \xrightarrow{\sim} k(\vec{\mathscr{S}}/\mathscr{F}_{\mathscr{N}})$ is finitely represented, so is $\mathscr{S} \xrightarrow{\sim} k^1_{\mathscr{N}}(\vec{\mathscr{S}})$.

2) Now we suppose *$\mathscr{S}$ finite and finitely represented.* By induction on $\dim \mathscr{S} = \sum_{x,y \in \mathscr{S}} \dim \mathscr{S}(x, y)$, we then prove that $\vec{\mathscr{S}}$ is finitely represented: Indeed, by induction hypothesis, we know that $\vec{\mathscr{S}}/\vec{\mathscr{I}}$ (13.5) is finitely represented

for each ideal $\mathscr{I} \neq 0$ of $\mathscr{S}$. By 13.10, it follows that $\mathscr{S} \xrightarrow{\sim} k^f(\vec{\mathscr{S}})$ for some cocycle f. By Theorem 13.9, it thus remains to prove that $\vec{\mathscr{S}}$ is zigzag-free.

Now each zigzag $C\colon \mathscr{P}\mathbb{A}_\infty \to \vec{\mathscr{S}}$ naturally induces a k-functor $C'\colon k(\mathscr{P}\mathbb{A}_\infty) \to k^f(\vec{\mathscr{S}})$. As in the case $f = 1$ considered in Lemma 13.8, the induced morphism of functors $C'(x, y)\colon k(\mathscr{P}\mathbb{A}_\infty)(x, y) \to k^f(\vec{\mathscr{S}})(C'x, C'y)$ admits a retraction. Since $\mathscr{P}\mathbb{A}_\infty$ is not finitely represented, we infer as in 13.8 that $\mathscr{S} \xrightarrow{\sim} k^f(\vec{\mathscr{S}})$ is not finitely represented: contradiction.

3) Suppose $\mathscr{S}$ *infinite and finitely represented*. By part 2), $\vec{\mathscr{S}'}$ is finitely represented for each finite full subspectroid $\mathscr{S}'$ of $\mathscr{S}$. By 13.10 it again follows that $\mathscr{S} \xrightarrow{\sim} k^f(\vec{\mathscr{S}})$ for some f. We then conclude as in part 2).

4) Suppose that $\vec{\mathscr{S}}$ is *minimal representation-infinite*. Then $\vec{\mathscr{S}}/\vec{\mathscr{I}}$ is finitely represented for each ideal $\mathscr{I} \neq 0$ of $\mathscr{S}$. By part 1), it follows that $\mathscr{S}/\mathscr{I}$ is finitely represented. By 2) and 3), $\mathscr{S}$ is not finitely represented.

5) Conversely, if $\mathscr{S}$ is minimal representation-infinite, $\vec{\mathscr{S}}/\vec{\mathscr{I}}$ is finitely represented for each $\mathscr{I} \neq 0$ (part 2 and 3). By part 1), $\vec{\mathscr{S}}$ itself is not finitely represented.

The remaining statements follow from 13.9, 13.10, 13.14 and 13.16. $\surd$

Remark. For each $N \in \mathbb{N}$, there are only finitely many isoclasses of ray-categories having at most N non-zero morphisms, hence only finitely many isoclasses of spectroids $k(P)$ or $k^1_{\mathscr{N}}(P)$ with dimension $\leqslant N$. As a consequence, *there are only finitely many isoclasses of finitely represented algebras with dimension* $\leqslant N$.

13.18. Remarks and References

1. We simply say *finitely represented* instead of "locally representation-finite" [14, Bautista/Gabriel/Roiter/Salmerón, 1983/85].
2. See [79, Janusz, 1969] and [98, Kupisch, 1970]. *Cyclic* means "generated by one element".
3. The main results of section 13 are taken from [14, Bautista/Gabriel/Roiter/Salmerón, 1983/85], a good example of international cooperation in older times and a good example of convergent contributions of four authors and of common achievement. The historical information spread by Math. Rev. 87g: 16031 and by its Erratum is false and slanted respectively.
4. *stable* here means stable under composition (13.6).
5. [20, Bongartz, 1979/80].
6. loc. cit. (3).
7. [26, Bongartz, 1984/85]. For the first occurrence of non-standard algebras see [127, Riedtmann, 1983] and [128, Riedtmann, 1983].
8. The induction argument used in the present proof is due to K. Bongartz.

14. Finitely Represented Algebras

One of the main objects we had in view was a description of the finitely represented algebras (which are finite-dimensional and admit only finitely many isoclasses of indecomposable modules[1]). We devote this last section to an exposition of the attained results in the language developed to this end.

In Sect. 13, we reduced the study of finitely represented algebras over an algebraically closed field k to the investigation of finitely represented finite ray-categories. To be precise: Each exhaustive sequence $x_1, \ldots, x_s$ of objects of such a category P gives rise to a finite-dimensional "matrix-algebra"[2] $\bigoplus_{i,j} k(P)(x_i, x_j)$ (or to several "matrix-algebras" $\bigoplus_{i,j} k^1_{\mathcal{N}}(P)(x_i, x_j)$ if char $k = 2$). And these matrix-algebras coincide up to isomorphism with the finitely represented algebras.

Thus it remains for us to describe the finitely represented ray-categories. To this effect, we shall give full references but must renounce proofs.

14.1. Let us start from an arbitrary ray-category P and soar: We call *covering* of P a functor $F: P' \to P$ such that P' is a ray-category[3] and that the following two conditions are satisfied:

a) If μ' is a morphism of P', $F\mu' = 0$ is equivalent to $\mu' = 0$.

b) For each non-zero $\mu \in P(x, y)$ (resp. $\mu \in P(y, x)$) and each $x' \in P'$ such that $Fx' = x$, there is a unique $y' \in P'$ and a unique $\mu' \in P'(x', y')$ (resp. $\mu' \in P'(y', x')$) such that $F\mu' = \mu$.

The coverings of P are the objects of a category $/P$ in which a morphism from $F': P'' \to P$ to $F: P' \to P$ is determined by a functor $E: P'' \to P'$ such that $F' = FE$. The condition implies that E is a covering of P'. According to a point of view propagated by Grothendieck, *the fundamental group* $\Pi(P, p)$ of P at a point $p \in P$ is the group of automorphisms of the *fibre-functor* Φ_p which maps each object $F: P' \to P$ of $/P$ onto the set $\Phi_p(F) = F^{-1}(p) = \{x' \in P': Fx' = p\}$. The fundamental theorem[4] about coverings states that, *if P is "connected"*[4], *Φ_p induces an equivalence between $/P$ and the category formed by the $\Pi(P, p)$-sets*, i.e. the sets subjected to a left action of $\Pi(P, p)$. In particular, if P is connected, there is a *universal covering* $U: \tilde{P} \to P$ such that $\Phi_p(U) \xrightarrow{\sim} \Pi(P, p)$.

For the reader's convenience, we briefly recall the construction of $\tilde{P}$ and $\Pi(P, p)$: A *walk* of P is a formal composition $\alpha_n | \ldots | \alpha_2 | \alpha_1$, $n \geqslant 1$, of non-zero morphisms or of formal inverses μ^* of non-zero morphisms μ; the domain of α_1 is the *origin* of the walk, the range of α_n is its *terminus*. The *homotopy* is the smallest equivalence relation $\sim$ on the set of all walks such that: 0) $\mathbb{1}_x \sim \mathbb{1}_x^*$ for all x; 1) $\mu|\mu^* \sim \mathbb{1}_y$ and $\mu^*|\mu \sim \mathbb{1}_x$ for all non-zero $\mu \in P(x, y)$; 2) $\mu|\nu \sim \mu\nu$ and $\nu^*|\mu^* \sim (\mu\nu)^*$ for each composable pair (μ, ν) of morphisms of P such that $\mu\nu \neq 0$; 3) $v \sim v'$ implies $u|v \sim u|v'$ and $v|w \sim v'|w$ whenever the involved walks are composable. With these definitions, as in topology, $\Pi(P, p)$ is identified with the group formed by the homotopy classes of walks from p to p. The *universal cover* $\tilde{P}$ has as points the homotopy classes of walks with terminus p ....

Example 1. Figure 1 describes the universal covers of the dumb-bell (case $s = t = 3$ of 13.12), of the penny-farthing (case $n = 7$, $e(i) = 7$, $\forall i$) and of the diamond. In all 3 cases, the fundamental group has one free generator. The notation is the obvious one.

Example 2. Figure 2 describes the universal cover of the ray-category defined by the quiver

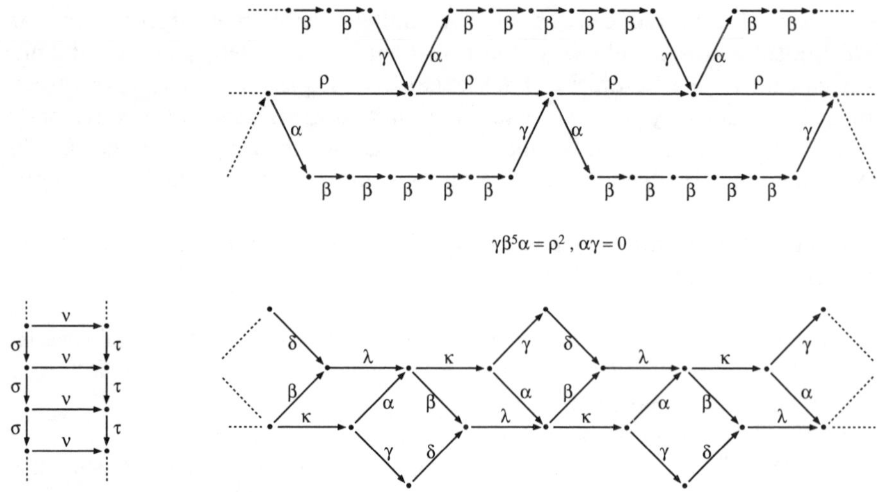

$\gamma\beta^5\alpha = \rho^2$, $\alpha\gamma = 0$

$\tau\nu = \nu\sigma$, $\sigma^3 = 0 = \tau^3$

$\beta\alpha = \delta\gamma$, $\alpha\kappa = \lambda\beta$, $\kappa\lambda = 0$

Fig. 1

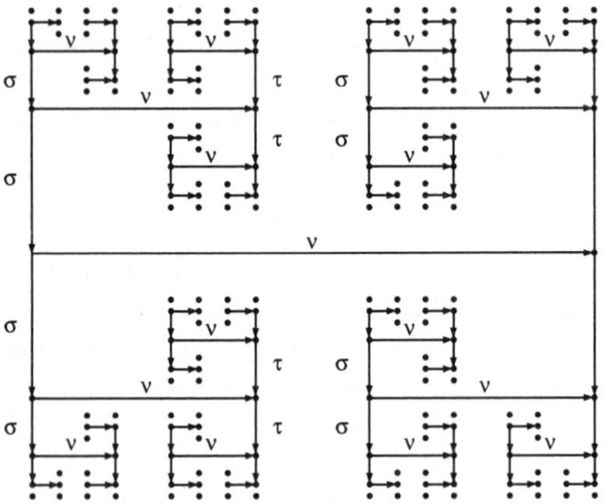

$\sigma^2 = \nu\sigma = \tau^3 = 0$

Fig. 2

$$\overset{\sigma}{\circlearrowright} y \xrightarrow{\nu} x \overset{\tau}{\circlearrowleft}$$

and the relations $\sigma^2 = \nu\sigma = \tau^3 = 0$. In this case, the fundamental group is a free non-commutative group with two generators.

Example 3. If P is the ray-category of a supportive arboresque spectroid or of one of the 4'299 critical arboresque spectroids represented by Fig. 11 of Sect. 10, the universal covering $\widetilde{P} \to P$ is an isomorphism.

As our examples show, the universal covering $U: \tilde{P} \to P$ of a ray-category P induces a *covering*[5] $Q_U: Q_{\tilde{P}} \to Q_P$ of the associated quivers. This is a general fact: If $F: P' \to P$ is an arbitrary covering of the ray-category P, a morphism μ of P' is irreducible if and only if so is $F\mu$. It follows that F induces a covering $Q_F: Q_{P'} \to Q_P$. If F is universal, Q_F may be universal as in Example 2, but in general it is not.

14.2. A ray-category P is called *simply connected* if it is connected and if each covering $F: P' \to P$ with connected domain P' is an isomorphism. It is equivalent to say that P is connected and that the universal covering of P is an isomorphism. Of course, if P is connected, the universal cover $\tilde{P}$ of P is always simply connected.

We further say that a subcategory K of a ray-category P is *convex* if it is full and connected and if all vertices x_i of all paths $x_0 \to x_1 \to \cdots \to x_n$ of Q_P with extremities x_0, x_n in K lie in K.

Theorem. *Let p be a "point" of a ray-category P and $\tilde{P}$ a universal cover of P.*
a) *P is zigzag-free if and only if so is $\tilde{P}$. Under these conditions, the fundamental group $\Pi(P, p)$ is free non-commutative.*
b) *If P is zigzag-free and simply connected, each convex subcategory is simply connected, the quiver Q_P admits only finitely many paths from a point $x \in P$ to a point $y \in P$, and all these paths have the same composition in P.*

This is a central statement in the theory of finitely represented algebras. The proof, of course, is essentially combinatorial.[5]

14.3. Let $F: P' \to P$ be a covering of the ray-category P. Our intention is to examine the relations between the modules over P and P'. We are primarily interested in the left adjoint $F_.$ of the *pull-up functor* $F^*: \operatorname{Mod} k(P) \to \operatorname{Mod} k(P')$, $M \mapsto M \circ F^{op}$. The *push-down* $F_.M'$ of a module $M' \in \operatorname{Mod} k(P')$ is defined by $(F_.M')(x) = \coprod_{Fx'=x} M'(x')$ (the summands are indexed by the points $x' \in P'$ "lying over" $x \in P$); the action of a non-zero morphism $\alpha \in P(x, y)$ is described by Fig. 3, where $\alpha' \in P'(x', y')$ denotes the unique morphism of P' with range y' which lies over α.

As we shall see, the relation between $\operatorname{Mod} k(P)$ and $\operatorname{Mod} k(P')$ is wholly transparent in a context, often met with in practice, which slightly generalizes universal coverings: To be precise, we call *Galois covering*[6] of P the pair formed

$$\coprod_{Fx'=x} M'(x') = (F_.M')(x) \xleftarrow{(F_.M')(\alpha)} (F_.M')(y') = \coprod_{Fy'=y} M'(y')$$

$$\big\uparrow \text{can} \qquad\qquad\qquad\qquad\qquad \big\uparrow \text{can}$$

$$M'(x') \xleftarrow{\quad M'(\alpha') \quad} M'(y')$$

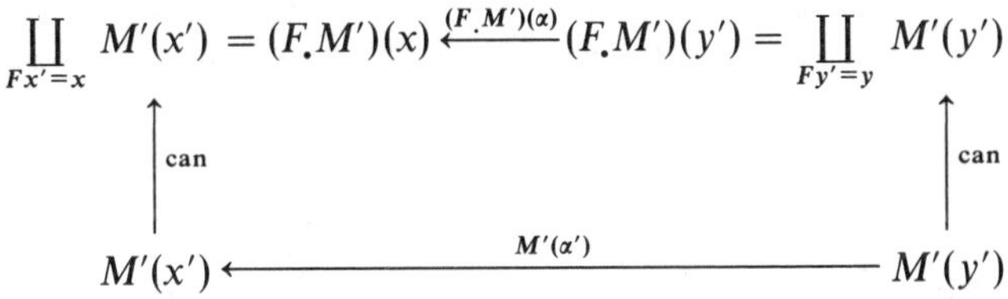

Fig. 3

by a covering $F: P' \to P$ such that $F^{-1}(q) \neq \emptyset$, $\forall q \in P$, and by a group G of automorphisms of F in $/P$ which acts freely and transitively on the fibres $F^{-1}(q)$. For instance, P' may be a direct sum of copies of P on which G acts freely and transitively. More interesting is the case where P is connected and $P' = \tilde{P}/N$ is the quotient of $\tilde{P}$ by a normal subgroup N of the fundamental group: The objects and non-zero morphisms of $\tilde{P}/N$ are the orbits of N in the sets of objects and of non-zero morphisms of $\tilde{P}$. The group acting on $\tilde{P}/N$ is identified with $\Pi(P, p)/N$. The quotient $\tilde{P}/N$ is a ray-category because it covers the ray-category P.

In the following lemma, ${}^gM' \in \operatorname{Mod} k(P')$ denotes the *translate* $x' \mapsto {}^gM'(x') = M'(g^{-1}x')$ of a module $M' \in \operatorname{Mod} k(P')$ by a group element $g \in G$.

Lemma.[7] *Let $F: P' \to P$ be a Galois covering with group G. Then we have $F_\bullet {}^gM' \tilde{\to} F_\bullet M'$ and $\coprod_{g \in G} {}^gM' \tilde{\to} F^\bullet F_\bullet M'$ canonically for all $M' \in \operatorname{Mod} k(P')$ and $g \in G$.* $\surd$

14.4. If $F: P' \to P$ is a covering of the ray-category P and M a finite-dimensional module over P, $F^\bullet M \in \operatorname{Mod} k(P')$ is pointwise finite but in general not finite-dimensional. By contrast, the push-down $F_\bullet M'$ of a finite-dimensional module M' over P' is always finite-dimensional. The question is whether $F_\bullet M'$ is indecomposable if so is M'.

Lemma.[7] *Let $F: P' \to P$ be a Galois covering with group G and $M' \in \operatorname{mod} k(P')$ an indecomposable module over P' such that ${}^gM' \not\tilde{\to} M'$ if $1 \neq g \in G$. Then $F_\bullet M'$ is indecomposable and each $L' \in \operatorname{mod} k(P')$ such that $F_\bullet L' \tilde{\to} F_\bullet M'$ is isomorphic to some ${}^gM'$, $g \in G$* $\surd$

Of course, the condition ${}^gM' \not\tilde{\to} M'$ is satisfied if g generates an infinite subgroup of G. (Since g leaves no point fixed, it cannot stabilize the finite support of M'). It is also fulfilled if the support of M' is finitely represented.[8]

Let us turn to the *representation-quivers* of P and P'. The action of G on $\operatorname{mod} k(P')$ clearly induces an action on the vertices of $\Gamma_{k(P')}$. (For each vertex $V \in \Gamma_{k(P')}$ and each $g \in G$, $gV \in \operatorname{ind} k(P')$ is the unique vertex isomorphic to ${}^gV$.) This action commutes with the translation (10.1). If $\mu(U, V)$ denotes the number of arrows from a vertex U to V, we further have $\mu(U, V) = \mu(gU, gV)$ for all $g \in G$.

Now denote by $\Gamma^1_{k(P')}$ *the union of the connected components of* $\Gamma_{k(P')}$ *which contain no vertex V such that $gV = V$ for some $g \in G$, $g \neq 1$.* The action of G on the vertices of $\Gamma^1_{k(P')}$ then extends to an action on the translation-quiver $\Gamma^1_{k(P')}$ and gives rise to a translation-quiver $\Gamma^1_{k(P')}/G$: The vertices of the quotient are the orbits $GV = \{gV: g \in G\}$ of the vertices V of $\Gamma^1_{k(P')}$; the number of arrows between two vertices is given by $\mu(GU, GV) = \sum_{g \in G} \mu(U, gV)$; finally, $\bar{\tau}(GU)$ is defined as $G\bar{\tau}(U)$ whenever U is not injective.

Theorem.[7] *Let P be a ray-category and $F: P' \to P$ a Galois covering with group G.*

a) *If $U \in \Gamma^1_{k(P')}$ is not injective, the push-down of an almost split sequence of* mod $k(P')$ *with tail U is an almost split sequence of* mod $k(P)$ *with tail $F_\bullet U$.*

b) *The push-down functor* $F_\bullet$: mod $k(P') \to$ mod $k(P)$ *induces an isomorphism of the translation-quiver* $\Gamma^1_{k(P')}/G$ *onto the union of some connected components of* $\Gamma_{k(P)}$.

c) *P is finitely represented if and only if so is P'. If these conditions are satisfied,* $\Gamma^1_{k(P')}$ *equals* $\Gamma_{k(P')}$ *and* $\Gamma_{k(P')}/G$ *is identified with* $\Gamma_{k(P)}$ $\surd$

As the following isomorphisms show, the push-down functor also relates the morphisms of mod $k(P)$ to those of mod $k(P')$:[7]

$$\operatorname{Hom}_{k(P)}(F_\bullet M', F_\bullet N') \xrightarrow{\sim} \operatorname{Hom}_{k(P')}(M', F^\bullet F_\bullet N') \xrightarrow{\sim} \operatorname{Hom}_{k(P')}\Big(M', \coprod_{g \in G} {}^gN'\Big)$$

$$\xrightarrow{\sim} \coprod_{g \in G} \operatorname{Hom}_{k(P')}(M', {}^gN'), \qquad \forall M', N' \in \text{mod } k(P').$$

14.5. Example 1. Let us determine the representation-quiver of the diamond D (13.12). Fig. 4 depicts D and the right half $[\tilde{D}$ of its universal cover (compare with 14.1, Fig. 1). To $[\tilde{D}$ we apply[9] the "knitting algorithm" (compare with 10.2 and 10.4): Fig. 5 reproduces the left end of the obtained postprojective component of $\Gamma_{k([\tilde{D})}$. The numbers ascribed to the vertices are the values of the dimension-function at the left end of $[\tilde{D}$. Fig. 6 describes the part of $\Gamma_{k([\tilde{D})}$ between two "incisions". The left incision coincides with the incision in Fig. 5. The right incision is obtained from the left one by translation. If V is a vertex on the left incision, V' its translate and d, d' the corresponding dimension-functions, we have $d'(p_0) = 0$ and $d'(p_{n+1}) = d(p_n)$ for $p = x, y, z$ or t and $n \in \mathbb{N}$. It follows that the postprojective component of $\Gamma_{k([\tilde{D})}$ is obtained by translating the piece of Fig. 6 indefinitely to the right and by gluing the obtained pieces together along the translated incisions. Accordingly, the dimension-functions are simply "shifted." In particular, if we fix a point p_n of $[\tilde{D}$, almost all postprojectives vanish at p_n. This implies that $[\tilde{D}$ is finitely represented and that $\Gamma_{k([\tilde{D})}$ coincides with its postprojective component.

Since $[\tilde{D}$ is finitely represented, so is $\tilde{D}$ itself. The representation-quiver $\Gamma_{k(\tilde{D})}$ is obtained by translating the pieces of Fig. 6 indefinitely to the right *and* to the left. Gluing all these pieces together gives rise to a stripe, infinite on both sides,

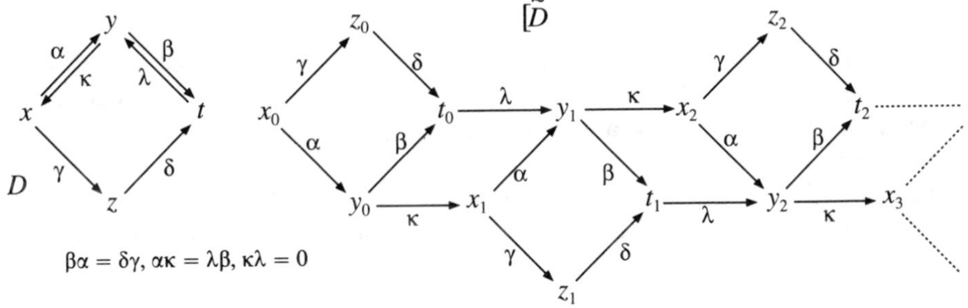

Fig. 4

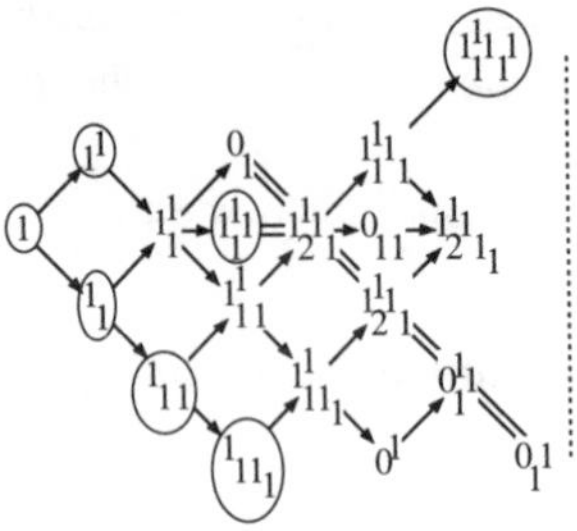

Fig. 5

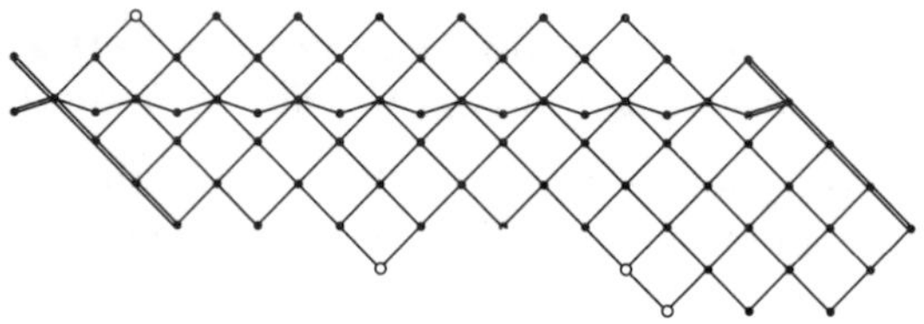

Fig. 6

on which an infinite cyclic group acts by shifting the constitutive pieces. The quotient of this stripe by the group action is identified with the representation-quiver $\Gamma_{k(D)}$ of $k(D)$; this quotient is of course also obtained by identification of both ends of Fig. 6, where the projective vertices are marked with ringlets.

By Theorem 14.6 below, ind $k(D) = \mathcal{I}$ mod $k(D)$ is isomorphic to the mesh-category $k_\tau(\Gamma_{k(D)})$. Thus it is completely determined by Fig. 6.

Example 2. Let $P = \mathscr{P}\tilde{A}_n$ be the ray-category attached to a quiver $\tilde{A}_n$ with shifting orientation (11.3, 13.6). Then $\tilde{P}$ is the ray-category attached to a linear quiver, infinite at both ends and with shifting orientation. The fundamental group is infinite cyclic. The representation-quiver $\Gamma_{k(\tilde{P})}$ has 4 connected components whose quotients by the fundamental group are the components Π, I, Γ_0 and Γ_∞ of 11.3. The remaining components Γ_λ of $\Gamma_{k(P)}$, $\lambda \in k\setminus\{0\}$, are not described by Theorem 14.4.[10]

14.6. Theorem.[11] *Let $\mathscr{S}$ be a finitely represented spectroid, ind $\mathscr{S} = \mathcal{I}$ mod $\mathscr{S}$ the category formed by representatives of the indecomposable finite-dimensional representations, $\Gamma_{\mathscr{S}}$ the representation-quiver of $\mathscr{S}$ and $k_\tau(\Gamma_{\mathscr{S}})$ the associated mesh-category* (10.3). *Then there is an isomorphism $k_\tau(\Gamma_{\mathscr{S}}) \tilde{\rightarrow}$ ind $\mathscr{S}$ which is the identity on the objects if and only if $\mathscr{S}$ is standard, i.e. isomorphic to the linearization $k(\vec{\mathscr{S}})$ of its ray-category $\vec{\mathscr{S}}$* $\surd$

The condition $\mathscr{S} \tilde{\rightarrow} k(\vec{\mathscr{S}})$ is satisfied in particular if char $k \neq 2$ (13.5).

14.7. A ray-category Σ is called *critical* if its linearization $k(\Sigma)$ is a critical arboresque spectroid (10.7). Accordingly, there are 3 countable families of critical spectroids, which are associated with the 3 first series listed in 10.7, Fig. 9 (the spectroids of the last series of Fig. 9 are not distributive). Besides these 3 series, there are 4'299 critical ray-categories which are associated with the spectroids of 10.7, Fig. 11. The entire list has been subjected to various verifications and is considered as secure.

Theorem.[12] *A connected ray-category P is finitely represented if and only if its universal cover $\tilde{P}$ is zigzag-free* (13.9) *and contains no critical convex subcategory.*

The theorem is regarded as a summit of the whole theory. It is powerful and surprisingly handy.[13]

Corollary. *A ray-category P is finitely represented if and only if there is no cleaving functor $C: \Sigma \to P$ where Σ is critical or equal to $\mathscr{P}\mathbb{A}_\infty$* (13.9).

By 13.8 we already know that the condition is necessary. To prove the sufficiency, we may suppose that P is connected and consider a universal covering $U: \tilde{P} \to P$. Like P, $\tilde{P}$ is zigzag-free, and each full subcategory Σ of $\tilde{P}$ gives rise to a cleaving functor $U|\Sigma: \Sigma \to P$. If P is finitely represented, it follows that Σ is not critical. Our theorem then implies that $\tilde{P}$ is finitely represented. $\surd$

Premonition of the existence of such a criterion lingered through the fifties. However, in contrast to the theorem, the corollary seems unwieldy.

Example 1. Let P be the ray-category defined by the quiver Q of Fig. 7 and the relations $\nu\sigma = \tau\nu$, $\sigma^3 = 0 = \tau^5$. Then it is easy to verify that $\tilde{P}$ (Fig. 7) is zigag-free. So let us search for a convex critical subcategory Σ. If Q_Σ is a tree, the convexity condition enforces linearity which contradicts criticality. If Q_Σ is not a tree, it contains a "square" of $Q_{\tilde{P}}$. If it contains only one, we examine the critical arboresque spectroids whose quivers contain 1 square (10.7: second series of Fig. 9 and families 11–35 of Fig. 11); we then notice that the convexity condition is only compatible with the families 12, 14, 17, 20 and 21; these 5 families now are excluded by the relations $\sigma^3 = 0 = \tau^5$. Finally, in the case of 2 squares, the convexity condition is only compatible with the families 85, 86 and 87 of 10.7, Fig. 11. These 3 families are excluded by our relations. As a consequence, P is finitely represented.

Example 2. Let now T be the ray-category defined by the same quiver as P (Fig. 7) and the relations $\nu\sigma = \tau\nu$, $0 = \sigma^3 = \tau^6 = \tau^2\nu$. In this case, $\tilde{P}$ contains a critical convex subcategory whose linearization belongs to the family 12 of 10.7, Fig. 11. Hence, P is not finitely represented.

14.8. Theorem.[14] *If a finite-dimensional algebra A over an algebraically closed field k is not finitely represented, there are infinitely many dimensions for which the number of isoclasses of indecomposable A-modules is infinite.*

Proof-sketch. Since A is modularly equivalent to the algebra of its spectroid, we are reduced to the corresponding statement for a *finite* spectroid $\mathscr{S}$. If $\mathscr{S}$ is not distributive, $\mathscr{S}$ admits a quotient $\mathscr{T}$ of a full subcategory whose quiver $Q_{\mathscr{T}}$ has one of the following forms: $\circlearrowright\cdot\circlearrowleft$, $\cdot\rightrightarrows\cdot$, $\circlearrowright\cdot\leftrightarrows\cdot\circlearrowleft$, $\sigma\circlearrowright\cdot\xrightarrow{\delta}\cdot\circlearrowleft\rho$. In the first

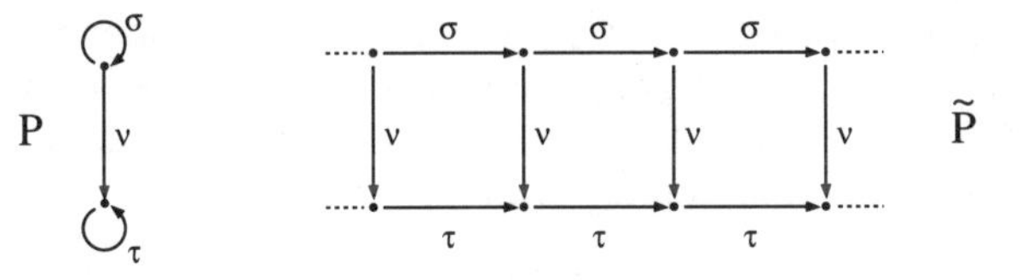

Fig. 7

three cases, we may further suppose that $\mathscr{R}_{\mathscr{T}}^2 = 0$; in the last case, we may suppose that $\mathscr{T}$ is defined by the relations $0 = \sigma^2 = \rho^2 = \rho\delta\sigma$. All four cases are easy and well-known.

Now, if $\mathscr{S}$ is distributive, we may assume that $\mathscr{S} \xrightarrow{\sim} k^f(\vec{\mathscr{S}})$ for some f (13.17a). Each zigzag of $\vec{\mathscr{S}}$ then provides a cleaving functor $C: \mathscr{P}\tilde{A}_n \to \vec{\mathscr{S}}$, where $\mathscr{P}\tilde{A}_n$ is the ray-category attached to a quiver of type $\tilde{A}_n$ with shifting orientation (11.3). The cleaving functor C extends to a linear functor $C': k\tilde{A}_n \to k^f(\vec{\mathscr{S}})$ which induces a restriction-functor $C^{\bullet}$: mod $k^f(\vec{\mathscr{S}}) \to$ mod $k\tilde{A}_n$ and its right adjoint $C_{\bullet}$: mod $k\tilde{A}_n \to$ mod $k^f(\vec{\mathscr{S}})$. Since the statement of our theorem holds for $k\tilde{A}_n$, it must hold for $k^f(\vec{\mathscr{S}}) \xrightarrow{\sim} \mathscr{S}$ (extend the argument of Lemma 13.8).

Finally, if $\mathscr{S}$ is distributive and $P = \vec{\mathscr{S}}$ zigzag-free, we have $\mathscr{S} \xrightarrow{\sim} k(P)$ and can apply Theorem 14.7. Supposing P connected, we obtain a critical convex subcategory Σ of $\tilde{P}$. By 12.9, Σ admits "enough" infinite families of indecomposable modules. Extending them to $\tilde{P}$ by 0 and pushing the extended modules down to P (14.4), we obtain the required infinite families for P. $\surd$

14.9. For mathematicians working in the area, the conjecture[15] leading to Theorem 14.8 has been a strong incentive for forty years. In contrast to its influence, the amount of its applications is surprisingly small. There is at least one, unexpected and unconjectured before its discovery.

Theorem.[16] *Let Alg_n denote the algebraic subvariety of* $\mathrm{Hom}_k(k^n \otimes_k k^n, k^n)$ *formed by the multiplication laws which are associative and admit a unit element* ($n \in \mathbb{N}$). *The multiplication laws defining a finitely represented algebra then form an open subvariety of Alg_n.*

Corollary. *Let V be an infinite irreducible algebraic subvariety of Alg_n formed by pairwise "non-isomorphic" multiplication laws. Then each point of V gives rise to an infinitely represented algebra-structure on k^n.*

14.10. Having thus brought our discourse to its natural conclusion, we cannot close it without at least mentioning important developments extending beyond its scope[17]. We must here content ourselves with reproducing parts of [CB] which round off the view afforded by Corollary 4.5 and Theorem 14.8.

Let $k[X]$ and $k\langle X, Y\rangle$ denote the free associative algebras generated by X and by X, Y respectively (the algebras of the quivers consisting of 1 and 2 loops!). A *finite-dimensional* algebra A is called *tame*[18] if it is not finitely represented and if, for each $d \in \mathbb{N}$, there are finitely many $k[X]$-A-bimodules M_i which are free of rank d over $k[X]$ and such that each indecomposable A-module of dimension d is isomorphic to $k[X]/(X - \lambda) \otimes_{k[X]} M_i$ for some i and some $\lambda \in k$. It is called *wild* if there is a $k\langle X, Y\rangle$-A-bimodule which is finitely free over $k\langle X, Y\rangle$ and such that the functor F: $\mathrm{mod}_{pf}\, k\langle X, Y\rangle \to \mathrm{mod}\, A$, $N \mapsto N \otimes_{k<X,Y>} M$ preserves indecomposability and heteromorphism[19].

Theorem 1.[20] *Each finite-dimensional algebra A is either wild or tame or finitely represented.*

Let us call *tube of circumference* $c \geqslant 1$ a translation-quiver which is isomorphic to the connected component Γ_0 of the representation-quiver Γ_Q of a quiver Q of type $\tilde{A}_{c-1}$ with steady orientation (11.2).

Theorem 2.[21] *The representation-quiver Γ_A of a tame finite-dimensional algebra A has at most countably many connected components which are not tubes of circumference* 1. *Among these components, only finitely many contain an indecomposable of prescribed dimension.*

14.11. Remarks and References

1. If an algebra A is finitely represented, each A-module is a (direct) sum of finite-dimensional indecomposables (see for instance [133, Ringel/Tachikawa, 1975]).
2. The multiplication obeys the law of matrix-multiplication.
3. It would be enough to require that P' has "zero-morphisms".
4. *Connected* means that the quiver Q_P (13.6) is connected. For finite coverings, A. Grothendieck's point of view is explained in [66, 1970].
5. Here we mean coverings in the topological-combinatorial sense. For a proof of the theorem, see [46, Fischbacher, 1981].
6. Through coverings of algebraic varieties, the notion is related to Galois extensions of fields of functions.
7. [61, Gabriel, 1980/81]. Compare with [65, Green, 1980/81].
8. [104, Martinez/de la Peña, 1983].
9. This method of sectioning $\tilde{D}$ is due to K. Bongartz.
10. For an investigation of the remaining components by means of coverings, see [40, Dowbor/Skowronski, 1987].
11. [27, Bongartz/Gabriel, 1980/82], [33, Bretscher/Gabriel, 1981/83].
12. [24, Bongartz, 1982/84], [25, Bongartz, 1984].
13. [45, Fischbacher, 1986], [47, Fischbacher/de la Peña, 1986].
14. [12, Bautista, 1984/85], [26, Bongartz, 1984/85].
15. The conjecture, known as second conjecture of Brauer-Thrall, is attributed to R. Brauer and R.M. Thrall in [78, Jans, 1954/57].
16. [57, Gabriel, 1975].
17. The proofs use the notion of a bocs (see [136, Roiter, 1980]).
18. Here we restrict the word *tame* to algebras which are not finitely represented.
19. *heteromorphism* = property of not being isomorphic!
20. [43, Drozd, 1980]; see also (21) and [150, Gabriel/Nazarova/Roiter/Sergejchuk/Vossieck, 1992].
21. [37, Crawley-Boevey, 1988].

Bibliography

[1] Assem I., Happel D.: Generalized tilted algebras of type A_n. Commun. Algebra *9* (1981), 2101–2125, Zbl.481.16009. Erratum, Commun. Algebra *10* (1982), 1475.

[2] Assem I., Skowronski A.: Iterated tilted algebras of type D_n. Preprint.

[3] Auslander M.: Representation dimension of Artin algebras. Queen Mary College Mathematics Notes, London (1971), Zbl.331.16026; see also Representation theory of Artin algebras I. Commun. Algebra, *1* (1974), 177–268, Zbl.285.16028; II, 269–310, Zbl.285.16029.

[4] Auslander M.: Applications of morphisms determined by objects. Proc. Conf. Represent. Theory, Philadelphia (1976), Lect. Notes Pure App. Math. *37* (1978), 245–327, Zbl.404.16007.

[5] Auslander M.: Isolated singularities and existence of almost split sequences (Ottawa 1984). Lect. Notes Math. *1178* (1986), 194–242 (notes by L. Unger), Zbl.633.13007.

[6] Auslander M., Platzeck M.I., Reiten I.: Coxeter functors without diagrams. Trans. Am. Math. Soc. *250* (1979), 1–46, Zbl.421.16016.

[7] Auslander M., Reiten I.: Stable equivalence of dualizing varieties I. Advances in Math. *12* (1974), 306–366, Zbl.285.16027.

[8] Auslander M., Reiten I.: Representation theory of Artin algebras. Commun. Algebra, III, *3* (1975), 239–294, Zbl.331.16027; IV, *5* (1977), 443–518, Zbl.396.16007; V, *5* (1977), 519–554, Zbl.396.16008; VI, *6* (1978), 257–300, Zbl.446.16027.

[9] Auslander M., Smalo S.O.: Almost split sequences in subcategories. J. Algebra *69* (1981), 426–454; Addendum *71* (1981), 592–594, Zbl.457.16017 and Zbl.474.16022.

[10] Bautista R.: Sections in Auslander-Reiten quivers (Ottawa 1979). Lect. Notes Math. *832* (1980), 74–96, Zbl.446.16029.

[11] Bautista R.: Sections in Auslander-Reiten components II. An. Inst. Mat. UNAM *20* (1980), 157–175, Zbl.532.16021.

[12] Bautista R.: On algebras of strongly unbounded representation type. (Ottawa 1984) Comment. Math. Helv. *60* (1985), 392–399, Zbl.584.16017.

[13] Bautista R., Brenner S.: Replication numbers for non-Dynkin sectional subgraphs in finite Auslander-Reiten quivers and some properties of Weyl roots. Proc. Lond. Math. Soc., III. ser. *47* (1983), 429–462, Zbl.505.16012.

[14] Bautista R., Gabriel P., Roiter A.V., Salmerón L.: Representation-finite algebras and multiplicative bases (Torun 1983). Invent. Math. *81* (1985), 217–285, Zbl.575.16012.

[15] Bautista R., Martinez R.: Representations of partially ordered sets and 1-Gorenstein Artin algebras. Proc. Conf. on Ring theory (Antwerp 1978). Lect. Notes Pure Appl. Math. *51* (1979) 385–433, Zbl.707.16003.

[16] Beilinson A.A.: Coherent sheaves on $\mathbb{P}^n$ and problems of linear algebra. Funkts. Anal. Prilozh. *12* (1978), 68–69, Zbl.402.14006. English transl.: Funct. Anal. Appl. *12* (1979) 214–216.

[17] Beilinson A.A., Bernstein I.N., Deligne P.: Faisceaux pervers. Astérisque *100* (1982), Zbl.536.14011.

[18] Bernstein I.N., Gelfand I.M., Gelfand S.I.: Algebraic bundles on $\mathbb{P}^n$ and problems of linear algebra. Funkts. Anal. Prilozh. *12* (1978), 66–67, Zbl.402.14005. English transl.: Funct. Anal. Appl. *12* (1979) 212–214.

[19] Bernstein I.N., Gelfand I.M., Ponomarev V.A.: Coxeter functors and Gabriel's theorem. Usp. Mat. Nauk *28* (1973) 19–33, Zbl.269.08001. English transl.: Russ. Math. Surv. *28* (1973), 17–32.

[20] Bongartz K.: Zykellose Algebren sind nicht zügellos (Ottawa 1979). Lect. Notes Math. *832* (1980), 97–102, Zbl.457.16020.

[21] Bongartz K.: Tilted Algebras. Lect. Notes Math. *903* (1981), 26–38, Zbl.478.16025.

[22] Bongartz K.: Treue einfach zusammenhängende Algebren (Oberwolfach 1981). Comment. Math. Helv. *57* (1982), 282–330, Zbl.502.16022.

[23] Bongartz K.: Algebras and quadratic forms. J. Lond. Math. Soc., II Ser. *28* (1983), 461–469, Zbl.532.16020.

[24] Bongartz K.: A criterion for finite representation type (Luminy 1982). Math. Ann. *269* (1984), 1–12, Zbl.552.16012.

[25] Bongartz K.: Critical simply connected algebras (Luminy 1982). Manuscr. Math. *46* (1984), 117–136, Zbl.537.16024.

[26] Bongartz K.: Indecomposable modules are standard (Essen 1984). Comment. Math. Helv. *60* (1985), 400–410, Zbl.591.16014.

[27] Bongartz K., Gabriel P.: Covering spaces in representation theory (Puebla 80). Invent. Math. *65* (1982), 331–378, Zbl.482.16026.

[28] Borel A.: Linear Algebraic Groups. Benjamin, New York, 1969, Zbl.186,332.

[29] Bourbaki N.: Groupes et algèbres de Lie. Chap. 4–6; Hermann, Paris, 1968, Zbl.186,330.

[30] Brauer R.: Math. Reviews *8* (1947), 561.

[31] Brenner S.: A combinatorial characterization of finite Auslander-Reiten quivers (Ottawa 1984). Lect. Notes Math. *1177* (1986), 13–49, Zbl.644.16017.

[32] Brenner S., Butler M.C.R.: Generalizations of the Bernstein-Gelfand-Ponomarev reflection functors (Ottawa 1979). Lect. Notes Math. *832* (1980), 103–169, Zbl.446.16031.

[33] Bretscher O., Gabriel P.: The standard form of a representation-finite algebra (Oberwolfach 1981). Bull. Soc. Math. France *111* (1983), 21–40, Zbl.527.16021.

[34] Bretscher O., Läser C., Riedtmann C.: Self-injective and simply connected algebras (Oberwolfach 1981). Manuscripta Math. *36* (1981), 253–307, Zbl.478.16024.

[35] Butler M.C.R.: The construction of almost split sequences I. Proc. Lond. Math. Soc., III. Ser. *40* (1980), 72–86, Zbl.443.16018.

[36] Cartan H., Eilenberg S.: Homological Algebra. Princeton Univ. Press, 1956, Zbl.75,243

[37] Crawley-Boevey W.W.: On tame algebras and bocses. Proc. Lond. Math. Soc., III. Ser. *56* (1988), 451–483, Zbl.661.16026.

[38] Dlab V., Ringel C.M.: Indecomposable representations of graphs and algebras (Bonn 1973). Mem. Am. Math. Soc. *173* (1976), Zbl.332.16015.

[39] Donovan P., Freislich M.R.: The representation theory of finite graphs and associated algebras. Carleton Lecture Notes 5, Ottawa (1973), 83 pp., Zbl.304.08006.

[40] Dowbor P., Skowronski A.: Galois coverings of representation-infinite algebras. Comment. Math. Helv. *62* (1987), 311–337. Zbl.628.16019.

[41] Dräxler P.: Aufrichtige gerichtete Ausnahmealgebren. Bayreuther Math. Schr. *29* (1989) see also Sur les algèbres exceptionnelles de Bongartz. Comptes-Rendus Acad. Sc., Paris, Sér. I, *311* (1990), 495–498.

[42] Drozd Ju.A.: Coxeter transformations and representations of posets. Funkt. Anal. Prilozh. *8* (1974), 34–42. English transl.: Funct. Anal. Appl. *8* (1974), 219–225, Zbl.356.06003.

[43] Drozd Ju.A.: Tame and wild matrix problems (Ottawa 1979). Lect. Notes Math. *832* (1980), 242–258, Zbl.457.16018.

[44] Dynkin E.: The structure of semi-simple Lie-algebras. Usp Mat. Nauk *2* (1947), 59–127. English transl.: Am. Math. Soc. Transl. *9* (1950), 328–469.

[45] Fischbacher U: The representation-finite algebras with at most 3 simple modules (Luminy 1982). Lect. Notes Math. *1177* (1986), 94–114, Zbl.626.16010.

[46] Fischbacher U.: Zur Kombinatorik der Algebren mit endlich vielen Idealen. J. Reine Angew. Math. *370* (1986), 192–213, Zbl.584.16018.

[47] Fischbacher U., de la Peña J.A.: Algorithms in representation theory of algebras. Lect. Notes Math. *1177* (1986), 115–134, Zbl.631.16010.

[48] Fitting H.: Über die direkten Produktzerlegungen einer Gruppe in direkt unzerlegbare Faktoren. Math. Z. *39* (1934), 16–30, Zbl.9,202.

[49] Freyd P.: Abelian categories. Harper and Row (1964), Zbl.121,21.

[50] Gabriel P., Sur les catégories abéliennes localement noethériennes et leurs applications aux algèbres étudiées par Dieudonné. Séminaire J.P. Serre, Collège de France (1960), 118 pp.

[51] Gabriel P.: Des catégories abéliennes (Paris 1961). Bull. Soc. Math. Fr. *90* (1962), 323–448, Zbl.201,356.

[52] Gabriel P.: Unzerlegbare Darstellungen I (Oberwolfach 1970). Manuscr. Math. *6* (1972), 71–103, Zbl.232.08001.

[53] Gabriel P.: Indecomposable representations II (Rome 1971). Symp. Math. *11* (1973), 81–104, Zbl.276.16001.

[54] Gabriel P.: Lectures in Representation Theory (Ottawa 1972). Manuscript.

[55] Gabriel P.: Représentations indécomposables des ensembles ordonnés (Ottawa 1972). Sémin. P. Dubreil, Paris (1972–73), 301–304, Zbl.329.16019.

[56] Gabriel P.: Degenerate bilinear forms. J. Algebra *31* (1974), 67–72, Zbl.282.15014.

[57] Gabriel P.: Finite representation type is open. Lect. Notes Math. *488* (1975), 132–155, Zbl.313.16034.

[58] Gabriel P., Riedtmann C.: Group representations without groups (Giessen 76, Oberwolfach 77). Comment. Math. Helv. *54* (1979), 240–287, Zbl.447.16023.

[59] Gabriel P.: Christine Riedtmann and the selfinjective algebras of finite representation type. Proc. Conf. on Ring Theory (Antwerp 1978), Lect. Notes Pure Appl. Math. *51* (1979) 453–458, Zbl.431.16008.

[60] Gabriel P.: Auslander-Reiten sequences and representation-finite algebras (Ottawa 1979). Lect. Notes Math. *831* (1980), 1–71, Zbl.445.16023.

[61] Gabriel P.: The universal cover of a representation-finite algebra (Puebla 80). Lect. Notes Math. *903* (1981), 68–105, Zbl.481.16008.

[62] Gelfand I.M., Ponomarev V.A.: Indecomposable representations of the Lorentz group. Usp. Mat. Nauk *23* (1968), 3–60. English transl.: Russ Math. Surv. *23* (1968) 1–58, Zbl.236.22012.

[63] Gelfand I.M., Ponomarev V.A.: Problems of linear algebra and classification of quadruples of subspaces in a finite-dimensional vector space. Colloq. Math. Soc. Janos Bolyai *5* (Tihany 1970), (1972) 163–237, Zbl.294.15002.

[64] Godement R.: Topologie algébrique et théorie des faisceaux. Hermann, Paris, 1958, Zbl.80,162.

[65] Green E.L.: Group-graded algebras and the zero relation problem (Puebla 1980). Lect. Notes Math. *903* (1981), 106–115, Zbl.507.16001.

[66] Grothendieck A.: Revêtements étales et groupe fondamental (Paris 1960). Lect. Notes Math. *224* (1970), 118–142, Zbl.224.14002.

[67] Grothendieck A.: Groupes de classes des catégories abéliennes et triangulées, complexes parfaits. Lect. Notes Math. *589* (1977), 351–371, Zbl.359.18013.

[68] Gruson L.: Simple coherent functors (Ottawa 1974). Lect. Notes Math. *488* (1975), 156–159, Zbl.318.18012.

[69] Happel D.: Composition factors for indecomposable modules. Proc. Am. Math. Soc. *86* (1982), 29–31, Zbl.498.16026.

[70] Happel D.: On the derived category of a finite-dimensional algebra. Comment. Math. Helv. *62* (1987), 339–389, Zbl.626.16008.

[71] Happel D., Ringel C.M.: Tilted algebras (Puebla 1980). Trans. Am. Math. Soc. *274* (1982), 399–443, Zbl.503.16024.

[72] Happel D., Vossieck D.: Minimal algebras of infinite representation type with preprojective component. Manuscripta Math. *42* (1983), 221–243, Zbl.516.16023.

[73] Harada M., Sai Y.: On categories of indecomposable modules I. Osaka J. Math. *7* (1970), 323–344, Zbl.248.18018.

[74] Hochschild G.: On the structure of algebras with nonzero radical. Bull. Am. Math. Soc. *53* (1947), 369–377, Zbl.32,8.

[75] von Höhne H.J.: On weakly positive unit forms. Comment. Math. Helv. *63* (1988), 312–336, Zbl.662.15014.

[76] Hughes D., Waschbüsch J.: Trivial extensions of tilted algebras. Proc. Lond. Math. Soc. *46* (1983), 347–364, Zbl.488.16021.

[77] Jacobson N.: The radical and semi-simplicity for arbitrary rings. Am. J. Math. *67* (1945), 300–320, Zbl.60,73.

[78] Jans J.P.: On the indecomposable representatons of Algebras (Thesis Univ. of Michigan 1954). Ann. Math., II. Ser., *56* (1957), 418–429, Zbl.79,52.

[79] Janusz G.: Indecomposable modules for finite groups. Ann. Math., II. Ser., *89* (1969), 209–241, Zbl.179,23.

[80] Jordan C.: Traité des substitutions et des équations algébriques, Paris 1870. Reprint Paris 1957, Zbl.78,12.

[81] Kac V.: Infinite root systems, representations of graphs and invariant theory (Oberwolfach 1979). Invent. Math. *56* (1980), 57–92, Zbl.427.17001.

[82] Kaplansky I.: Projectives modules. Ann Math., II. Ser., *68* (1958), 372–377, Zbl.83,258.

[83] Kapranov M.M.: The derived categories of coherent sheaves on grassmannians. Funkts. Anal. Prilozh. *17* (1983), 78–79. English transl.: Funct. Anal. Appl. *17* (1983) 145–146, Zbl.571.14007.

[84] Kapranov M.M.: The derived category of coherent sheaves on a quadric. Funkts. Anal. Prilozh. *20* (1986), 67. English transl.: Funct. Anal. Appl. *20* (1986) 141–142, Zbl.607.18004.

[85] Keller B.: Chain complexes and stable categories. Manus. Math. *67* (1990), 379–417.

[86] Keller B.: Derived categories and universal problems. Commun. Algebra. *19* (1991), 699–747.

[87] Keller B.: Algèbres héréditaires par morceaux de type D_n. Comptes-Rendus Acad. Sc., Paris, Sér. I, *312* (1991) 483–486.

[88] Keller B., Vossieck D.: Sous les catégories dérivées. Comptes-Rendus Acad. Sc. Paris, Sér. I, *305* (1987), 225–228, Zbl.628.18003.

[89] Keller B., Vossieck D.: Aisles in derived categories. Bull. Soc. Math. Belg., Ser. A *40* (1988), 239–253, Zbl.671.18003.

[90] Keller B., Vossieck D.: Dualité de Grothendieck-Roos et basculement. Comptes-Rendus Acad. Sc. Paris, Sér. I, *307* (1988), 543–546, Zbl.663.18005.

[91] Kleiner M.M.: Partially ordered sets of finite type. Preprint Inst. Math. Ukr. Ac. Sc. 71-2 (1971), 1–15. Zap. Nauchn. Semin. Leningr. Otd. Mat. Inst. Steklova *28* (1972), 32–41. English transl.: J. Sov. Math. *3* (1975), 607–615, Zbl.345.06001.

[92] Kleiner M.M.: On faithful representations of partially ordered sets of finite type. Preprint loc. cit., 16–22. Zap. Nauchn. Semin. Leningr. Otd. Mat. Inst. Steklova *28* (1972), 42–59. English transl.: J. Sov. Math. *3* (1975), 616–628, Zbl.345.06002.

[93] Kraft H.: Geometric methods in representation theory (Puebla 1980). Lect. Notes Math. *944* (1982), 181–258, Zbl.517.14016.

[94] Kraft H., Riedtmann C.: Geometry of representations of quivers (Durham 1985). In: Representations of Algebras, Lond. Math. Soc. Lect. Notes Ser. *116* (1986), 109–145, Zbl.632.16019.

[95] Kronecker L.: Algebraische Reduktion der Scharen bilinearer Formen, Sitzungsber. Akad. Berlin (1890), 1225–1237, Jbuch. 22, 169.

[96] Kruglyak S.A.: Representations of algebras with zero square radical. Preprint Inst. Math. Ukr. Ac. Sc. 71-4 (1971), 1–12. Zap. Nauchn. Semin. Leningr. Otd. Mat. Inst. Steklova *28* (1972), 60–68, Zbl.323.16012. English transl.: J. Sov. Math. *3* (1975) 629–635.

[97] Krull W.: Über verallgemeinerte endliche abelsche Gruppen. Math. Z. *23* (1925), 161–196, Jbuch 51, 166.

[98] Kupisch H.: Unzerlegbare Moduln endlicher Gruppen mit zyklischer p-Sylow Gruppe. Math. Z. *108* (1969), 77–104, Zbl.188,90.

[99] MacLane S.: Homology. Springer-Verlag, New York Berlin Heidelberg, 1963, Zbl.133,265.

[100] MacLane S.: Categories for the Working Mathematician. Springer-Verlag, New York Berlin Heidelberg, 1971, Zbl.232.18001.

[101] Matlis E.: Injective modules over noetherian rings. Pac. J. Math. *8* (1958): 511–528, Zbl.84,266.

[102] Marmaridis N.: Reflection functors (Ottawa 1979). Lect. Notes Math. *832* (1980), 382–395, Zbl.446.16030.

[103] Martinez R.: Algebras stably equivalent to *l*-hereditary (Ottawa 1979). Lect. Notes Math. *832* (1980), 396–431, Zbl.443.16019.

[104] Martinez R., de la Peña J.A.: Automorphisms of representation-finite algebras. Invent. Math. *72* (1983), 359–362, Zbl.491.16028.

[105] Mitchell B., Theory of Categories. Academic Press, 1965, Zbl.136,6.

[106] Morita K.: Duality for modules and its applications to the theory of rings with minimum condition. Sc. Rep. Tokyo Kyoiku Daigaku, *6* (1958), 83–142, Zbl.80,257.

[107] Müller W.: Unzerlegbare Moduln über artinschen Ringen (Bonn 1973). Math. Z. *137* (1974), 197–226, Zbl.268.16022.

[108] Nazarova L.A.: Integral representations of Klein's four-group. Dokl. Akad. Nauk SSSR *140* (1961), 1011–1014. English transl.: Sov. Math., Dokl. *2* (1961) 1304–1308, Zbl.106,26. Representations of the local ring of a curve with 4 branches. Izv. Akad. Nauk SSSR, Ser. Math. *31* (1967), 1361–1378 English transl.: Math. USSR, Izv. *1* (1967) 1305–1321, Zbl.222.16028.

[109] Nazarova L.A.: Representations of quivers of infinite type. Izv. Akad. Nauk SSSR, Ser. Mat. *37* (1973), 752–791, Zbl.298.15012. English transl.: Math. USSR, Izv. 7 (1973) 749–792.

[110] Nazarova L.A., Roiter A.V.: Representations of partially ordered sets. Zap. Nauchn. Semin. Leningr. Otd. Mat. Inst. Steklova *28* (1972), 5–31. English transl.: J. Sov. Math. *3* (1975), 585–606, Zbl.336.16031.

[111] Nazarova L.A., Roiter A.V.: Representations and forms of weakly completed posets (Russian). In: Linear Algebra and Theory of representations, Ukr. Akad. Nauk (1983), 19–54, Zbl.566,16017.

[112] Nazarova L.A., Roiter A.V.: Representations of bipartite completed posets. Comment. Math. Helv. *63* (1988), 498–526, Zbl.671,18002.

[113] Nazarova L.A., Roiter A.V.: Representations of a biinvolutive poset (Russian). Preprint Inst. Math. Ukr. Ac. Sc. 91-34 (1991), 1–31.

[114] Nazarova L.A., Roiter A.V., Gabriel P.: Représentations indécomposables: un algorithme. Comptes-Rendus Acad. Sc., Paris, Sér. I, *307* (1988), 701–706. Zbl.662.16019.

[115] Nazarova L.A., Zavadski A.G.: Partially ordered sets of finite growth. Funkts. Anal. Prilozh. *16* (1982), 72–73. English transl.: Funct. Anal. Appl. *16* (1982), 135–137, Zbl.499.16021.

[116] Nesbitt C.: On the regular representations of algebras. Ann. Math., II. Ser., *39* (1938), 634–658, Zbl.19,102.

[117] Nesbitt C., Scott W.M.: Some remarks on algebras over an algebraically closed field. Ann. Math., II. Ser. *44* (1943), 534–553, Zbl.60,82.

[118] Ovsienko S.A.: Weakly positive integral quadratic forms (Russian). In: Schurian Matrix problems and quadratic forms. Ukr. Akad. Nauk (1978), 3–17.

[119] Ovsienko S.A.: Boundedness of roots of weakly positive integral quadratic forms (Russian). In: Representations and quadratic forms, Ukr. Akad. Nauk *155* (1979), 106–123, Zbl.449.15018.

[120] Parthasarathy R. *t*-structures dans la catégorie dérivée associée aux représentations d'un carquois. Comptes-Rendus Acad. Sc., Paris, Sér. I, *304* (1987), 355–357, Zbl.613.16012.

[121] de la Peña J.A., Takane M.: Spectral properties of Coxeter transformations and applications. Arch. Math. *55* (1990), 120–134, Zbl.687.16017.

[122] de la Peña J.A., Simson D.: Preinjective modules, reflection functors, quadratic forms and Auslander-Reiten sequences. Trans. Am. Math. Soc. *329* (1992), 733–753. See also: Simson D.: Two-peak posets of finite prinjective type. Preprint 1990, 13 pp.

[123] Perlis S.: A characterization of the radical of an algebra. Bull. Am. Math. Soc. *48* (1942), 128–132.

[124] Reiten I., Van den Bergh M.: Two-dimensional tame and maximal orders of finite representation-type. Mem. Am. Math. Soc. *408* (1989), 72 pp., Zbl.677.16002.

[125] Rickard J.: Morita theory for derived categories. J. Lond. Math. Soc., II. Ser. *39* (1989), 436–456, Zbl.642.16034.

[126] Riedtmann C.: Algebren, Darstellungsköcher, Überlagerungen und zurück (Ottawa 1979). Comment. Math. Helv. *55* (1980), 199–224, Zbl.444.16018.

[127] Riedtmann C.: Representation-finite selfinjective algebras of class D_n. Compos. Math. *49* (1983), 231–282, Zbl.514.16019.

[128] Riedtmann C.: Many algebras with the same Auslander-Reiten quiver. Bull. Lond. Math. Soc. *15* (1983), 43–47, Zbl.487.16021.

[129] Ringel C.M.: Finite-dimensional hereditary algebras of wild representation type. Math. Z. *161* (1978), 235–255, Zbl.415.16023.

[130] Ringel C.M.: The rational invariants of tame quivers. Invent. Math. *58* (1980), 217–239, Zbl.433.15009.

[131] Ringel C.M.: Bricks in hereditary length categories. Result. Math. *6* (1983), 64–70, Zbl.526.16023.

[132] Ringel C.M.: Tame algebras and integral quadratic forms. Lect. Notes Math. *1099* (1984), Zbl.546.16013.

[133] Ringel C.M., Tachikawa H.: QF3-rings. J. Reine Angew. Math. *272* (1975), 49–72, Zbl.318.16006.

[134] Roiter A.V.: The unboundedness of the dimension of the indecomposable representations of algebras that have an infinite number of indecomposable representations. Izv. Akad. Nauk SSSR, Ser. Mat. *32* (1968), 1275–1282, Zbl.167,310. English transl.: Math. USSR, Izv. *2* (1968), 1223–1230.

[135] Roiter A.V.: The roots of integral quadratic forms. Tr. Mat. Inst. Steklova *148* (1978), 201–210. English transl.: Proc. Steklov Math. Inst. *148* (1980) 207–217, Zbl.443.10020.

[136] Roiter A.V.: Matrix problems and representations of bocses (Ottawa 1979). Lect. Notes Math. *831* (1980), 288–324, Zbl.473.18006.

[137] Schmidt O.: Unendliche Gruppen mit endlichen Ketten. Math. Z. *29* (1929), 34–41, Jbuch. 54,148.

[138] Sergej chuk V.V.: Classification problems for systems of forms and of linear maps. Izv. Akad. Nauk SSSR, Ser. Mat. *51* (1987), 1170–1190, Zbl.653.15006. English transl.: Math. USSR, Izv. *31* (1988) 481–501.

[139] Serre J-P.: Faisceaux algébriques cohérents. Ann. Math., II Ser. *61* (1955), 197–278, Zbl.67,162.

[140] Shafarevich I.R.: Foundations of Algebraic Geometry. Nauka, Moscow, 1972, 567 pp. English transl.: Springer-Verlag, New York Berlin Heidelberg, 1974, 439 pp, Zbl.258.14001.

[141] Shafarevich I.R.: Algebra I. Encyclopaedia of Mathematical Sciences 11. Viniti, Moscow, 1986. English transl.: Springer-Verlag, Berlin Heidelberg New York, 1990, 258 pp.

[142] Springer T.A.: Linear Algebraic Groups. Birkhäuser, 1980, 304 pp., Zbl.453.14022.

[143] Tannenbaum A.: Invariance and System Theory. Lect. Notes Math. *845* (1981), Zbl.456.93001.

[144] Thrall R.M.: On ahdir algebras (preliminary report). Bull. Am. Soc. *53* (1947), Abstract 22, 49–50.

[145] Verdier J.L.: Catégories dérivées, état 0 (Paris 1961) Lect. Notes Math. *569* (1977), 262–311, Zbl.407.18008.

[146] Vossieck D.: Représentations de bifoncteurs et interprétation en termes de modules. Comptes-Rendus Acad. Sc. Paris, Sér. I *307* (1988), 713–716, Zbl.661.16025.

[147] Weierstrass K.: Zur Theorie der bilinearen und quadratischen Formen. Monatsber. Akad. Berlin (1868) 310–338, Jbuch.1,54.

[148] Yoshii T.: On algebras of bounded representation type. Osaka Math. J. *8* (1956), 51–105, Zbl.70,31.

[149] Zavadski A.G.: Representations of partially ordered sets of finite growth. Preprint, Kiev (1983), 1–76.

[150] Gabriel, P., Nazarova L.A., Roiter A.V., Sergejchuk V.V., Vossieck D.: Tame and wild subspace problems, to appear in Ukrainian Mathematical Journal.

A bibliography from 1984 to 1990 can be found in Ringel C.M., Recent Advances in the Representation Theory of Finite Dimensional Algebras, 141–192, in Representation Theory of Finite Groups and Finite-Dimensional Algebras (Eds. G.O. Michler and C.M. Ringel), Progress in Math. Vol. 85, Birkhäuser, 1991.

Subject Index

List of Symbols

In the following list, the symbols $\square$ and $\circ$ denote variables (categories, functors, quivers, sets ...)

Latin alphabet

Greek alphabet

Special signs